PRACTICAL GUIDE TO EXPERIMENTAL DESIGN

PRACTICAL GUIDE TO EXPERIMENTAL DESIGN

NORMAND L. FRIGON

DAVID MATHEWS

John Wiley & Sons, Inc.

New York • Chichester • Brisbane • Toronto • Singapore • Weinheim

This text is printed on acid-free paper.

This publication is designed to provide accurate and authoritative information in regard to the subject matter covered. It is sold with the understanding that the publisher is not engaged in rendering legal, accounting, or other professional services. If legal advice or other expert assistance is required, the services of a competent professional person should be sought.

Library of Congress Cataloging-in-Publication Data

Frigon, Normand L.

Practical guide to experimental design / Normand L. Frigon, David Mathews.

p. cm.

Includes index.

ISBN 0-471-13919-X (cloth : alk. paper)

1. Experimental design 2. Research, Industrial—Statistical methods. I. Mathews, David, 1940- . II. Title.

T57.37.F75 1997

658.4'033—dc20 96-23728

Printed in the United States of America

10 9 8 7 6 5 4 3 2 1

CONTENTS

PREFACE

Today all businesses are faced with growing competition—on national, international, regional, and local levels. Add to this competitive environment the rapidly growing technology base and the very short life cycle of products, processes, and services. The pressure on business to make precise, fact-based decisions rapidly is imperative to success. An important decision-making tool available to business, industry, and government is Design of Experiments (DOE).

The purpose of this book is to bring the basic ability to perform designed experiments to all decision makers: engineering, technical, service, and business. We have specifically structured the text to build the expertise and procedures necessary to perform, evaluate and make decisions using designed experiments. The examples are integrated into the text to provide a clear, concise, and comprehensible approach to the training, planning, application, and implementation of DOE. This book is all about making fact-based decisions using a common sense approach.

We start by building an understanding of the historical perspective of DOE in Chapter 1, where we present the several approaches to performing DOE and provide the overall understanding of processes, procedures, and methodologies. The book then proceeds in a very structured and deliberate way to the planning and management necessary to successfully accomplish a DOE, the basic statistical understanding needed to comprehend variance and how variance is used to make decisions, and the application of the statistical underpinnings of DOE. We then introduce the analytical tools of Analysis of Variance (ANOVA), full and fractional factorial DOE, and the use of DOE as a fault isolation and failure analysis tool. In the last part of the book we provide case studies, examples, and exercises from research, design, and development; industrial processes and products; management and services; and small-business applications.

Each new concept in the book is illustrated and presented step-by-step. Our experience from many years of management, consulting, and training is that this step-by-step approach is an important part of the application and implementation of DOE and should be carefully studied. The book is primarily intended as a practical applications guide and also contains enough material for either two semesters or three quarters of work.

Our objective in writing this book is to offer a single, comprehensive guide to businesses at all levels, small and large, for the practical use of DOE in business decisions. We have integrated a coherent, well-structured strategy for any business to achieve world-class competitive status. This is done by understanding the dynamics of the implementation of DOE in a business environment where planning and management are key elements for the successful experimental design. This approach is best described by the DOE implementation model that forms the basis for this book.

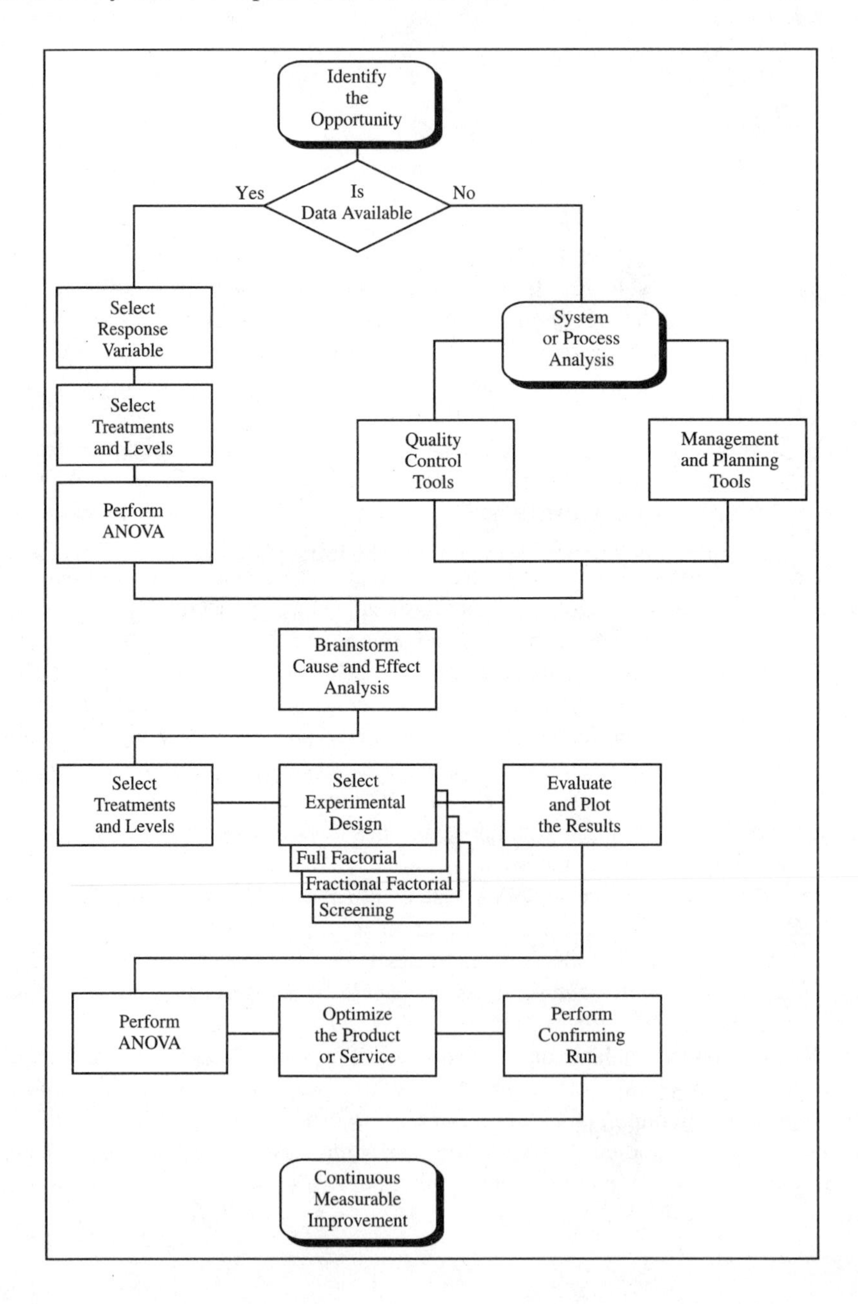

1

INTRODUCTION: GETTING STARTED

To survive and thrive in today's globally competitive environment, your products, services, and processes must be on target the first time, every time. The only way to accomplish this is through a comprehensive quality and reliability management strategy that optimizes your products and services during the design and development stages and continuously improves them throughout their life cycle. The most direct route to world-class competitiveness for your products and services is through variability reduction. Reducing the variability of your products, services, and process has a direct and immediate effect on your market competitiveness and profit. This variability reduction decreases defect rates, improves yields, lowers scrap rates, reduces rework, expands market potential, and reduces warranty costs. A critical tool in accomplishing variability reduction is the Statistical Design of Experiments. This practical guide discusses how to bring designed experiments to your activity at a practical operational level. To accomplish this, there are a few elements you must understand about designing, planning, and executing a designed experiment:

- The Improvement Process
- The Design of Experiments Model
- Getting Started
- Basic Statistical Understanding
- Making Intelligent Choices
- Analysis of Variance
- Full Factorial Experiments
- Fractional Factorial Experiments
- Fault Isolation and Failure Analysis

THE IMPROVEMENT PROCESS

The final decision for change is always a business decision. The business leader must, therefore, have confidence in the information used to make improvement decisions. The quality of the information used to make these decisions is very dependent on the structure of the improvement process, the validity of experimental design, and the analytical methods used. Businesspeople also must be able to address questions of experimental design and analysis in making the decisions to allocate resources and in designing the programs for improvement.

Here we will integrate the technical tools of Process Analysis and Product and Process Control with the new tools of Analysis of Variance (ANOVA) and Design of Experiments (DOE). These new tools will allow us to structure continuous measurable improvement efforts for new and existing products and processes. Figure 1.1 provides the structured approach to Continuous Measurable Improvement (CMI).

All improvement programs that are designed to achieve world-class competition must start with an in-depth understanding of the voice of the customer. It is crucial to know what is expected in the marketplace and to establish product and process parameters accordingly. The seven management tools can then be used to translate the voice of the customer to a quality function deployment, putting the marketplace language into the technical terms that engineers, analysts, and managers use to design a process. By evaluating the resulting process, you can find the optimum parameters for the services or products and their associated processes.

The process is stabilized using process control tools and is further improved with process analysis. For these existing processes we recommend ANOVA to evaluate the existing data to find candidates for further improvement, and Quality Function Deployment (QFD) and DOE to improve the process. The loss function enables you to determine whether the improvement to the process will result in a cost-effective change. Here we will explore the application and implementation of the improvement tools and how they fit into continuous measurable improvement.

We will present new tools for the planning, management, and evaluation of the improvement process. First, we will examine ANOVA, which is an advanced statistical technique that partitions variation and enables us to make decisions concerning its significance. We will review the implementation and application of ANOVA.

We will then examine DOE. This section will include the classical methods applied to Full Factorial, Fractional Factorial, and Screening experiments. We will then present the Taguchi Method. Additionally, we will present the Loss Function as a tool used to determine the economic feasibility of recommended changes and improvements.

THE DESIGN OF EXPERIMENTS MODEL

The DOE model in Figure 1.2 applies to designed experiments for new product and process development and for existing processes. This process integrates many tools and techniques and introduces the use of ANOVA and the design of experiments. As

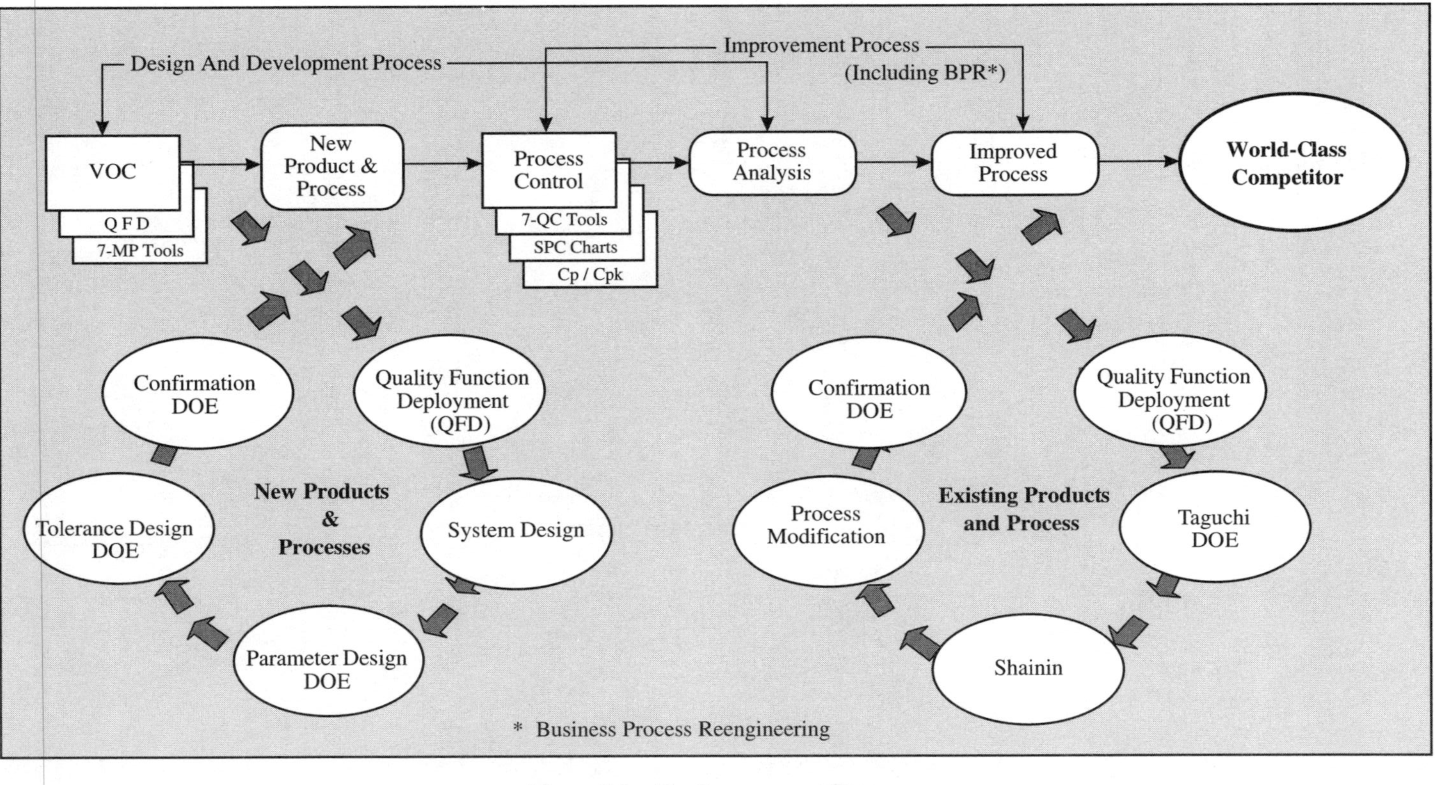

Figure 1.1 The Improvement Process

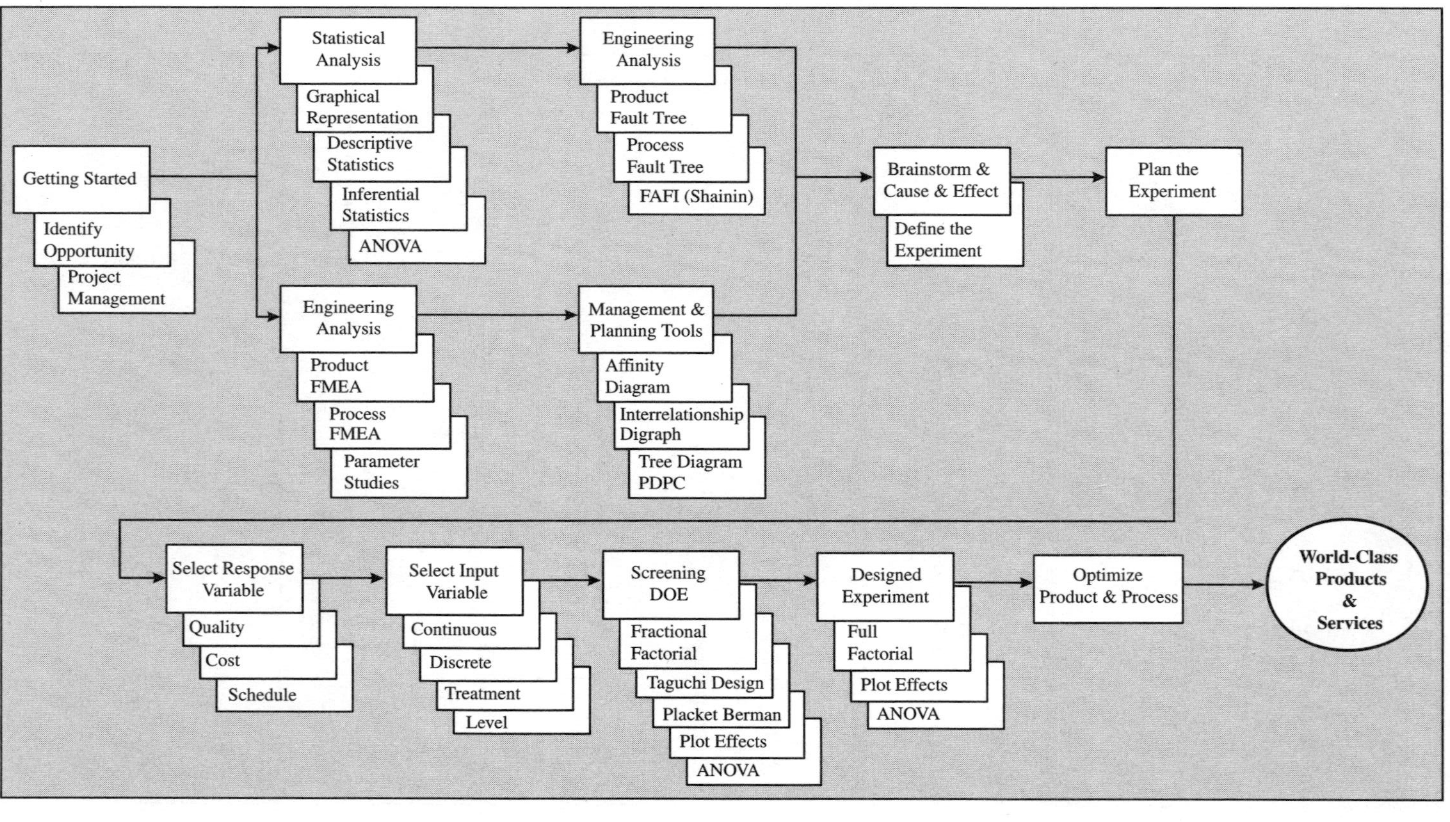

Figure 1.2 The DOE Process

with all of our technical tools, we must first define the process in order to gain a clear understanding of what we want to improve. The process and product must be understood at a very detailed level to determine all the treatments (input variables) that affect the product's quality characteristics (output variables). We need to know what these parameters are, how they were set, what relevance they have to customer needs, and how we can affect the end product. This understanding of the process provides us with the information to identify improvement opportunities.

If an improvement opportunity relates to an existing process, we can use available data. These data must be of sufficient quality and quantity to determine which response variables and treatments can be subjected to ANOVA evaluation to further our improvement process. The availability of this data can significantly reduce the number of experimental runs that will be required by providing a priori information to reduce the number of screening runs required.

If no data are available concerning the process or product, we must use the seven Management Tools and seven Quality Tools to develop sufficient information for a brainstorming session by an integrated team of subject experts to select treatments, levels, and response variables. The next step in the DOE process is to select the appropriate experimental design.

The selection of an experimental design is dependent upon the number of treatments that are to be evaluated, the level of data that is required, and the cost associated with accomplishing the experimental runs. Because full factorial experiments are the most comprehensive, they are also the most costly and time consuming to run. Fractional factorials provide information only on a limited scope of data and are less costly; screening experiments (including Taguchi designs) also provide a limited scope of data, but can provide some information concerning interactions. When many treatments are being considered, we recommend that you first perform screening experiments to identify the significant contributors to variance and then use fractional factorials or full factorials as refining experiments.

We evaluate the results of these experiments using two distinct methods. First, the data are evaluated for level effects and graphed for visual analysis. Second, we subject the data to ANOVA to determine whether the effects are significant, which treatments are the most significant, and the percent contribution to overall variance from the treatments and their levels. We will review these methods as we develop full factorial designed experiments.

Several DOE methods have been developed by various factions, each proposing its particular model as the best or only solution. It is not our purpose to support or detract from any of these approaches; that conflict is academic—we will focus on applications. The following approaches to DOE apply most directly to management, industrial, and administrative processes of concern to business.

DOE- AND FACT-BASED DECISION MAKING

As Dr. Margaret Wheatly has indicated in her landmark work on the chaos theory, all systems that are in disorder are seeking order and all systems in order are tending to

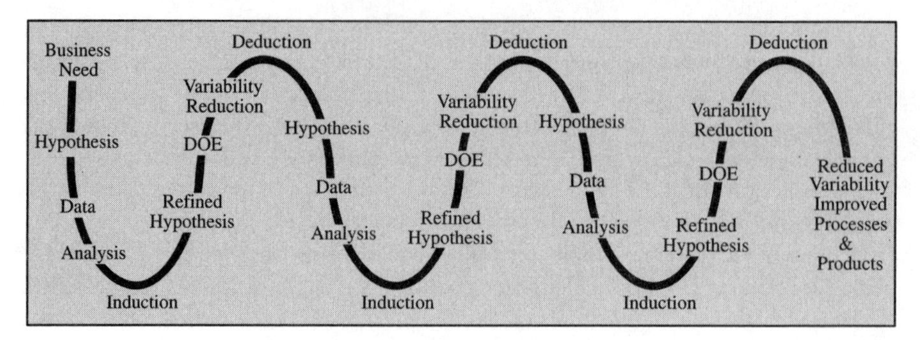

Figure 1.3 The Scientific Method and Fact-Based Business Decision Making Using DOE

disorder. This natural state of change exists within all business processes. Existing processes can be made more orderly and uniform, thereby reducing their variation. Existing processes become "out of control" with time and change. This nature of our business processes lends itself to the scientific method of data, hypothesis, induction, and deduction on an iterative basis. Figure 1.3 reflects this cycle and demonstrates how the scientific method fits the DOE model and the realities of the current volatile business environment.

Figure 1.3 emphasizes the interactive nature of process improvement using DOE. Although the exact nature of the chaos of business, its variability, and changes in environment are inexact, data and information can bring order to that chaos. This view of the scientific method of solving business problems is not intended to provide a unique approach to solving problems. Rather, it is a process that will take all analysts along the same generic path regardless of the tools selected to do the job. This process would work with a number of tools presented in this book. DOE in the induction-deduction cycle could be replaced with ANOVA, Chi-Square, Nonparametrics, or any other analytical tool. The important point is that all of these tools fit into a cycle and a process, just as the DOE process uses many of the statistical methods within its model. The convergence of these powerful tools and techniques harnesses the natural variability of products and services and brings order to chaos. The strategy of this book follows very closely the understanding of the scientific approach to problem solving.

First we consider how to make the decisions leading to a successfully designed experiment and whether there is any way to make fact-based decisions along the way. This helps you identify the opportunity for improvement, define your experiment, and plan the experiment as if it were any other business project.

Chapter 2 provides a basic understanding of the statistical underpinnings leading to DOE. These are the basic statistical building blocks that will provide you with an understanding of the several different types of data you will be dealing with in preparing for and executing your designed experiment. Some of the basic statistical concepts of Chapter 2 are:

- Knowing and understanding variation
- Maintaining records of your data

- Sampling
- Data validation
- Descriptive statistics
- Graphical representation of data
- Discrete distributions
- Continuous distributions

In Chapter 3 we begin the process of using statistics to make decisions. These are the first tools to be used in preparing to perform a Designed Experiment. They allow you to make rational comparisons of your existing data and information. The statistical tools you will use to evaluate business (industrial, process, product, service, and administrative) data are:

- Chi-square (χ^2)
- Confidence intervals
- Student's distribution and test
- F distribution and test
- Signed Rank Test
- Rank Sum Test
- Kruskal-Wallis Test
- Friedman Test
- Contingency Tables

ANOVA is an effective analysis tool that allows for the simultaneous comparison of populations to determine if they are identical or if they are significantly different. It is an important resource for the evaluation of existing data and as a tool to be used in conjunction with a Designed Experiment. Chapter 4 discusses how ANOVA can determine whether the variance we see in our data is significant and measures that significance. The following four approaches to ANOVA are provided:

- One-Way ANOVA
- Two-Way ANOVA
- Linear Contrasts
- Multivariant ANOVA

Chapter 5 covers Full Factorial Experiments, which provide a comprehensive analysis of all treatments, their levels, and interactions. Evaluating full factorial experiments includes:

- The analysis of the effects of the treatments and treatment levels
- Graphing these effects
- Performing an ANOVA to determine the significance of the treatments
- Determining the percent contribution of the variance at each level

Fractional Factorial Experiments are the most frequently used experimental designs. As the number of treatments to be evaluated in a full factorial experiment grows large, the cost in time, resources, and budget also grows. The number of runs required to complete a full factorial repeat or replicate quickly, thus outgrowing the resources of most experimenters. Fractional factorial experiments in Chapter 6 provide a means to economically perform experiments under these conditions. Several key points contribute to our ability to fractionalize and use fractional factorial experiments:

- When there are many variables to be evaluated
- In a sequence of experiments used to refine and identify significant factors
- To optimize systems where there is little or no direct concern about interactions

Chapter 8, Fault Isolation and Failure Analysis, presents the troubleshooting tools. Personal preferences may vary, but one useful pair of definitions considers *fault isolation* to be the localizing of a problem and *failure analysis* to be the determination of exactly what went wrong and how it happened. To accomplish Fault Isolation and Failure Analysis (FIFA), it is important to use a structured approach. Some of the tools used for FIFA are:

- Pareto Analysis
- Component Search
- Multi-Vari
- Autopsy
- Paired Alternatives
- Cochran's Q
- Failure Modes and Effects Analysis
- Fault Tree Analysis

After using all of these analytical techniques, there may still be some unanswered questions. We have provided additional analytical techniques in Chapter 7.

The final two chapters of the book are Industrial Case Studies (Chapter 9) and Nonindustrial Case Studies (Chapter 10). These case studies are drawn fron actual improvement projects that have used these tools. They are provided to give the reader a sense of reality for the applicability of these tools. They demonstrate clearly that Design of Experiments works in improving and optimizing products and services.

GETTING STARTED

Getting started is always difficult because there does not seem to be any starting place. How do you really make the decisions leading to conducting a successful designed experiment, and how do you make fact-based decisions along the way? In this chapter we will review the following basic concepts, which are also shown in Figure 1.4.

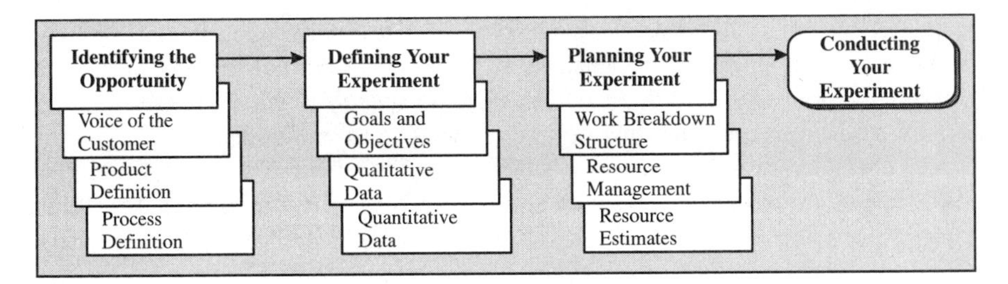

Figure 1.4 Getting Started

- Identifying the Opportunity
- Defining Your Experiment
- Planning Your Experiment

IDENTIFYING THE OPPORTUNITY

Knowing and understanding your customer, products, and processes is the first step toward performing a successful designed experiment. This will preclude the tyranny of assumptions that so often causes you to carry out a useless experiment. This is a basic search for the facts concerning your customer's needs and requirements, your product, or your process. To accomplish this you must use a structured approach to becoming intimate with these important factors. Yes, you must be intimate with your customer, product, or process before you can truly understand all of the factors affecting its performance. You accomplish that by knowing and understanding your customer; knowing and understanding your product; and/or knowing and understanding your process.

Knowing and Understanding Your Customer

Often when describing a product or service for design, development, or improvement, the voice of the customer is not fully heard or understood beyond a marketing perspective. Customers describe products and services in terms of the attributes of their performance. This definition is too narrow for the purpose of planning a designed experiment. In preparing for a designed experiment, therefore, we must include a more comprehensive customer perspective that can be translated into measurable characteristics for a designed experiment. This information, derived from the customers and the marketplace, is referred to as the Voice of the Customer (VOC), and it is an important factor in determining what you ultimately will decide to measure in your designed experiment.

The VOC in identifying opportunities for improvement may include attributes of the entire customer experience that are necessary to achieve customer satisfaction. There are three types of customer satisfaction. The first, basic expectations, includes those product and service features that the customer takes for granted. These are items that customers do not ask the supplier about, but which, if they are omitted, will result in extreme customer dissatisfaction. The second type of product or service attribute includes the items that the customer will specify as wants. These are often written descriptions of a product or service. They are usually expressed as performance characteristics that are requested in addition to the basic expectations. The third type of product and service feature encompasses the attributes that the customer did not request and does not know about (the engine will run 100,000 miles!). These are the characteristics that excite or "wow" the customers.

Your challenge in knowing and understanding your customer is to determine the DOE metrics that are your customer's basic expectations, wants ,and delights. Then you must determine how they can be measured as output variables or controlled as input variables to your experiment. This information can then help identify the opportunity and define your experiment. This information is best organized using a Voice of the Customer Table as shown in Table 1.1.

The Voice of the Customer Table (VOCT) is used to collate data about the customers and their expectations about the customer experience. Using this information and the table you have constructed, you now can determine metrics for your designed experiment. Information from the voice of the customer most frequently is used as the quality characteristic, output variable, or uncontrolled variables of your experiment. This information can also be used to determine environmental inputs to create robust designs. But remember that these metrics, like all metrics used for control or improvement, should have some relationship to quality, cost, and schedule.

Knowing and Understanding Your Product

We have decided to make the first phase of product definition the voice of the customer. This approach makes the customer your first consideration in conducting a designed experiment. The VOC will add the relevance of the marketplace to your efforts and ensure that you will have a result that is relevant to the product's or service's final use. Now you will be prepared to review the specifications and drawings of the product from this new perspective. What can I measure that will increase the satisfaction of the customer? How can I use the specifications to gain knowledge of the product, its assemblies, and components? What will the drawings tell me, and can I determine key measures from the drawings?

Product and service specifications come from many sources. They may be as simple as the written description of a service or as complex as a government specification for an aircraft. Although these two types of specifications are very diverse, they serve the same purpose—they provide a written record of the requirements of a system as it is designed and/or produced. This document provides you with information concerning the systems output characteristics and how they can be measured, the system requirements, its subassemblies and components. Often the composition of materials is specified, as are the specific tests required to accept the material. To perform a successful DOE, the specifications must be reviewed for the specific measures of output and input and their relationship to product, process, or service performance.

Customer Satisfaction	#	Attribute	Metric	Parameter
Expectations	1	Safety	Critical Failure	λ 0.001
	2	Reliability	MTBF	1 Year
	3	Durability	MTBM	25K Miles
	4	Economy	MPG	16-18
	5	Heater/AC	Temperature	70% /–10%
	6	Radio	Mono	Quality
Desires	1	Leg Room	Space	Cubic Feet
	2	Trunk Room	Space	Cubic Feet
	3	Hi Reliability	MTBF	1.5 Years
	4	Durability	MTBM	50K Miles
	5	Colors	Number	Range
	6	Radio w/Tape	Stereo	Quality
	7	Recline Seats	Range of Movement	Comfort
	8	Warranty	Time	36 Months
Delights	1	Keyless Entry	Reliability	λ 0.001
	2	Digital Instruments	Brightness	Visual
	3	Radio w/Tape	Stereo	Quality
	4	Power Seats	Range of Movement	Comfort
	5	Power Windows	Electric	Convenience
	6	Hi Reliability	MTBF	2 Years
	7	Durability	MTBM	100K Miles
	8	Warranty	Time	72 Months

Table 1.1 Voice of the Customer Table

Knowing and Understanding Your Process

To know and understand your processes you must use a structured approach to process analysis. This methodology is based on the principle that all activities can be viewed as processes that are systematic, repetitive series of actions to develop or produce products or services.

There are three types of processes; industrial, administrative, and managerial. Figure 1.5 shows that all three have the same characteristics: input >>>> process >>>>> output. These recursive characteristics enable us to apply the six phases of process analysis (Figure 1.6) to know and understand any process.

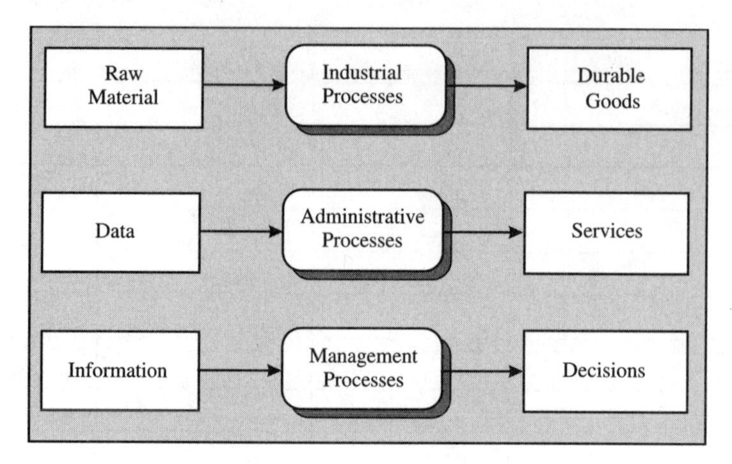

Figure 1.5 Three Types of Processes

Industrial Processes. Industrial processes come to mind immediately when we think of process analysis. These are the processes that produce things. The inputs to industrial processes are raw materials. These raw materials can be in the form of basic materials, such as iron ore, steel, and coal; subassemblies, such as computer boards and engine parts; or equipment for rework or repair, such as engines or automobiles that require overhaul or aircraft that require upgrade and modification. Industrial processes lend themselves most easily to the technical resources for process improvement. Processes such as repairing, rebuilding, or upgrading things are also industrial processes. In those cases, the items to be repaired, rebuilt, or upgraded, together with the new parts, rework kits, or upgrades, are the raw materials of the process.

Administrative Processes. Administrative processes are those that frustrate customers (internal and external) most frequently. Administrative processes produce the paper, data, and information that other processes use. They also produce products used directly by the customers, such as tax returns, pay checks, reports, and data. These processes include some of the most complex and bureaucratic challenges in the pursuit of world-class competitiveness. The improvement of administrative processes affects all other processes. Special attention must be paid to the dilatory effect of inefficient and ineffective administrative processes on quality, cost, schedule, personnel morale, team spirit, management processes, and industrial processes.

Management Processes. Management processes are the structured means by which businesses and individuals make key decisions. It is very important to understand that management is the process of using data to make decisions. That process works best when properly accomplished in organizations with a Management 2000 infrastructure. This structured, quantifiable approach to decision making ensures fact-based decisions. These fact-based decisions are supported by quantifiable data derived from the application of the technical tools.

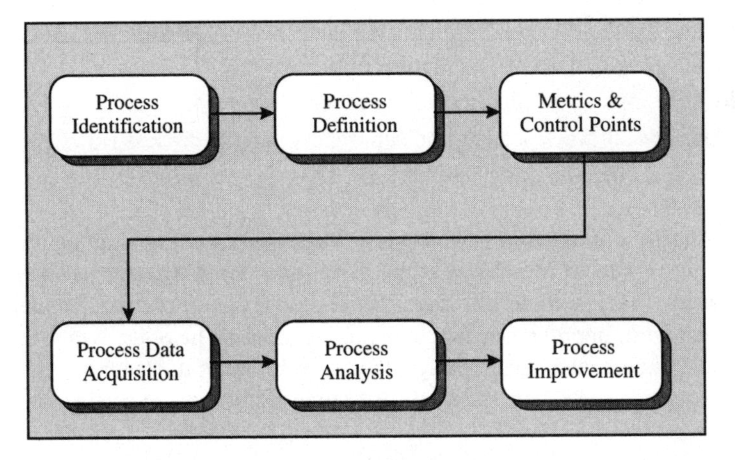

Figure 1.6 Six Phases of Process Analysis

Process Analysis. The six phases of process analysis, presented in Figure 1.6, provide a structured method of identifying and describing the elements of this transformation. The detailed understanding provided by a formal process analysis is a required precursor for the use of process data as technical resources for a designed experiment. You must first understand a process, its functional elements, work activities, and measurable parameters before applying any of that information to a designed experiment.

Process Identification. Selecting process metrics for use in a designed experiment begins with identifying the process we are going to analyze and determining who is specifically responsible for the process and who has the authority to change it. To properly identify a process we must determine its initial input, its work activities, and its output. We must also identify process ownership. Every element of an organization can identify dozens of activities that are performed, including many that are important. To implement continuous measurable improvement, however, we must select those activities that are critical to the mission of the organization, business, and individuals. These, then, are the processes to be managed and improved. Resources for improvement activities are always limited; therefore, we must have a means of prioritizing those activities. The following two steps and their associated actions provide a structured method for determining process identification and the associated process ownership.

1. Acquire and review all process documentation.
2. Determine process ownership.

Process Definition. The following steps provide a structured method to complete the process definition and provide the basis for completing the flowcharts. In steps 1–4 you develop lists of information that are used in step 5 to create the flow chart.

1. Define the internal elements and boundaries of the process.
2. Identify the outputs and customers of the process.
3. Identify inputs and suppliers of the process.
4. Identify each work activity and subprocess and the process flows.
5. Flowchart the process.

Process Metrics and Control Points. After identifying and defining the process, the process analysis team must review the goals and objectives for its analysis and improvement. Each goal must be concise and measurable, stressing one theme, such as cost reduction, improvement of quality, or schedule. The objectives must then be established, documented, and clearly understood by all of the team members before the next step can be taken successfully. Some questions that help clarify goals include:

- Is our objective to reduce defects?
- Is our objective to improve our process or activity?
- Is our objective to reduce the queue?
- Is our objective to improve the throughput?
- Is our objective to improve our timeliness?
- Is our objective to reduce the cost?

Once the goals and objectives for the process analysis have been clearly and concisely written and agreed upon, the required control points and metrics can be identified.

Metrics are the measurements necessary to monitor the selected process and to determine if it is satisfying the requirements. These measurements should be based on customer requirements; they should be easy to understand and specific. In selecting control points, remember that they must be tied to results of critical operations and that they are at process decision points. The measures of process quality, cost, and schedule (Q$S) are of prime concern when measuring processes. We will next describe the step-by-step procedure to determine process metrics and control points.

Performing Process Measurement. The following five steps and their actions provide a structured method for determining the process metrics and critical process elements to be measured:

1. Review process output requirements.
2. Review process input requirements.
3. Review all process elements.
4. Define the measures of the process elements.
5. Define measures and control points.

Define the measures of the critical process elements: The measures selected for the process inputs, outputs, and each critical process element should relate to the goals and objectives for the process analysis. Determine the measure of each critical element and control point by its direct relationship to the goals and objectives of the process analysis.

- Select a metric to calculate the measure of each selected element.
  - Quality (Q)
  - Cost ($)
  - Schedule
- Design a statistic to be used in measuring the selected element
  - Reject rate
  - Warranty returns
  - Cost of scrap
  - Time to complete a work activity

Process Data Acquisition. After we have identified the process critical measures and control points, we must collect data to facilitate the analysis of these points. Determining whether the process is meeting customer needs and expectations requires measurable data. These data are also essential to determine if the process is in control and to ensure the success of improvement efforts. The following three steps and the associated actions describe the procedure to perform data acquisition.

1. Determine data media.
2. Specify the scope of the data acquisition requirement.
3. Gather data from the critical elements/control points.

Process Analysis. It is imperative that our activities be consistent with the stated objective for the designed experiment. Therefore, after prioritizing opportunities, you must base decisions for action on their fiscal, quality, and/or schedule impact. The following three steps and associated actions will aid in completing this phase of process analysis.

1. Analyze data using statistical techniques
2. Compare performance and requirements
3. Perform further analysis of selected elements

DEFINING YOUR EXPERIMENT

Now you should clearly understand the VOC and how it integrates with your products, services, and processes. This information is a vital element in the process of defining your experiment. Now we will define what you are going to accomplish in your designed experiment and which treatments you will use to evaluate your experimental output. DOE is often a resource-intensive activity; therefore, DOE should never be performed without a clear understanding of how it will affect the organization. It is important at this point in the DOE process to identify the vision of your organization and how performing this designed experiment will enhance that vision.

What are you attempting to accomplish with your designed experiment? On the surface this may seem to be a simple question—improvement, of course! However, as you become more intimate with your product, service, or process at lower and lower levels, it becomes less apparent exactly how any specific designed experiment will help your organization achieve its vision. You can maintain the proper perspective by establishing goals and objectives for your designed experiment and correlating those to the organization's direction to achieve its vision. Once these goals and objectives have been established, quantitative data and qualitative data that exist within your organization can be used to select specific experimental inputs (treatments) and output (product or service characteristics).

Goals and Objectives

Based on achieving your vision, you must now establish goals and objectives for your designed experiment. Goals are the overall desired result you must accomplish to achieve your vision. Objectives delineate exactly what needs to be accomplished to achieve a specific goal. Your designed experiments should be constructed to meet specific objectives. These all build upon the same foundation of achieving the organization's vision, as indicated in Figure 1.7. DOE is used like many other management and technical tools to accomplish specific objectives in support of achieving your goals and vision.

Vision. Every business pursuit begins with a vision, and with the hope that what is an idea today will ultimately become a reality. Each vision is a unique idea of a possible future accomplishment. It is what you and your organization will be when the vision is reached. This vision can be what the entrepreneur visualized for a new enterprise, the future course of an existing company, the strategic plan for a division of a company, or even a personal vision of what you hope to achieve. Opportunities to achieve your vision will often include taking advantage of technical advances, changes in the marketplace, innovations in products and processes, and changes to the existing business method.

In order to achieve a vision, the resources of the organization must focus on that vision. They need to understand the vision and believe in the possibility of achieving it. The vision must be founded on business, economic, and technical possibilities and must be attainable from each of these perspectives. Using designed experiments can and will translate these possibilities into realities.

Goals and Objectives. Establishing goals and objectives ensure that there is a structured way to achieve your vision. Each goal you establish represents a specific accomplishment that must be attained to achieve your vision. Each goal is an element that is needed to develop the company's strategic vision. These goals and objectives are cascaded to each organization to develop organizational specific short-, intermediate-, and long-term objectives. The objectives are the "How" of accomplishing the

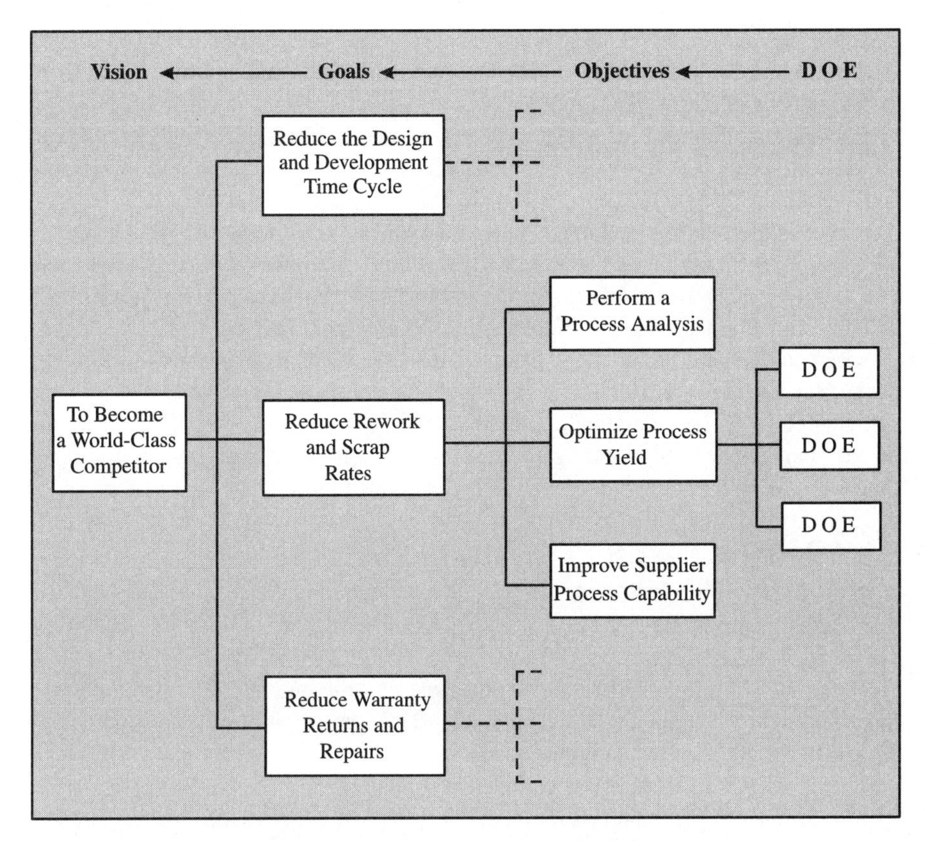

Figure 1.7 Vision Goals and Objectives

goals. Each objective is a specific action to be taken, such as "Performing a process analysis" or "Optimizing process yield."

One of the major obstacles to successful implementation of DOE is the misalignment of the planned DOE with the organization's vision, goals, and objectives. At times designed experiments are planned and executed without regard to their relationships to achieving specific goals and objectives. This can lead to suboptimizing a process or product, thereby not supporting the achievement of the organization's vision. The strength of defining your experiment in relation to the organization's vision, goals, and objectives is that you can avoid this suboptimization and focus your limited resources on accomplishing effective designed experiments.

Qualitative Data

When using quantitative data, the emphasis is on the collection, analysis, and interpretation of numerical data. In these activities, we count or measure things; we use the relationships of numbers, and the characteristics of mathematical distributions, to turn the data into information for decision making. In business planning processes, however, we must deal with large amounts of language data, or qualitative data, such as ideas, opinions, perceptions, or desires. The usual statistical methods do not work in the analysis of language data.

Recognizing the need for new tools and techniques, in 1972 the Society of Quality Control Technique Development, sponsored by the Japanese Federation of Science and Technology, began looking for ways to enhance planning activities. This exhaustive investigation culminated in a proposal for Seven New Quality Control Tools. In 1979, Shigeru Mizuno wrote a book entitled *Management for Quality Improvement, The Seven New QC Tools*. GOAL/QPC had this book translated into English in 1983 and the English translation was published by Productivity Press in 1988.

GOAL/QPC and the American Supplier Institute began studying and teaching the Seven New QC Tools in 1983. Since that time, many others have refined the tools and contributed to their understanding. Some, like GOAL/QPC, refer to them as the Seven Management and Planning Tools (7 M&P Tools). The management and planning tools we will review are:

- Affinity Diagram
- Interrelationship Matrix
- Relationship Matrix

These tools are flexible in their application, and the output from one can be the input for another. In this manner we are able to leverage their power, thus increasing their value in all planning activities.

Affinity Diagram. The first task, then, is to distill the qualitative information into key ideas or common themes. The Affinity Diagram is a very effective tool for achieving this result. It organizes language data into groupings and determines the key ideas or common themes. The results can then be used for further analysis in the planning process.

The development of an Affinity Diagram is a three-step process. This is a creative task, requiring analysis of ideas, association of common thoughts, and determination of patterns from large amounts of data. An individual can develop the Affinity Diagram, but because it is a creative process, often a team is more effective. The three steps for developing an Affinity Diagram are:

1. Group data.
2. Select grouping titles.
3. Refine groupings.

Step 1. Group Data Begin by collating the ideas, opinions, perceptions, desires, or issues as individual data elements. Write each one on an individual piece of paper, such as a Post-it note. Then arrange them on a flat surface such as a wall, white board, or window. The objective is to cluster the ideas together in logical associations. Some people like to write tentative titles and cluster the ideas under them. We do not recommend this, because it can stifle creativity. Instead let the associations and patterns drive the title.

During this step, it is important that every member of the team participate. Begin by placing the data elements on the flat surface without regard to association. Then direct the team to arrange the notes in logical patterns. The team does this together, without discussion or evaluation of the choices. Each person is allowed to move the Post-it notes around at will. This may seem chaotic, but set a time limit (e.g., 15 minutes), encourage participation, and soon order and agreement will come out of the chaos (Figure 1.8).

Step 2. Select Grouping Titles The next step is to decide on a title for each grouping. The title needs to represent an action that reflects the main idea or theme of the grouping. The titles, therefore, need to be complete sentences, stated as actions.

In some instances, determining the title requires a compromise among the ideas in the grouping. Keep in mind that, at this point in developing the Affinity Diagram, the

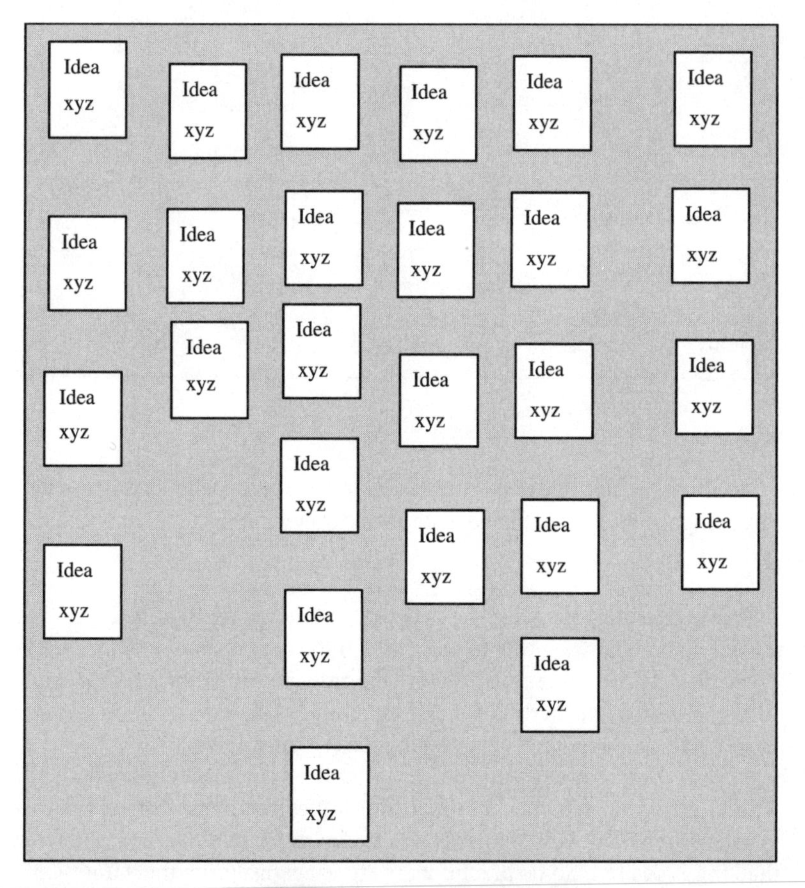

Figure 1.8 Associate Ideas

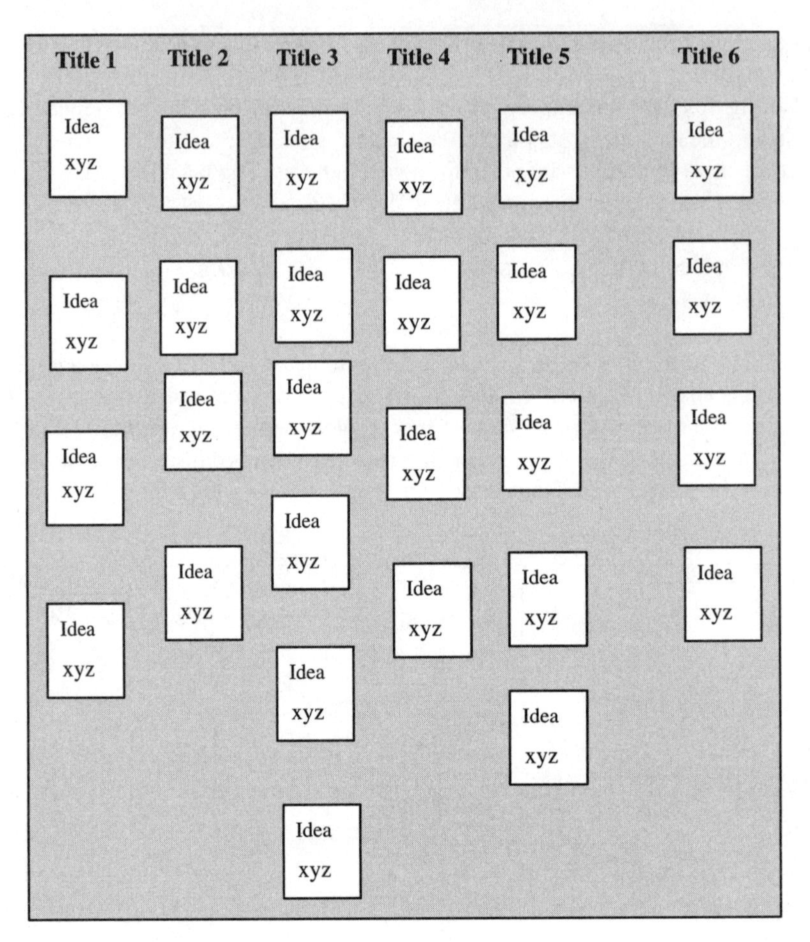

Figure 1.9 Title the Groupings

title is important because it defines the action to be taken. Avoid evaluating the ideas in the groupings at this point; the next step will further clarify the issues and the titles (Figure 1.9).

Step 3. Refine Groupings After the groupings have appropriate titles, it is time to review each item under each title to see if it still fits or if it should be included under a different title. At the same time, review the titles to ascertain whether any of the groupings can be consolidated. The resulting Affinity Diagram will bring order to the original collection of apparently unrelated ideas (Figure 1.10).

The Affinity Diagram is a very powerful tool for dealing with language data. It quickly and effectively distills the data into logical patterns that reveal common themes and associations. It is also effective in dealing with old issues when new creative thinking is desired. Some practitioners argue against using the Affinity Diagram when the situation is simple and requires quick results.

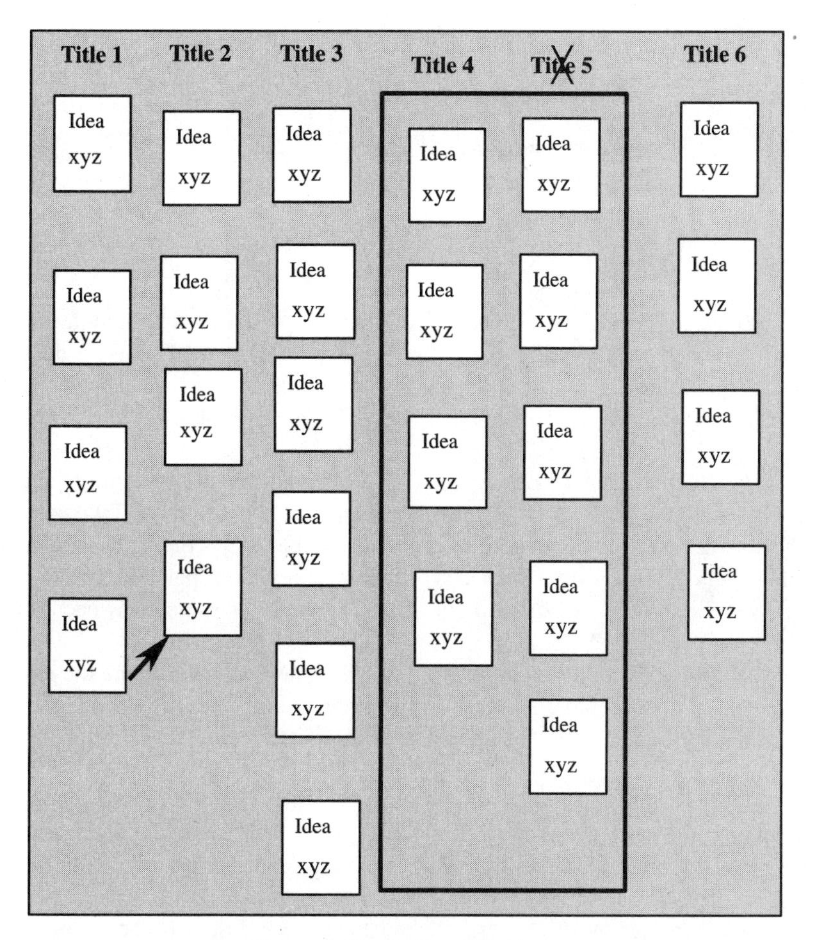

Figure 1.10 Refine Groupings

Interrelationship Matrix. The relationships among language data elements are not linear and they are often multidirectional. In other words, an idea or issue can affect more than one other idea or issue, and the magnitudes of these effects can vary. Additionally, the relationships are often hidden or not clearly understood.

The Interrelationship Digraph is an effective tool for understanding the relationships among ideas and for mapping the sequential connections between them. The usual input for the Interrelationship Digraph is the results of an Affinity Diagram. The Interrelationship Digraph can, however, be used to analyze a set of actions or ideas without first developing an Affinity Diagram.

We use the information developed from the Interrelationship Digraph to establish priorities and to determine optimum sequencing of actions. Frequently, teams develop an Affinity Diagram, skip the Interrelationship Digraph, and go on to use another tool to develop their plan. This is a big mistake. The Interrelationship Digraph always provides an important understanding of the data you are analyzing.

Building an Interrelationship Diagraph consists of five steps:

1. Develop the matrix of issues.
2. Determine causal relationships.
3. Mark the causal relationships.
4. Sum the interrelationships.
5. Set priority for the issues.

Step 1. Develop the Matrix of Issues The first step in developing the Interrelationship Digraph using the Achieving the Competitive Edge method is to develop a matrix of the issues. Enter each issue on the horizontal and vertical axes. Add a total column and a total row to the matrix, as shown in Figure 1.11.

Step 2. Determine Causal Relationships The second step in developing the matrix is to determine the causal relationships between each pair of issues. Take each issue on the vertical axis and compare it to each of the other issues on the horizontal axis. For this method, the question is: "Does the horizontal issue depend on, or is it caused by, the vertical issue?" The question needs to be worded the same way each time.

Step 3. Mark the Causal Relationships For this method we evaluate the extent of each causal or dependency relationship: strong, medium, weak, or none. Figure 1.12 is an example of a matrix with symbols added.

Step 4. Sum the Interrelationships After determining all the relationships, score them in both the vertical and horizontal axes. Place the totals in the appropriate column or row. Figure 1.13 shows the matrix with the scores added.

Step 5. Set Priorities for the Issues Review the results of Step 4. The issues with the largest sum totals have the greatest impact on the other issues. In this matrix this corresponds to the row totals at the bottom of the matrix. These are the "critical few"

Depend →	A	B	C	D	E	Total
Issue A						
Issue B						
Issue C						
Issue D						
Issue E						
Total						

Figure 1.11 Interrelationship Matrix

Depend →	A	B	C	D	E	Total
Issue A						
Issue B	⊙			⊙		
Issue C	△	⊙		△		
Issue D	⊙				○	
Issue E	⊙		⊙			
Total						
⊙ Strong=9		○ Medium=3		△ Weak=1		

Figure 1.12 Matrix of Issues with Causal Relationships

issues. Solving these problems, implementing these actions, or providing these services will have the greatest influence on the problem, goal, or customer requirement in your designed experiment.

The totals in the column, on the left side of the matrix, reflect the issues that are affected by the other issues. In this case the highest total indicates that an issue is most affected by the other issues. It is, therefore, the most dependent.

You can use the insight provided by this evaluation to prioritize actions, or to determine the issues necessary for further planning your designed experiment.

Relationship Matrix. The Matrix Diagram is a tool for organizing language data (ideas, opinions, perceptions, desires, and issues) so that they can be compared to one another. The matrix diagram reveals the relationships among ideas and visually

Depend →	A	B	C	D	E	Total
Issue A						0
Issue B	⊙			⊙		18
Issue C	△	⊙		△		11
Issue D	⊙				○	12
Issue E	⊙		⊙			18
Total	28	9	9	10	3	59
⊙ Strong=9		○ Medium=3		△ Weak=1		

Figure 1.13 Matrix of Issues with Scores

demonstrates the influence each element has on every other element. The construction of a Matrix Diagram is a four-step process:

1. Select the matrix elements.
2. Select the matrix format.
3. Complete the matrix headings.
4. Determine relationships or responsibilities.

Step 1. Select the Matrix Elements The matrix elements fall into categories, the vertical axes of which are the "whats" and the horizontal axes of which are the "hows." You can derive these elements from a new or previously developed brainstorming session, Affinity Diagram, or Interrelationship Digraph.

Step 2. Select the Matrix Format As stated previously, the matrix format depends on the number of sets of data to be analyzed. The most common format of the relationship matrix is demonstrated in Figure 1.14.

Step 3. Complete the Matrix Headings After the language data are collected, sorted, and divided into sets, and the matrix format is selected, fill in the headings of the matrix.

Step 4. Determine Relationships Examine each of the interconnecting nodes in the matrix and determine if there is a relationship. As in the Achieving the Competitive Edge method for developing an Interrelationship Digraph, evaluate the relationships and mark the matrix accordingly (Figure 1.15). At this point, sum the rows and columns and interpret the matrix. In the matrix in Figure 1.15, the row on the bottom indicates the priority of actions that must be taken to accomplish your goal. The col-

	Action 1	Action 2	Action 3	Action 4	Action 5	Action 6	Total
Issue A							
Issue B							
Issue C							
Issue D							
Issue E							
Total							

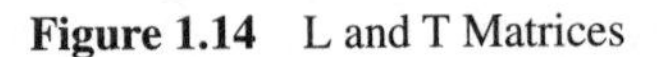

Figure 1.14 L and T Matrices

	Action 1	Action 2	Action 3	Action 4	Action 5	Action 6	Total
Issue A	⊙	○		⊙		○	24
Issue B	○	○	△		○	△	11
Issue C	○			⊙	△		13
Issue D		△	⊙	○		○	16
Issue E	⊙			⊙	○	⊙	30
Total	24	7	10	28	7	16	94

Figure 1.15 Completed L Matrix

umn on the right indicates the dependency of the issues to be accomplished. The highest priority action is action 3 and the most interdependent issue is issue E.

Quantitative Data

Over the years, statistical methods have become prevalent throughout business, industry, and science. With the availability of advanced, automated systems that collect, tabulate, and analyze data, the practical application of these quantitative methods continues to grow. Statistics today play a major role in all phases of modern business and are indispensable in performing an effective designed experiment. These tools are essential in the evaluation of your input and output variables for designed experiments. We will review the following quantitative methods, which will assist you in getting started. Additional basic statistical tools will be fully discussed in Chapter 3.

- Cause-and-Effect Analysis
- Data Tables
- Histograms
- Pareto analysis

Cause-and-Effect Analysis. After identifying a goal for your designed experiment, it is necessary to determine the causes that are effecting its accomplishment. The cause-and-effect relationship is at times obscure. A considerable amount of analysis often is required to determine the specific cause or causes affecting the problem.

Cause-and-effect analysis uses diagramming techniques to identify the relationship between an effect and its causes. Cause-and-effect diagrams are also known as fishbone diagrams. Figure 1.16 demonstrates the basic fishbone diagram. The problem box contains the problem statement or goal being evaluated for cause and effect. The prime arrow functions as the foundation for the major categories. The six basic cate-

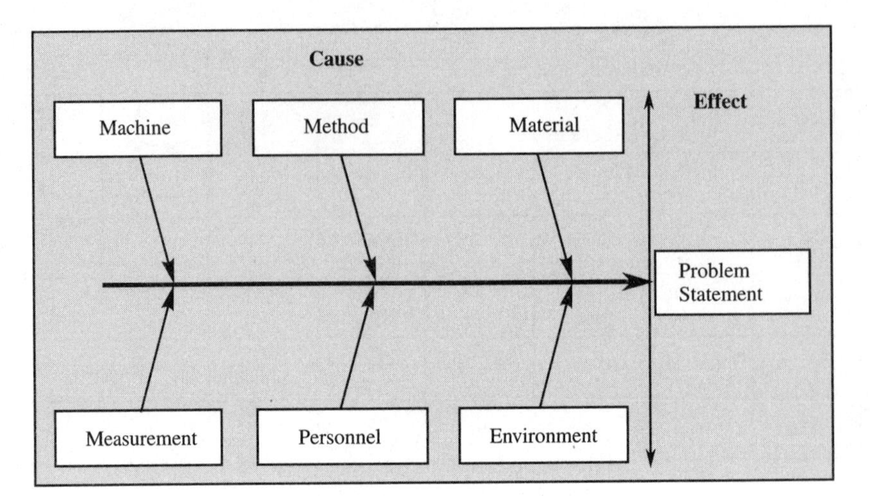

Figure 1.16 Cause-and-Effect Diagram

gories for the primary causes of the problems most frequently are personnel, method, materials, machinery, measurements, and environment. Other categories may be specified based on the needs of the analysis.

When you have identified the major causes contributing to the problem, then you can determine the causes related to each of the major categories. Using your cross-functional team in a brainstorming session, identify the possible causes related to each of the categories, as shown in Figure 1.17. Based on the cause-and-effect analysis of the problem and the determination of causes contributing to each major category, identify potential input variables for your designed experiment.

Data Tables. Data Tables, or data arrays, provide a systematic method for collecting and displaying data and in most cases are forms designed for that purpose. These

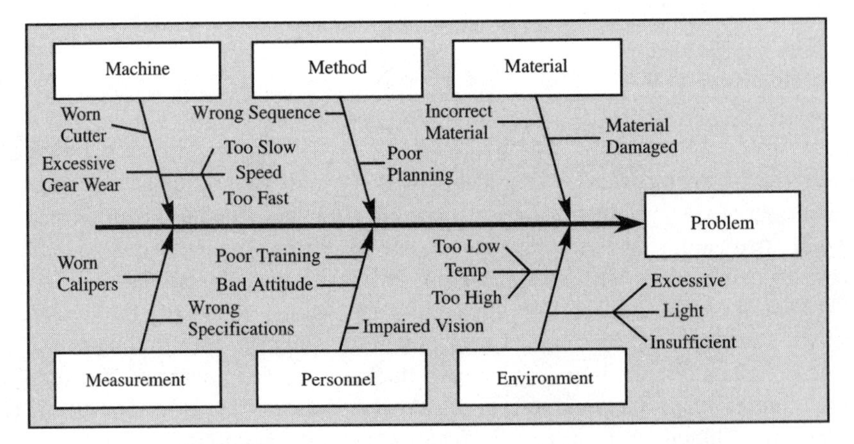

Figure 1.17 Completed Cause-and-Effect Diagram

tables are used most frequently where data are available from automated media. They provide a consistent, effective, and economical approach to gathering data, organizing it for analysis, and displaying it for preliminary review. Data tables sometimes take the form of manual check sheets where automated data are unnecessary or unavailable. Data tables and check sheets should be designed to minimize the need for complicated entries. Simple-to-understand, straightforward tables are a key to successful data gathering.

The effective use of data tables requires that certain decisions be made concerning the data. Determine why you are collecting the data and what specific analysis and analytical methods you will use. Identify what type of data are needed to perform the analysis. Determine whether the data are currently available, and from what source, and on what media. Identify where, when, and how the data will be collected, and by whom. Clearly define the data categories as you design your data table. Identify what data reduction tools you will use (automated, manual, programs) and how you will organize the data. Finally, consider what actions are contemplated. Will the data collected support the planned conclusions and recommendations?

Table 1.2 is an example of an attribute (pass/fail) data table for a material inspection function. From this simple check sheet, several data points become apparent. The total number of defects is 34. The highest number of defects is from Supplier A, and the most frequent defect is incorrect test documentation. We can subject this data to further analysis by using Pareto analysis, control charts, and other statistical tools.

Bar Charts. A bar chart is a graphical representation of data as a frequency distribution. This tool is valuable in evaluating both attribute (pass/fail) and variable (measurement) data. Bar charts offer a quick look at the data at a single point in time; they do not display variance or trends over time. A bar chart displays how the cumulative data look *today*. It is useful in understanding the relative frequencies (percentages) or frequency (numbers) of the data and how these data are distributed.

Defect	Shift 30-Day Period			
	1	2	3	Total
Run-up Test	15	10	30	55
System Integration	25	20	30	75
Motor Static Test	45	20	0	65
Motor Integration	5	50	60	115
Totals	90	100	120	310

Table 1.2 Check Sheet for Material Receipt and Inspection

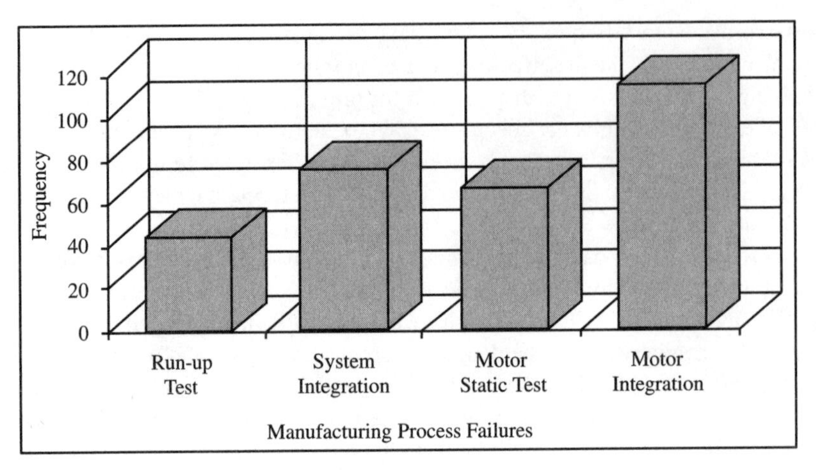

Figure 1.18 Bar Chart

Bar Charts for Attribute Data Bar charts for attribute data are easy to construct from data tables and check sheets. The bar chart in Figure 1.18 is based on the data in Figure 1.20. A bar chart of this data graphically demonstrates the relationships and frequencies of defects found in the manufacturing process. The frequency of occurrence appears on the vertical (Y) axis and the attribute elements are on the horizontal (X) axis.

Pareto Analysis. There are three uses and types of Pareto analysis. The basic Pareto analysis identifies the vital few contributors that account for most quality problems in any system. The comparative Pareto analysis focuses on any number of program options or actions. The weighted Pareto analysis gives a measure of significance to factors that may not appear significant at first, such as cost, time, and criticality. Pareto analysis can be used to evaluate the following:

Quality	Cost	Schedule
Failures	Repair	Cycle Times
Warranty Returns	Warranty	Throughput
Conformance	Labor	Delivery Times
Frequency of Repairs	Facilities	Milestones
Rejects	Disability	Deadlines
Reliability	Compensation	Shipping Schedules
Mean Time to Failure	Rates of Returns	Receipt Schedules
Frequency of Failures	Unit Costs	Process Times

Pareto analysis is performed to determine the significant few and the insignificant many. This Pareto analysis arranges the bar chart data by frequency of occurrence and provides the cumulative percentages of occurrence as indicated in Figure 1.19. The

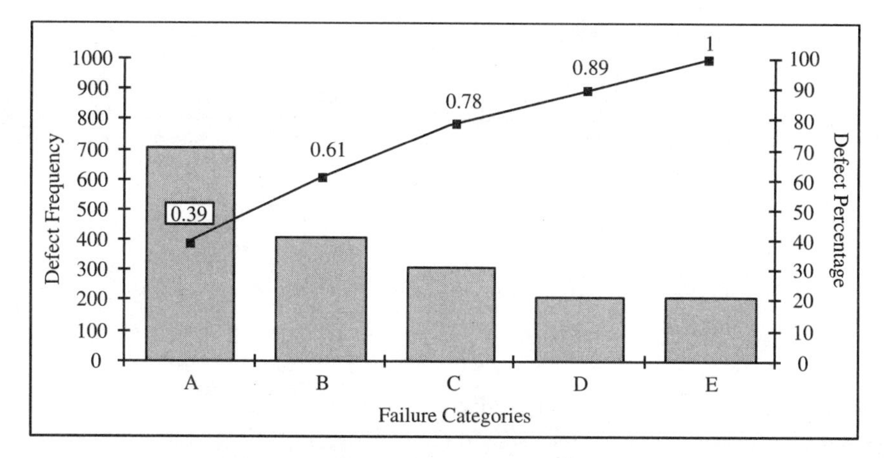

Figure 1.19 Basic Pareto Analysis for Frequency of Failures

discrete frequency is given on the left side of the chart and the percentage is given on the right side of the chart. In the basic Pareto analysis for frequency of failure, Category A is the most frequently occurring, accounting for 39% of all failures; Failure Category B is the next most frequently failing; and the cumulative percentage of failures for Categories A and B is 61%. Using this information, it becomes easy to understand where to concentrate corrective action efforts.

PLANNING YOUR EXPERIMENT

Now that you understand how to identify the opportunity and define your experiment, it is time to turn our attention to planning. This is a very necessary step in getting your experiment started and one that frequently falls prey to the tyranny of assumptions. "I have been doing this job for years and know what I want to do, how long it will take, and how much it will cost." This is a false statement. Unless you have accomplished planning at least to the level we recommend here, your experiment will take longer than planned, cost more than budgeted, and may fail. The following are the most basic necessities for you to plan your designed experiment.

- Work Breakdown Structure
- Time Lines
- Resource Estimates

Work Breakdown Structure

Establishing a Work Breakdown Structure (WBS) is the first step in planning your designed experiment. The WBS provides a framework to accomplish this. It is used to lay out the functional areas of your designed experiment and estimate the costs of

Level	Function	Description
1	Project Level	Total design and development project consisting of all functional areas.
2	Functional Level	Describes the design-and-development requirements for each functional area of the project.
3	Task Level	Describes the individual task elements to be performed in each functional area.

Table 1.3 Work Breakdown Structure Levels

facilities, materials, equipment, and labor. The WBS breaks down the overall designed experiment from its initial definition so that the tasks can be better defined on the basis of work elements and activities. The WBS provides an outline that can be monitored, planned, and managed throughout the design-and-development process. The task numbering system established in the WBS will carry through to the process flow chart and the Gantt chart, thus providing a basis for all the management tools for your designed experiment.

We will use the 3-level WBS shown in Table 1.3 to fully describe a design-and-development project. Depending on the complexity of your project, you can modify

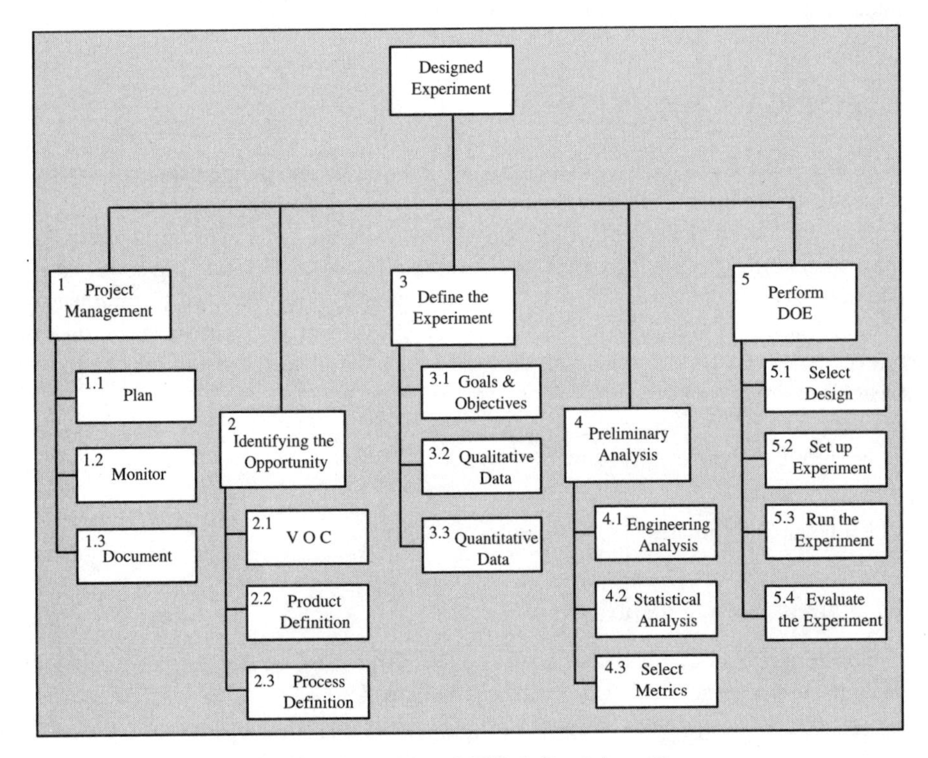

Figure 1.20 Typical Level 1 Work Breakdown Structure

the number of levels in any WBS to coincide with your organization's needs and the scope of the project.

The first level of the WBS relates to the integrated effort at the highest level and should not be related solely to any specific department or function. At this level, the WBS is used to describe the overall design-and-development project. Figure 1.20 demonstrates a typical Level 1 Work Breakdown Structure.

The functional level of the WBS provides a breakdown of the project level into homogeneous functional areas and provides the basis for assignment of project elements to specific departments or individuals (depending on the size of your organization). Figure 1.21 demonstrates a typical Level 2 Work Breakdown structure.

At the task level of the WBS, you will describe the work to be accomplished within each functional subtask of the Level 2 WBS. Be cautious when doing this; you should be describing homogeneous work elements, not minute jobs to be done by specific individuals. That would make this level of the WBS very large and unwieldy; remember, the purpose of the WBS is to provide a basis for making decisions. Figure 1.22 is an example of a Level 3 WBS.

As indicated in each of the three typical WBSs, a numbering system has been applied at each level. At the highest level, the numbering system corresponds to each functional level of the project; at the second level, these numbers are allocated down to the homogeneous work elements of the individual functional elements; and in the lowest level, the numbers represent individual task elements. This same numbering system is then used in the project management Gantt chart for estimating costs and expenses. You will also see this numbering system used in process flowcharting later

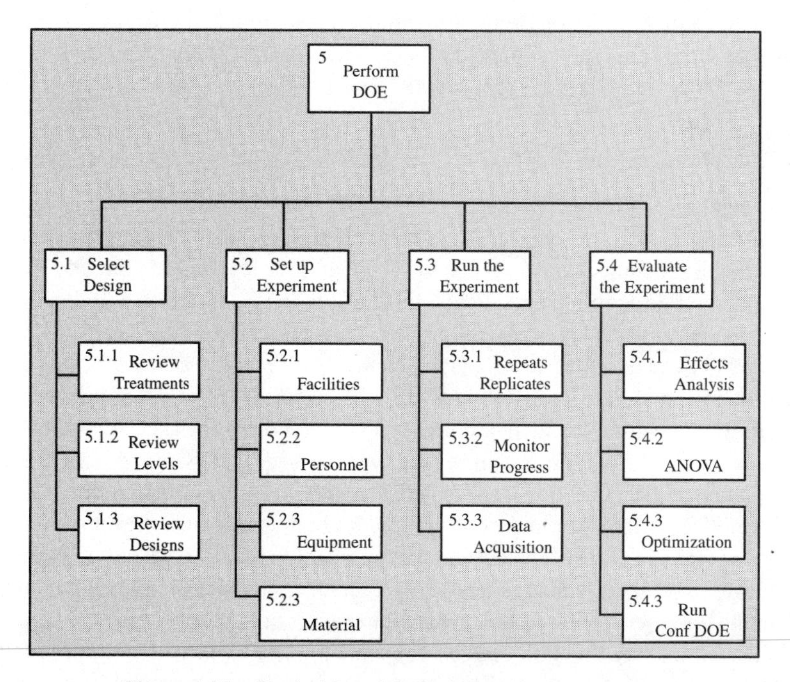

Figure 1.21 Typical Level 2 Work Breakdown Structure

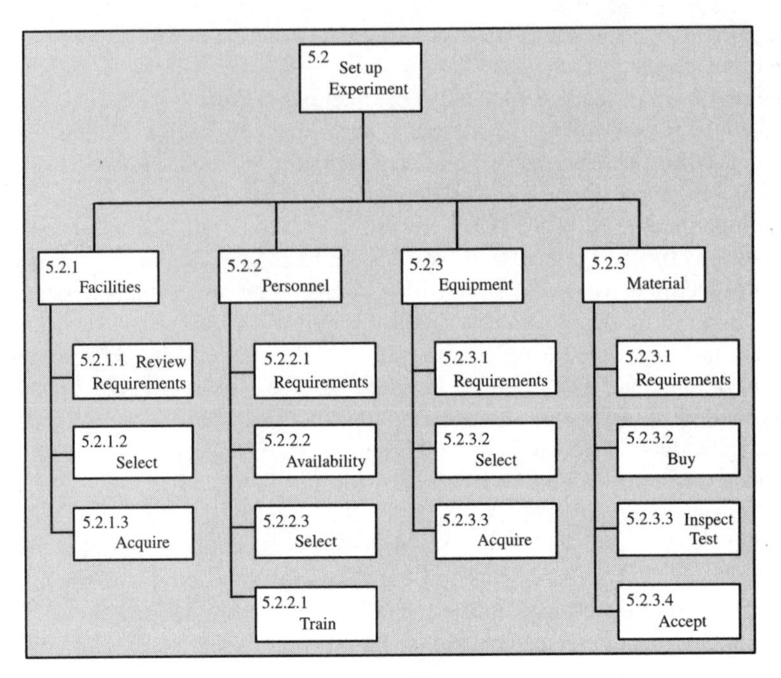

Figure 1.22 Typical Level 3 Work Breakdown Structure

in the book. This system provides an easy reference to coordinate the several elements in any management function (budgeting, milestone charts, and process analysis) and correlating these functions with the common measures of Quality, Cost and Schedule (Q$S).

Resource Management

The management of resources is necessary to plan, monitor, and execute any project. In performing a designed experiment, an excellent tool for managing resources and maintaining a schedule is the Gantt chart. The Gantt chart is derived directly from the WBS and can be accomplished at each level of the WBS. Here we will demonstrate the Gantt chart at level 1, for overall project management.

The WBS is transformed using a Gantt chart. As is demonstrated in Figure 1.23, the same numbering system used in the WBS is used to identify the tasks in the Gantt chart. To this structure, we then add the start and end dates for each subordinate task (milestones). This provides us with a form we can use to estimate the time required and sequence of each task to be performed, the start and end dates, and the milestones for accomplishing the designed experiment. Please note that in this example the time allotted for the execution of the designed experiment is three weeks out of a total time of eight weeks—this is typical of a properly executed plan. You should allot more time to planning and preparation of the designed experiment than to its execution.

In this Gantt chart, the dark time bars represent the summary of the times required to complete the tasks. The gray lines represent individual subordinate tasks in each

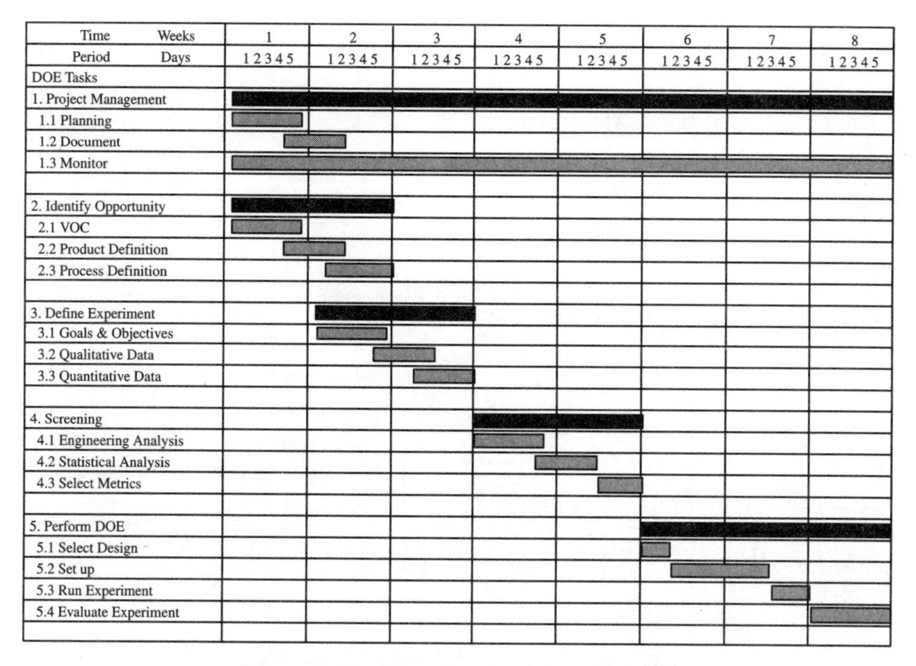

Figure 1.23 Gantt Chart Based on the WBS

functional area for which individual time estimates can be made for start date and end date. In this example, task 3, Define Experiment, will require eight weeks to complete.

Resource Estimates. The project cost estimate is very important, since it forms the basis for all of our financial analysis. It is necessary to generate this cost estimate to determine whether you have the finances available to complete the designed experiment. The cost estimate, coupled with the project schedule (Gantt chart), also allows you to monitor the project and to control costs and progress based on this initial estimate. There are many methods for estimating project costs. We will use a basic method based on the WBS and Gantt chart. Most project management software packages provide the capability to perform this evaluation quickly and easily.

First, use the WBS and its associated Gantt chart to obtain estimates of the scope of the project and the time and work hours it will require to complete the project. Use this knowledge to prepare a cost estimate based on:

- The cost performance of past projects
- Inputs from each functional area defined in the WBS and Gantt chart
- The appropriate allowance for risk and contingencies

These cost estimates can then be applied to the Gantt chart as indicated in Figure 1.24. The costs here have been estimated based on the weekly labor utilization with

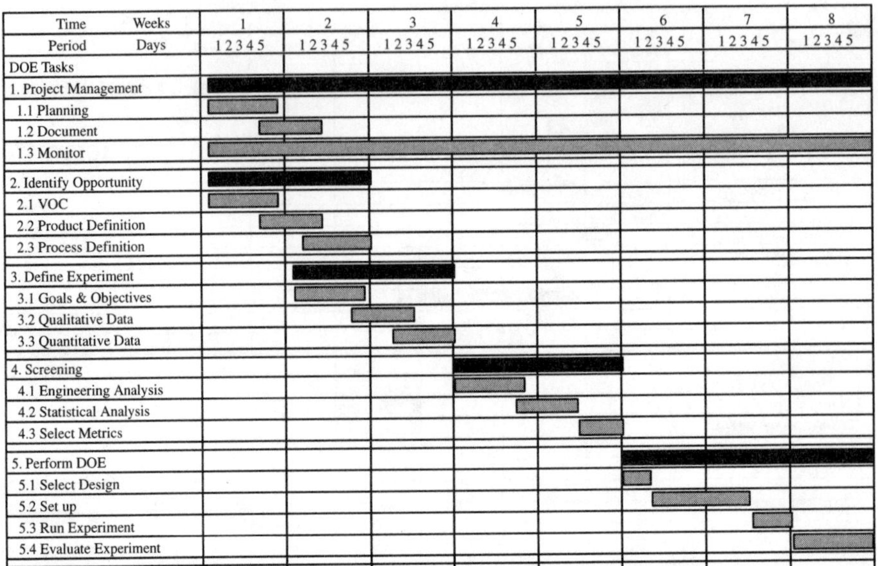

Time Period	Weeks 1	2	3	4	5	6	7	8
Days	1 2 3 4 5	1 2 3 4 5	1 2 3 4 5	1 2 3 4 5	1 2 3 4 5	1 2 3 4 5	1 2 3 4 5	1 2 3 4 5
DOE Tasks								
1. Project Management								
1.1 Planning								
1.2 Document								
1.3 Monitor								
2. Identify Opportunity								
2.1 VOC								
2.2 Product Definition								
2.3 Process Definition								
3. Define Experiment								
3.1 Goals & Objectives								
3.2 Qualitative Data								
3.3 Quantitative Data								
4. Screening								
4.1 Engineering Analysis								
4.2 Statistical Analysis								
4.3 Select Metrics								
5. Perform DOE								
5.1 Select Design								
5.2 Set up								
5.3 Run Experiment								
5.4 Evaluate Experiment								
Cost Estimate K$								
Labor	0.50	2.75	2.00	3.25	3.00	4.50	5.00	3.00
Facilities	0.00	0.00	0.00	0.00	0.00	0.75	0.75	0.75
Equipment	0.00	0.00	0.00	0.50	0.00	1.00	0.00	0.00
Material	0.00	0.00	0.00	0.25	0.00	2.50	0.00	0.00
Total	0.50	2.75	2.00	4.00	3.00	8.75	5.75	3.75
Cumulative	0.50	3.25	5.25	9.25	12.25	21.00	26.75	30.50

Figure 1.24 Development Project Cost Estimate

any appropriate multipliers (remember, labor costs you more that simple salary), the costs of facilities, equipment, and material. The costs for each week of the project are provided as well as the cumulative costs.

In this cost estimate, the total project costs are estimated to be $30,500 over an eight-week period. The project starts in Week One with a cost of $500 and ends on Week Sixteen with a cost of $3,750. The highest costs are incurred during Week Six. This cost peak is due to the acquisition of equipment and materials during that period. The proportion of costs over this period of time are graphically demonstrated in Figure 1.25.

WHAT HAVE YOU LEARNED?

- Identifying the opportunity for improvement using designed experiments is based on a broad range of available information. This starts with listening to the voice of the customer and knowing and understanding your products and processes.
  - Voice of the Customer: Information from the voice of the customer can come from many sources—marketing, the service department, warranty returns,

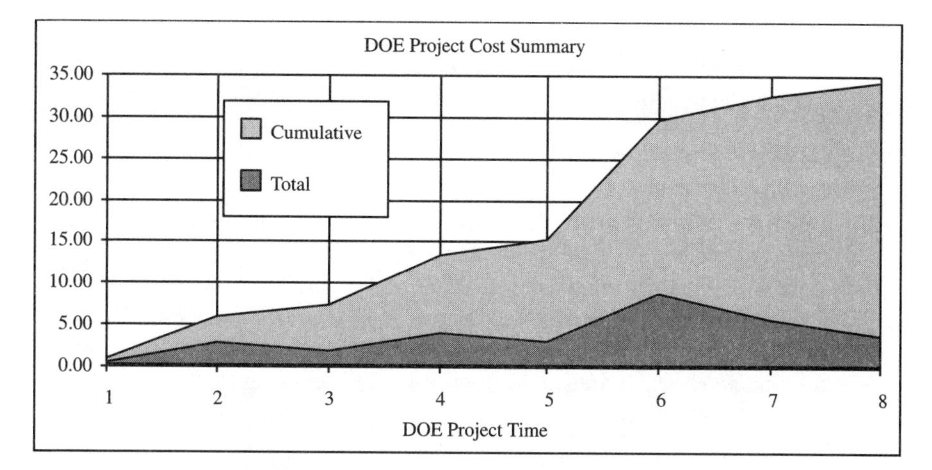

Figure 1.25 Landscape Graph of Project Costs

and customer surveys. All this information can be consolidated and translated into DOE metrics using the Voice of the Customer Table.

- Product definition: Drawings and specifications are the best source of detailed and technical information concerning a specific system. You should never assume that you know everything about a system; always review the specifications and drawings carefully. This is also a good source of metrics for your DOE.
- Process Definition: The process that produces your product or service has a very significant effect upon the end result. Knowing and understanding this process, its capabilities, and its limitations will provide a source of metrics for your DOE as well as an indication of what the controllable and uncontrollable environments are.

- The definition of a designed experiment is much more than simply picking one of the many performed arrays that are available. The basis for performing a designed experiment begins with the knowledge of what you are trying to accomplish at the highest level. Establishing this vision for your project will provide the goals and objectives you need to guide you through the rest of the process. Based on this vision, you now can use the available quantitative and qualitative data to refine and select the input and output elements of your experiment.
  - Goals and Objectives: Establishing the vision for your product or service will provide a path for improvement. Where do your realistically want this product to be? Looking at this vision on the horizon will give you another indication of what the appropriate metrics are. Your DOE(s) should always have this vision as the ultimate goal.
  - Qualitative Data: This deals with large amounts of language data such as ideas, opinions, perceptions, or desires. The usual statistical methods do not

work in the analysis of language data. There are three tools that deal best with translating qualitative data to fact-based information:

    - Affinity Diagram
    - Interrelationship Matrix
    - Relationship Matrix

  - Quantitative Data: Over the years, statistical methods have become prevalent throughout business, industry, and science. With the availability of advanced, automated systems that collect, tabulate, and analyze data, the practical application of these quantitative methods continues to grow. Statistics today play a major role in all phases of modern business and are indispensable in performing an effective designed experiment. The following tools are essential in the evaluation of your input and output variables for designed experiments.
    - Cause-and-Effect Analysis
    - Data Tables
    - Histograms
    - Pareto analysis

- Managing your experiment is much like managing any other project, and many of the same tools can and should be used. This is also an important part of your designed experiment because DOE projects fail more frequently as a result of lack of resources than they do for any technical concern. You need to ascertain the following:
  - Work Breakdown Structure
  - Resource Management
  - Resource Estimates

TERMINOLOGY

Process: A value-added action or work activity that has an input and an output. Typically the input to a process is raw materials and the output is a finished product. All processes have this same recursive nature: input>>>>>>>>work activity>>>>>>>>output.

Product: Something that is produced by a process. It is the result of the value-added work activity during the production. Products can be materials or services.

Design of Experiments: A fact-based decision making tool that is based on the variability of all products and services. It measures that variability and provides a structured method for planning, executing, and analyzing products and services.

Fact-Based Decision Making: The scientific method of induction and deduction as it is applied to business decisions. Fact-based decision making takes the subjectiveness and controversy out of making business decisions by quantifying the facts and presenting them in a rational way.

Industrial Processes: Processes that produce things. The inputs to industrial processes are raw materials, which can be in the form of basic materials such as iron ore, steel, and coal; subassemblies, such as computer boards and engine parts; or equipment for rework or repair, such as engines or automobiles that require overhaul or aircraft requiring upgrade and modification.

Administrative Processes: Produce the paper, data and information that other processes use. They also produce many of the services we use on a daily basis. Administrative processes are the processes that most frequently frustrate customers and employees.

Management Processes: The structured means by which businesses and individuals make key decisions. It is very important that we clearly understand that management is the process of using data to make decisions.

Process Analysis: Provides a structured method of identifying and describing the elements of any type of process. The detailed understanding provided by a formal process analysis is a required precursor for the use of process data as technical resources for a designed experiment.

Vision: Expresses the hope that what is an idea today will become a reality. Each vision is a unique idea of a possible future accomplishment. It is what you and your organization will be when the vision is reached.

Goal: An element that is needed to develop the company's strategic vision. Goals and objectives are cascaded to each organization to develop organizational specific short-, intermediate-, and long-term objectives.

Objectives: The "How" of accomplishing goals. Each objective is a specific action to be taken, such as "performing a process analysis" or "optimizing process yield."

EXERCISES

The following exercises can be performed using your organization as the model or they can be accomplished constructively as part of a class group.

1. Using your own organization as a model, establish the vision, goals, and objectives as outlined in this chapter.

2. Select one of your customers, services, or products and complete a voice of the customer table.

3. Conduct a process analysis through flowcharting of your company process or a process selected during class.

4. Select a team from your organization or the class. Based on the process analysis you just completed, select a single process element:

- Conduct a brainstorming session concerning what affects this process element.
- Use an affinity diagram to distill the results of the brainstorming session.

- Perform an Interrelationship Matrix Analysis.
- Perform a Relationship Matrix Analysis.

5. Decide on an improvement project based on the results achieved from Exercises 3 and 4 and select an opportunity for improvement. Using that improvement goal, plan a complete designed experiment.

- Make a work breakdown structure.
- Produce a Gantt chart and resource estimate for the improvement project.

2

BASIC CONCEPTS IN STATISTICS

Most decision making rests on comparisons. These comparisons form the foundation of fact-based decision making. The properly drawn comparison, using the tools that are presented here, will quantify your decisions. These decisions will apply directly to your existing data and also provide the basis for selecting characteristics and treatments for your designed experiments. These comparisons are made using statistical tools and visual tools such as charts and graphs. The purpose of this chapter is twofold:

- To provide tools that will assist you in evaluating existing data, to make decisions using that data, and to prepare for conducting designed experiments.
- To provide an understanding of basic statistical principles, especially variance, that will make designed experiments much more comprehensible.

This chapter also helps to identify and eliminate bad data, reduce the effect of nuisance factors (e.g., weather conditions) when taking data, determine which factors are having a significant influence on your process or product (e.g., machine tolerance variations or differences in personnel capabilities), and compare the importance of these factors. While the emphasis is on providing a single "complete set" of techniques that will be useful in most situations, a few other graphical techniques are presented which will ease the transfer of information from other organizations. This chapter will guide you through the following methods of making intelligent comparisons:

- Data Acquisition
    - Types of Data
    - Natural Variation in Data
    - Tips for Planning Tests

- Descriptive Statistics
    - Mean and Variance
    - Median and Range
    - Mode
    - Standard Deviation
- Graphical Representations
    - Histograms
    - Pareto Analysis
    - Scatter Plots
    - Dot Plots
    - Box-and-Whisker Plots
    - High-Low Plots
- Data Distributions
    - Discrete
    - Continuous

In a perfect world, saws would cut pieces exactly the same length every time, customer responses would be identical and would never vary, and statistical tests would be unnecessary. Comparisons would be easy. Perfection aside, one can imagine a world in which variation would be small and would form identical, symmetric patterns. Comparisons would be only a little harder, and statistical test results would always be unambiguous. In the real world, however, data are messy, missing, and skewed. Statistics become the tool of the pragmatist. The need to identify (a) a "best estimate" or "central tendency" and (b) a measure of how the data are scattered becomes clear. Only then can meaningful comparisons be made. Figure 2.1 illustrates the problem graphically.

DATA ACQUISITION

Types of Data

Definitions for several different types of data will be given here to assist you as you read the literature and to ensure that you use the proper terminology when reporting your own results. When expressed as numbers for computational purposes, all data necessarily consist of discrete (individual) points. However, the term *discrete* is also used to refer to the *variable* that is being measured. It means a variable that can only take on certain values. Quite often, discrete data are the result of arbitrarily or subjectively assigning a *rank* (number in an increasing sequence) to properties of items, such as hat sizes or intensity of responses in surveys. In the latter procedure, vaguely defined intervals with names like fair, moderately, and improved are assigned numbers. Such data are called *ordinal* data. Two-state (*dichotomous*) data such as yes/no, on/off, good/bad could have the number 1 assigned to one of the states and the num-

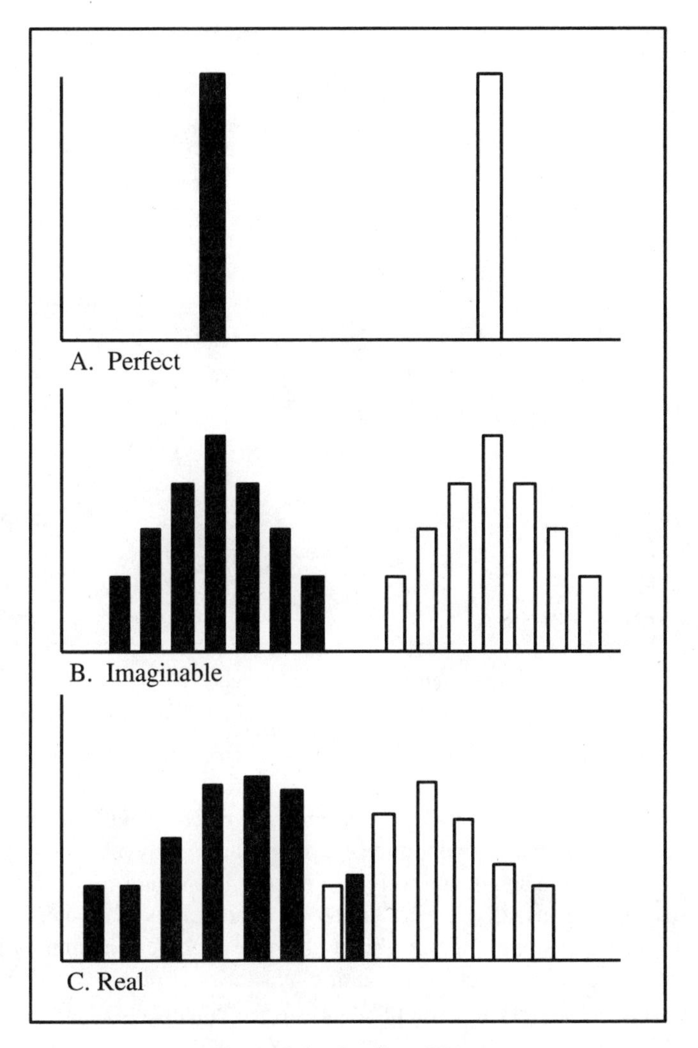

Figure 2.1 Quality of Data

ber 0 assigned to the other, with the analyst deciding whether 1 means good or bad based on some external criterion. If the data categories have no intrinsic rank, such as nationality (Latvian, Bolivian, Tibetan), the data are called *nominal*. Figure 2.2 provides a summary of these types of data.

Multistate data simply have more numbers involved. For example, qualitative responses from a customer, such as terrible, very bad, bad, okay, good, very good, excellent, could be made quantitative by ranking them from 1 to 7 for use in certain statistical tests. It is important to recognize that the 6 being used to represent very good is not a quantity but only a "place marker" that indicates it lies somewhere between good (5) and excellent (7) *on a nonuniform scale*. Notice also that the multistate "good" conveys more information than the dichotomous "good," because it is

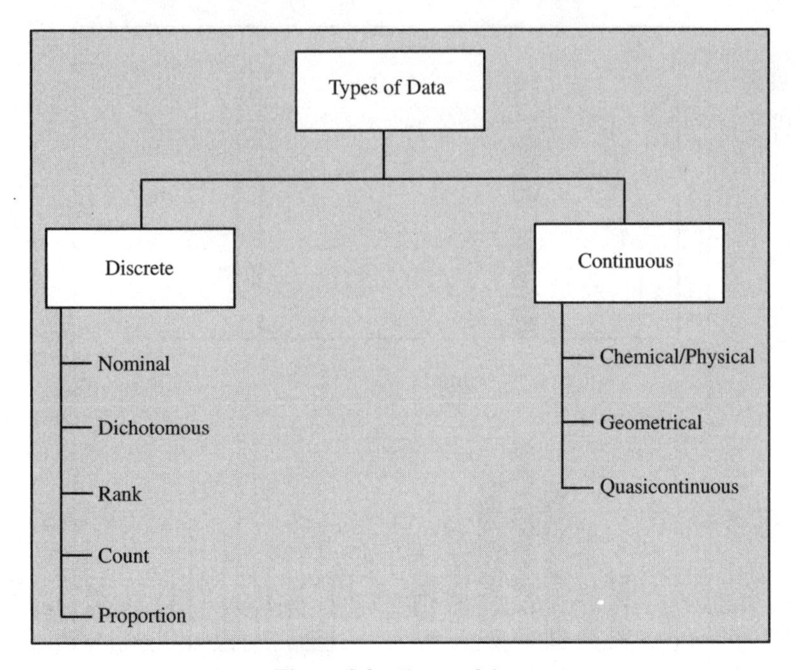

Figure 2.2 Types of data

based on a finer scale of resolution. The latter was achieved by using more categories for the same properties.

In contrast to a discrete variable, a *continuous variable* is one that can take on all possible values. Practically, a continuous variable can usually take on only a limited range. For instance, the height and weight of a human being are continuous variables. Continuous variables usually refer to the physical properties of an object. Everyday examples are dimensions and weight, with more technical examples involving mechanical, electrical, optical, or thermal characteristics of some raw material or item produced. When very large numbers of measurements are analyzed, it is often convenient to approximate a discrete variable by a continuous one. This will be discussed later.

Understanding Variation

In order to understand the quality of your data, so you know how much to trust them, you must understand how to deal with variations in your data. To do this, you must understand how the sources of this variation influence your data, so you can recognize them. To be specific, some of these sources are differences in human capabilities, impurities in raw materials, electrical noise on signals from instruments, measurement and recording errors, improper control settings (mechanical or electrical). Three types of error will be considered (Figure 2.3).

Obviously, if you make a single, one-of-a-kind item, statistics won't help you find a mistake. However, if you make a number of such items, comparing the items

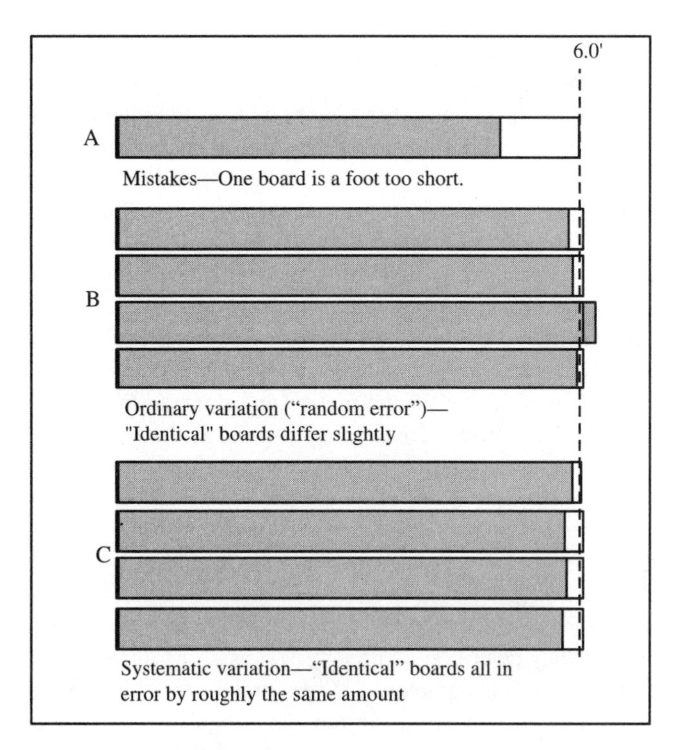

Figure 2.3 Types of Variation

will allow you to find a single one that is wrong. If the error is only slightly different from the ordinary variation, statistical comparison techniques may be your only recourse.

No measuring instrument is ever perfect. If it always gets *exactly* the same answer for a given measurement, maybe it isn't sensitive enough! Consider the meter in Figure 2.4. If the zero adjustment is not set perfectly, then every reading will be "off" by a certain amount. This is a systematic error. The magnitude of the difference between

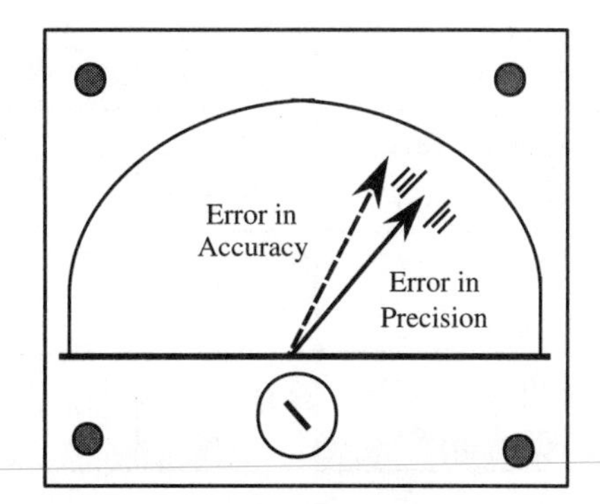

Figure 2.4 Typical Meter

the true value and the measured value is the *accuracy* of the meter, which can be improved by calibration. If a number of readings of the same true value are made, then friction in the bearings and other factors will lead to slight differences in the readings. The needle or pointer indicating the value may also waver around a central point when a single reading is being taken. The maximum extent of such differences from an average value provides a measure of the *precision* of the instrument. Calibration won't improve precision; it is built into the design of the meter. Similar types of imperfect readings occur with digital meters; people tend to believe every digit on a meter unless the numbers keep changing.

The quantity measured by the meter just discussed (voltage, flow rate, pressure, etc.) is called a *random variable* because its value can vary, and a particular allowable value can not be predicted. If the meter can only read up to 10 volts, then 23 volts is not an allowable value. The important point is that exact prediction of the next reading should not be possible, even if one reading is much more likely than another.

Considerations in Taking Data

Suppose your company makes sockets for outdoor lighting, and you need to determine how quickly they deteriorate with age. You ask for samples and are given: (a) 15-year-old sockets taken from houses along a cool, northern beach; (b) 10-year-old sockets taken from a mild (medium warmth and humidity), inland valley; and (c) 5-year-old sockets taken from a hot, dry southern desert. You can assume that the sockets are identical when they are made, so you have samples from *two* ages at each location. Now you can compare the effects of these locations, and others like them, on the sockets. If you could not assume that the sockets were identical at the start, then there would be no way to estimate deterioration with age, using just the data you would have. Notice that you do not have samples of the same age at two different locations, which would also be a help. Table 2.1 summarizes your data.

You must establish a metric, or scale of measurement, for the quality (or, conversely, the amount of deterioration) of the sockets so you can estimate an amount of

	Age (years)			
	0	5	10	15
Northern beach (cool, damp)	√			√
Inland valley (mild)	√		√	
Southern desert (hot, dry)	√	√		

Table 2.1 Data Available

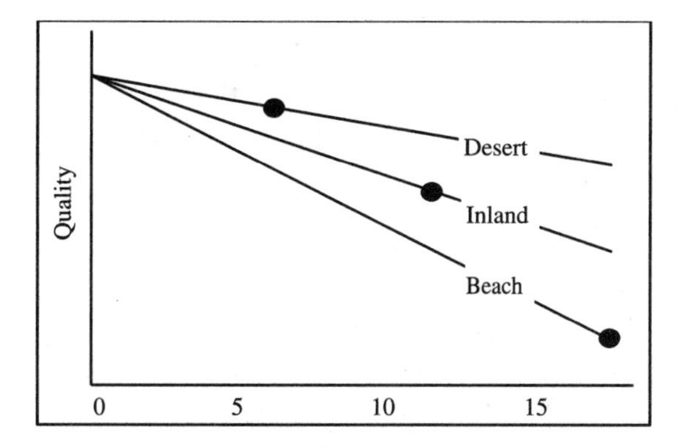

Figure 2.5 Deterioration with Age and Location

change over a given amount of time. With a metric, you can plot a graph like that in Figure 2.5 and thereby distinguish between the effects of age and location.

Notice that the effects of dampness and salt are mixed together, or *confounded*, with those of temperature. There may also be *interactions*, since the same concentration of salty water in air could cause significantly greater corrosion at high temperature than at low. In gathering data, always ask yourself, "Are there any environmental or other extraneous factors that could influence the results that I want to compare?" While it may be obvious to you that the damp, salty air is more important than the temperature in this problem, such experience may be lacking in other cases. You have enough data to compare the effects of the climate at these locations, and that may satisfy your needs completely, but you can only generalize to other locations exactly like these. Neither the temperature nor the humidity in the inland valley matches that of one of the other two locations, so you can't use this data to break the confounding of temperature and humidity and generalize to other locations. If you could also take a sample of 15-year-old sockets from a hot, southern beach or a dry northern region (preferably from both places), then you could do a factorial analysis (Chapters 4 and 5) that would allow you to get the answers you want. You must still watch carefully for other variables with effects that might not be negligible, such as air pollution. This is especially true if the other variables do not occur with equal intensity at all locations.

If you take data in sequence, be aware that the result of your next measurement may depend on the one you just took. That is, the results of the test may not be *independent.* For example, the act of testing a specimen may heat up the test fixture, so that continued rapid testing may cause a gradual bias in the results. Even if the items were actually identical, the third sample tested would respond more like the second than like the seventeenth. Similarly, a gradual upward drift in a setting of a dial will produce data that are not independent and therefore not random. This should be familiar to those who use control charts.

One common instance at which a truly random variable can become nonrandom is during warm-up of machinery or equipment. For example, signal amplitude from an amplifier could increase as the power supply warms up (Figure 2. 6). In short, warm-

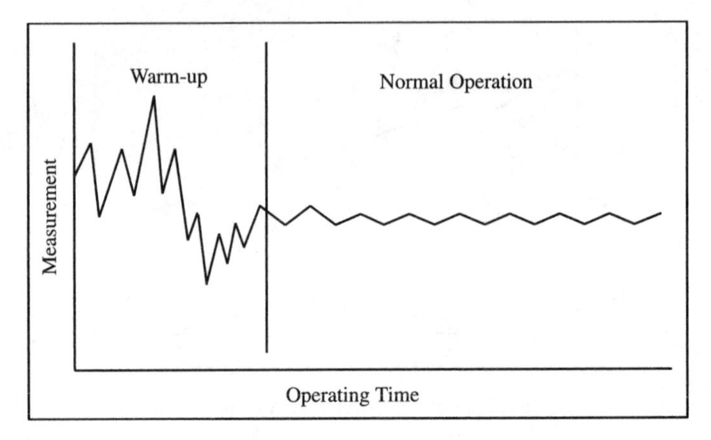

Figure 2.6 Start-up Effects

up is not a good time to take data that describes typical performance. Not allowing time for warm-up may be the source of a large part of your errors. Workers also need time to develop a rhythm and become focused, just as an athlete needs to warm up.

Lack of independence can arise in other data. People who take surveys know very well that they must not let a group of people hear each other's answers. The person answering will be guarded and the others will be influenced. This explains the rules governing discussions by jury members, who must not let others influence them before they form an opinion.

Sampling with Replacement

If a product is not consumed, it can be used for more than one test. For example, if hammers of different designs were to be evaluated by a group of carpenters, each carpenter could try all the hammer samples. Adjacent segments of the same piece of rubber, metal, or plastic can be used in tests of material properties in a research experiment. In several industries, the customers are effectively test items (as in medical studies). If the treatment has no long-term effect on the customer (e.g., haircut, suit pressed), then the same subjects can be used in repeated measurements after a suitable time delay. Thus, a service company can have options, and their associated advantages and disadvantages, that differ greatly from those of a company that produces a product.

There is a danger in using the same customer to rate items, such as Ford, GMC, and Chrysler rental cars, for example. The problem here is memory—decisions are not only subjective but conditional. Is the driver predisposed to prefer one car over another, based on previous experience with the brand or comments by friends? Will a Ford get the same rating if it is the first car tested as it would following the test of a Chrysler or a GMC car? Bias in the "test instruments" is a major problem when the same people are asked to make subjective evaluations on competing products or services; therefore, this practice must be avoided if at all possible.

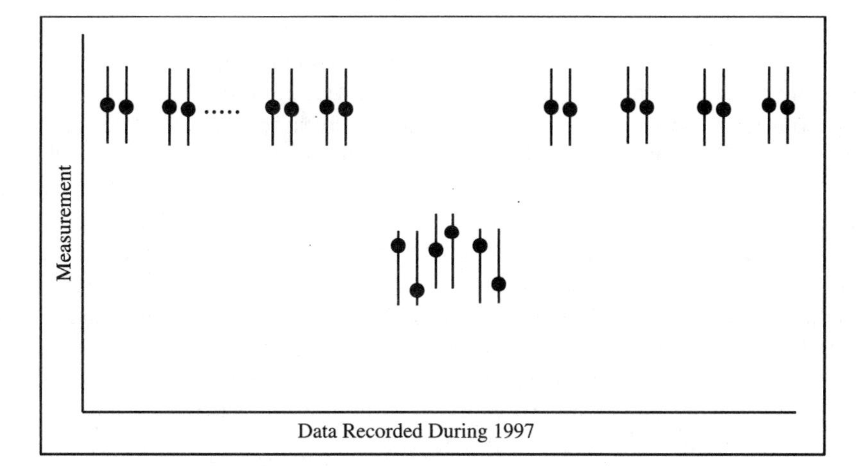

Figure 2.7 Mixing Special and Common Cause Variation Data

Keeping Records

Everyone knows it is important to keep copies of raw data as well as the results of analysis, but there is another point to keep in mind if you analyze your data statistically. Suppose that it is your job to investigate carburetor problems for a car company. Suddenly, bad carburetors start to show up, so you test as many this week as you normally test all year, and you select them from the line known to be having the problem. You fix the problem, but what happens to your company's database of periodic spot checks on carburetors if you dump in this special data? Suddenly, the database says that half of your cars are bad this year. This is illustrated graphically in Figure 2.7. If the purpose of the database is to provide data on typical performance of your product, you should not include data on special tests. You should keep a separate database with the results of special investigations, so you have a record of the symptoms associated with various failure modes. "Typical performance" data should be based on random selection. The special tests were limited to a group of items that were specifically selected because they were known to be different from the rest of the group—*inhomogeneous*—with respect to carburetor quality. Similarly, if you are doing a trend analysis of typical insurance claims by homeowners, you should obviously omit the time period after an earthquake or other natural disaster. For other analyses, however, that data might be the most important.

Planning Experiments

Most of the statistical techniques for design of experiments described in this book have their roots in the work of Fisher [1]. Fisher stressed three concepts: replication, randomization, and local control. Replicating is not *repetition*, which is simply doing the same test over and over again.

True *replication* requires repeating the same test under the influence of a variety of external (e.g., environmental) conditions. To do a true replication of an electronic measurement, you would want to turn off all power; disconnect and reconnect the instruments, probes, signal sources, and power sources; set all dials to zero and then back; allow different warm-up times for each replication; and you'd want to do this for different room temperatures each time. Obviously, in practice, these extreme measures are seldom carried out. The danger is that in simply splitting up measurements at different signal levels into groups and doing them in an arbitrary order, you will miss some important source of experimental error. The experimenter must decide on the appropriate level of replication, but it is wise to keep the dangers in mind. The purpose of repetition should be to assess the optimum intrinsic precision of the test (instrumentation, procedures, etc.), while the purpose of replication is to allow the full range of external sources of variability, including those of the test itself, to exert their influences on the outcome of the test.

Randomization is very important both from a theoretical point of view (since the classical statistical techniques don't work properly without it) and from a practical point of view (since the conclusions may not be valid beyond your test conditions without it). Most of the classical comparison techniques assume very restrictive conditions that are often not valid in practice. In particular, the error in a given measurement must be independent of the error in the measurement immediately preceding it. Fortunately, the statistics of randomization can be well approximated by exactly the same functions used in the classical tests. You can think of this as (loosely) saying that randomization allows the use of classical tests to be extended. Generation of random numbers for sample selection will be described later.

Local control is jargon for the rules you use to set levels of factors in your experiment. For example, in a test of five dyes, you might insist that each of the five be tested on the same four pieces of cloth. This restriction of how other factors (here, the variability of the cloth) can affect your experiment is called *blocking*. In a sense, a *designed experiment* is simply the multiple, systematic use of blocking, combined with replication and randomization. The critical factor is careful planning toward a specific goal. The latter always depends on going from the needs of the customers (internal as well as external), to requirements (specifications) on the product, to minimizing variability while optimizing performance and yield.

As an example of the importance of a well-planned experiment, a group of technicians were sent out to inspect 1,000 steel assemblies. The lead technician told them, among many other things, to indicate whether there was heavy, moderate, light, or no rust. They reported that 591 assemblies had no rust, 222 had light rust, 14 had moderate rust, and 173 had heavy rust. When asked, some of the technicians said that no rust meant none at all, while others thought it meant only negligible rust. All agreed on one thing: They only wrote down moderate when they could not decide between light and heavy! Defining a scale of measurement, and explaining it clearly, is at least as important in analysis of categorical data as it is in analysis of quantitative data.

If two sets of measurements both depend on a common factor, they may appear to be directly related when they are not. A famous example is that of plotting the number of insane people versus the number of computer users. This yields a smooth curve on a graph because both increase with population size. It may be very difficult to disprove such an apparent causal relationship, especially in this example.

Random Sampling

For our purposes, it is sufficient to select your samples for a given set of conditions by using the random number generator on a computer or pocket calculator, a table of random numbers, or some mechanical means like dice or very well shuffled playing cards. Note that random numbers will be used here only for ordering data, not for generating the data itself. For example, to generate the order of testing a series of n runs (sets of identical conditions, to be described later), use the "rand" or "random" function on your computer or calculator to generate a random number, multiply it by $n + 1$, and take the integer part of it (to the left of the decimal). Do this until you have n different numbers; that is, keep a number only the first time it appears, throwing away any repeats. See Figure 2.8. Details on random number generation with a computer program can be found in Cheney and Kinkaid [2].

To use a table of random numbers, make sure that it is a table of *uniformly* random numbers, pick a column, use the rightmost digit(s) from each number, and use them in sequence. If you need numbers from 1 to 80, take two digits and ignore any numbers greater than 80. If you run out of numbers in that column, go on to another. The table method of generating random numbers is demonstrated in Table 2.2.

These two techniques generate numbers that are equally likely to appear, because they use uniformly random numbers. The same effect can be obtained by using playing cards, but the rules for large sample sizes can be complicated and the deck must be very well shuffled. As an example of the possibilities with playing cards, suppose you needed to make ten runs with samples from three different groups. You could remove the face cards and the spades, use each of the remaining suits to represent a group, shuffle the deck well, and draw cards to determine the order of testing.

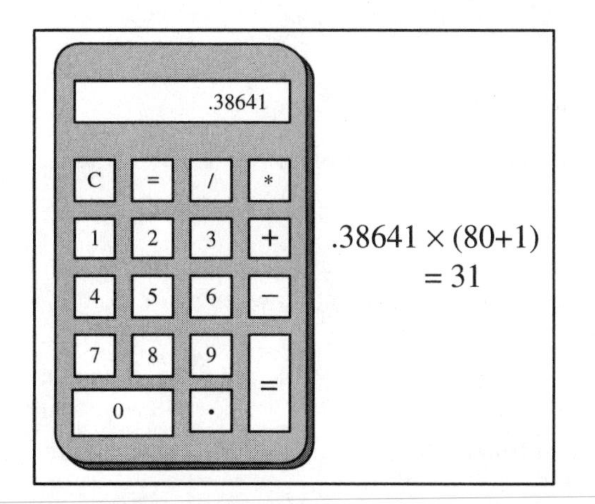

Figure 2.8 Generating Random Numbers Using a Calculator

054**63**	63
102**81**	Too Big
988**72**	72
648**09**	09
151**29**	29
831**72**	Repeat

Table 2.2 Generating Random Numbers Using a Table

Notice that the order of testing is randomized. Presumably, you can't tell ahead of time how a given test specimen will perform, but you might suspect that two specimens that were next to each other as they came out of the box or the oven (or as they walked in the door!) might behave similarly. Randomization separates them as far as test conditions are concerned—both the conditions you impose and those imposed by the environment.

Data Validation

Wherever possible, rough checks should be done while the data are being taken. Do a "sanity check" on your data—can the internal diameter of that tube really be 5 meters and not 5 millimeters? Is it reasonable that the 87th data point has a value of 416 when the first 86 all fell between 4.00 and 5.00? Often, errors can be caught and tests repeated or readings corrected on the spot. The analysis techniques given here are only as good as the data, so checking your data is extremely important. Statistical tests for the quality of your data will be given. A general rule: Always plot the data if you possibly can. If you have access to a computer spreadsheet with graphics capability, you should at least plot the sequence of points in the order they were taken. If feasible, take a laptop computer out into the field and enter the data as you take them.

Check List for Your Test

Since you know best how to obtain the data you need, only the following check list is included:

- Plan your experiment carefully so that it includes only the factors you really want, excludes environmental factors whenever possible, and tests a homogeneous population.
- Select the instrumentation, survey, or sampling technique so that it meets the needs of the test.
- Calibrate the instrumentation properly, select a good control group, or find some appropriate reference level.

- Save your raw data, including relevant comments and observations, even after the analysis is complete. Keep it for at least one year after the report is published, longer if appropriate.
- Do validation tests on your data.

Statistical Software Packages

If you are doing much with data analysis, you are already using a computer or a programmable pocket calculator. This chapter shows you what calculations to do, or what calculations your software is doing for you. Even if computer software is doing everything for you, you still need to understand the process. If the computer indicates that the average weight of the members of your staff is 178.459326 pounds, do you really believe that the data was that accurate or precise, and do you really need to know any closer than "about 178"? This chapter will help you answer the first part of the question; the second part is up to you! "Design of Experiments" is your tool, not your replacement.

You are strongly encouraged to use statistical software, but you should work a few problems with the techniques given here and compare the results to the answers the program generates. If it calculates values that can be looked up in a table, check it against a standard table. Be especially careful with the *t* table, as noted later. Modern electronics allows you to make measurements, send a digital output directly into a computer, use a statistical software package, and paste the answers directly into a report—all with a single stroke of a key. Properly set up and maintained, such a system can be extremely valuable. It can also turn out nonsense at an incredible rate if you don't know exactly what each step is doing and are not monitoring it.

DESCRIPTIVE STATISTICS

Sample Data Set

In order to simplify the explanation of the techniques to be used here, a specific set of data will be used as an example. As you will probably want to do with your own data, the raw data has been *coded* (statistician's jargon for "scaled") for ease of analysis. For example, .518376 became 51.83. This eliminated apparently insignificant digits that were actually beyond the precision of the instrument. These data are shown in Table 2.3. The numbers were taken exactly one minute apart. They are recorded down the page, starting from the left, in the order taken. These same numbers are placed in ascending order and recorded the same way in Table 2.4. Although this is not a necessary step for most of the analyses, particularly if you are using statistical software, it will be a convenient reference in the explanations that follow.

If you order the data as in Table 2.4, you can also check the first and last points to see whether they are obviously different from their adjacent points. A point that is significantly different from the rest is called an *outlier* and is usually the result of exper-

51.83	46.87	48.12	49.02	51.87
47.61	54.86	54.89	47.70	48.28
54.28	52.14	49.67	55.01	47.58
57.35	64.43	46.67	52.41	62.67
55.11	46.45	47.55	51.11	54.16
55.42	52.53	39.91	55.10	41.40
49.13	55.03	41.93	46.27	55.83
48.74	57.62	48.13	61.60	44.84
51.57	55.60	47.93	45.83	44.47
42.34	44.57	52.04	45.27	51.89
51.48	61.97	54.13	49.89	46.17
43.31	48.38	49.32	52.03	51.05
51.68	49.44	47.51	53.14	44.85
56.00	48.14	53.09	53.69	47.13
50.42	52.47	45.92	48.00	52.81
41.35	68.54	46.62	54.01	46.99
49.98	45.64	47.42	40.65	50.98
54.70	53.16	56.54	53.44	43.24
60.68	50.61	49.96	58.18	50.79
55.99	42.15	45.73	54.41	50.14

Table 2.3 Coded Data

imental error. Quantitative tests to determine whether this difference is statistically significant are given later.

Before anything else was done, the data were checked by graphing them in the order taken; the result is shown in Figure 2.9 The data vary randomly around a constant value, with no ramps, sudden jumps or "steps" in the local average, gradual drifts high or low, or single enormous values.

39.91	46.45	49.02	51.89	54.89
40.65	46.62	49.13	52.03	55.01
41.35	46.67	49.32	52.04	55.03
41.40	46.87	49.44	52.14	55.10
41.93	46.99	49.67	52.41	55.11
42.15	47.13	49.89	52.47	55.42
42.34	47.42	49.96	52.53	55.60
43.24	47.51	49.98	52.81	55.83
43.31	47.55	50.14	53.09	55.99
44.47	47.58	50.42	53.14	56.00
44.57	47.61	50.61	53.16	56.54
44.84	47.70	50.79	53.44	57.35
44.85	47.93	50.98	53.69	57.62
45.27	48.00	51.05	54.01	58.18
45.64	48.12	51.11	54.13	60.68
45.73	48.13	51.48	54.16	61.60
45.83	48.14	51.57	54.28	61.97
45.92	48.28	51.68	54.41	62.67
46.17	48.38	51.83	54.70	64.43
46.27	48.74	51.87	54.86	68.54

Table 2.4 Coded Data, Ordered

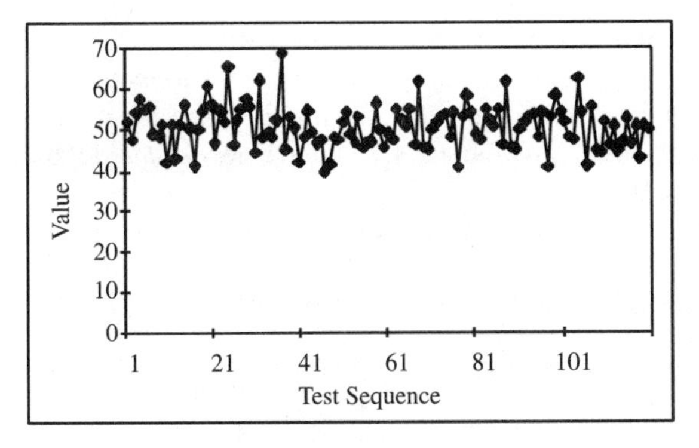

Figure 2.9 Test Sequence

You can get a good overview of your data with the dot diagram, as shown in Figure 2.10. It will be convenient to refer back to this figure in studying the techniques that follow. Notice that the numbers were rounded before plotting; this avoids having a separate dot for each value of "yield." It should also be mentioned that a bar of height equal to the highest dot could be used, making a histogram, but a standard histogram typically would have fewer bars and the heights would be divided by the total number of points, so that the vertical axis becomes frequency in place of count.

Median and Range

Suppose you have the data in Table 2.3 and want to calculate a number that describes its central point and a number that describes the amount of scatter around that point. Two common choices are: (1) the median $\tilde{y}$ for the central point and the range R for the amount of scatter, and (2) the mean $\bar{y}$ for the central point and the standard

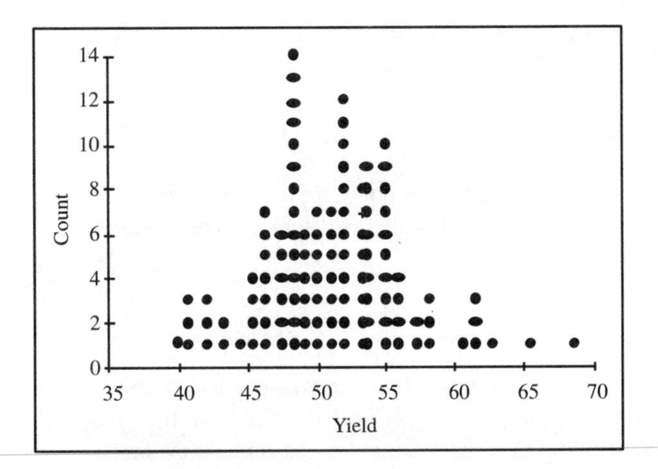

Figure 2.10 Dot Plot of Sample Values of y

statistic. Statistics that will be described here include the median, range, each rank in an ordered set of data, mean, standard deviation, and variance.

The *median* is the "middle" number. To find it, place the numbers in ascending order according to size, as in Table 2.4. You have now *ordered* the numbers; if you give each a number (The first is 1, the second is 2, . . .), you have also ranked them. If the *sample size* (how many numbers you have) is odd, the median is the middle number. If the sample size is even, the median is the average of the middle two numbers. The "elegant" way of writing this is;

$$\tilde{y} = \begin{cases} y_{(n+1)/2} & n \text{ odd} \\ \left[y_{(n/2)} + y_{(n/2)+1}\right]/2 & n \text{ even} \end{cases} \tag{2.1}$$

In the case of duplicate readings, or *ties*, give all the numbers in the tie the same rank, the average of the ranks. Thus, the ranks for 39.91, 40.65, 41.35, 41.40, 41.93, . . . ,68.54 from the data set are 1,2, 3, 4, 5, . . . , 100. Since n is even, we can now calculate y from Eq. (2.1) as follows:

- Since there are 100 values, the appropriate equation to use is Eq. (2.1) with n even.
- Using 100 values, the two center values are data element numbers 50 and 51.
- The next step is to find the average of these two values using Eq. (2.1) for n even, as indicated here.

$$\tilde{y} = \left[y_{(n/2)} + y_{(n/2)+1}\right]/2 = \left[y_{(100/2)} + y_{100/2+1}\right]/\ 2$$

$$= \left[y_{50} + y_{51}\right]/\ 2 = [50.42 + 50.61]/\ 2 \approx 50.52$$

The difference in the largest and smallest numbers is the *range*; for the data in Table 2.4, it is

$$R = y_{max} - y_{min} = y_{100} - y_1 = 68.54 - 39.91 = 28.63$$

Be sure to order the data before doing these calculations. The median and range are handy for making estimates with small sample sizes, where having to order the data is not a problem.

Mean and Standard Deviation

The mean and standard deviation are two more measures of central tendency and dispersion. They are the most commonly used measures for these purposes. The mean is a central tendency measure representing the arithmetic average of a set of observations. The standard deviation is a measure of dispersion representing the root mean

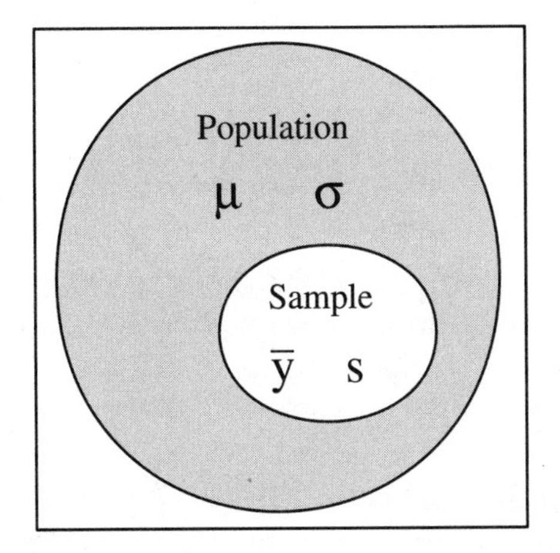

Figure 2.11 Population and Sample Relationship

square (RMS) distance between the mean and the individual observations, measured in the same units as the mean.

There are two means that you will use in performing your evaluations. One is the mean of the population, which is called mu (μ). The other is an estimate of the population mean based on your sample, which is called $\bar{y}$. There are also two measures of dispersion. One is a measure of the population dispersion, which is called sigma (σ). The other is an estimate of the population dispersion based on your sample, called s. Figure 2.11 demonstrates the relationship between populations and samples, and population parameters (Greek letters) and sample statistics (Roman letters).

Mean. To avoid confusion with the generalized use of the word "average" to signify typical or majority behavior, the more specific technical term "mean" will be used here. To calculate the *mean*, use Eq. (2.2). In this equation, sum the sample values y_i and divide by the sample size, n.

$$\bar{y} = \frac{\sum_{i=1}^{n} y_i}{n} \tag{2.2}$$

Using this equation, the mean can be calculated in two steps:

1. Sum all the data elements.
2. Divide the sum of the data elements by the number of data points.

These two steps are demonstrated in Table 2.5, using data from Table 2.3.

Standard Deviation. The standard deviation is a measure of dispersion used to indicate how close to the mean your observations actually are. This is an important mea-

51.83	46.87	48.12	49.02	51.87
47.61	54.86	54.89	47.70	48.28
54.28	52.14	49.67	55.01	47.58
57.35	64.43	46.67	52.41	62.67
55.11	46.45	47.55	51.11	54.16
55.42	52.53	39.91	55.10	41.40
49.13	55.03	41.93	46.27	55.83
48.74	57.62	48.13	61.60	44.84
51.57	55.60	47.93	45.83	44.47
42.34	44.57	52.04	45.27	51.89
51.48	61.97	54.13	49.89	46.17
43.31	48.38	49.32	52.03	51.05
51.68	49.44	47.51	53.14	44.85
56.00	48.14	53.09	53.69	47.13
50.42	52.47	45.92	48.00	52.81
41.35	68.54	46.62	54.01	46.99
49.98	45.64	47.42	40.65	50.98
54.70	53.16	56.54	53.44	43.24
60.68	50.61	49.96	58.18	50.79
55.99	42.15	45.73	54.41	50.14
$\bar{y} = \frac{\sum_{i=1}^{n} y_i}{n} = \frac{5066.55}{100} = 50.67$			$\sum y$	5066.55
			$\sum y_i / n$	50.67

Table 2.5 Calculating the Mean

sure when estimating product or process performance characteristics such as yield, cost, reliability, or time. Calculate the standard deviation using Eq. (2.3).

$$s = \sqrt{s^2} = \sqrt{\frac{SS}{n-1}} \tag{2.3}$$

Using this equation, the standard deviation can be calculated in three steps:

1. Calculate the Sum of the Squares (*SS*).
2. Calculate the variance (s^2).
3. Calculate the square root of the variance.

Step 1. Calculate the Sum of the Squares The sum of the squares *SS* is the sum of the squares of the distance from each point to the mean $(y_i - \bar{y})$ calculated more simply and accurately as the sum of the squares of the individual values of y (y_i) minus the square of the sum of the individual values divided by n. This is the form we use for the rest of the book, beginning with Eq. (2.4).

$$SS = \sum_{i-1}^{n} y_i^2 - n\bar{y}^2 = \sum_{i=1}^{n} y_i^2 - \frac{\left(\sum y_i\right)^2}{n} \tag{2.4}$$

Using this equation and applying it to the sample data set yields the sum of the squares displayed as a spreadsheet in Table 2.6.

y	y^2	y	y^2	y	y^2	y	y^2	y	y^2
51.83	2686.35	46.87	2196.80	48.12	2315.53	49.02	2402.96	51.87	2690.497
47.61	2266.71	54.86	3009.62	54.89	3012.91	47.70	2275.29	48.28	2330.958
54.28	2946.32	52.14	2718.58	49.67	2467.11	55.01	3026.10	47.58	2263.856
57.35	3289.02	64.43	4151.22	46.67	2178.09	52.41	2746.81	62.67	3927.529
55.11	3037.11	46.45	2157.60	47.55	2261.00	51.11	2612.23	54.16	2933.306
55.42	3071.38	52.53	2759.40	39.91	1592.81	55.10	3036.01	41.40	1713.96
49.13	2413.76	55.03	3028.30	41.93	1758.12	46.27	2140.91	55.83	3116.989
48.74	2375.59	57.62	3320.06	48.13	2316.50	61.60	3794.56	44.84	2010.626
51.57	2659.46	55.60	3091.36	47.93	2297.28	45.83	2100.39	44.47	1977.581
42.34	1792.68	44.57	1986.48	52.04	2708.16	45.27	2049.37	51.89	2692.572
51.48	2650.19	61.97	3840.28	54.13	2930.06	49.89	2489.01	46.17	2131.669
43.31	1875.76	48.38	2340.62	49.32	2432.46	52.03	2707.12	51.05	2606.102
51.68	2670.82	49.44	2444.31	47.51	2257.20	53.14	2823.86	44.85	2011.523
56.00	3136.00	48.14	2317.46	53.09	2818.55	53.69	2882.62	47.13	2221.237
50.42	2542.18	52.47	2753.10	45.92	2108.65	48.00	2304.00	52.81	2788.896
41.35	1709.82	68.54	4697.73	46.62	2173.42	54.01	2917.08	46.99	2208.06
49.98	2498.00	45.64	2083.01	47.42	2248.66	40.65	1652.42	50.98	2598.96
54.70	2992.09	53.16	2825.99	56.54	3196.77	53.44	2855.83	43.24	1869.698
60.68	3682.06	50.61	2561.37	49.96	2496.00	58.18	3384.91	50.79	2579.624
55.99	3134.88	42.15	1776.62	45.73	2091.23	54.41	2960.45	50.14	2514.02

$\sum y = 5066.55$ $\quad \sum y^2 = 259500.24$

$$SS = \sum y^2 - \frac{\left(\sum y\right)^2}{n} =$$

$\left(\sum y\right)^2 = 25669928.90$

$$259{,}500.24 - \frac{25{,}669{,}928.90}{100} = 2800.95$$

$\frac{\left(\sum y\right)^2}{n} = 256699.29$

2800.95

Table 2.6 Calculating the Sum of the Squares

$$SS = \sum y^2 - \frac{\left(\sum y\right)^2}{n} =$$

$$259{,}500.24 - \frac{25{,}669{,}928.90}{100} = 2800.95$$

Step 2. Calculate the Variance The variance is the average squared distance between the mean and each observation in your data set. Calculating the mean is an intermediate step in determining the standard deviation. The variance is calculated using Eq. (2.5) and the sum of the squares from step 1.

$$s^2 = \frac{SS}{df} = \frac{SS}{n-1} \tag{2.5}$$

The variance is the sum of the squares we have just calculated divided by the degrees of freedom. The degrees of freedom are the number of opportunities for our data to vary within the sample size minus the correction for the number of parameters that are estimates.

Applying the data we have calculated in Table 2.6, we can determine the variance as:

$$s^2 = \frac{SS}{n-1} = \frac{2800.95}{99} = 28.29$$

Step 3. Calculate the Standard Deviation The final step is to take the calculation of variance in the previous step and apply it to Eq. (2.4) as indicated here. This is the standard deviation of your sample data set.

$$s = \sqrt{s^2} = \sqrt{28.29} = 5.31$$

The advantage of using a computer should now be obvious! Note: When statisticians talk about the "sum of the squares," they usually mean *SS*, that is, the sum of the squares *of the differences from the data mean.*

Properties of Sums

If you add the same amount c to every number in a sample, then the mean and median change by that amount. Mathematically, the numerator in the equation becomes:

$$\sum_{i=1}^{n}(y_i + c) = \sum_{i-1}^{n} y_i + nc$$

For example: Subtracting .1 (adding $-.1$) from each number shifts the mean to 40.68 and the median to 40.52. The *pattern* (relative spacing of points) hasn't changed; only the scale has changed—the origin, or zero point, has shifted. The same is true for multiplication. Again, multiplying each y_i by .1 doesn't change the pattern, but it does change the scale. The mean becomes 5.068, etc. If a set of voltage readings is changed from volts to microvolts, their mean and median change to microvolts. If all the readings lie between 1.0012 and 1.0015 volts, subtracting 1.0000 from each number and then multiplying by 10,000 before doing the calculations (and converting the final result at the end) can be very convenient.

Adding the same amount to every number in a sample does not change the range, variance, and standard deviation. They are measuring the spread about a point that merely has been shifted. If you multiply every number by the same amount, then the range and standard deviation are multiplied by that amount, and the variance is multiplied by the square of that amount (since it is the square of the standard deviation).

Mode

A *mode* is a peak in a frequency distribution. The major mode is the highest peak; any others are minor modes. If more than one mode appears, it is almost always a good idea to find out why. It is usually an indication that your data are not homogeneous. Like the mean and median, the mode can be used as a measure of central tendency. Note, however, that the way you choose to "bin" your data in making the frequency distribution can shift the mode.

REPORTING RESULTS

Numerical Precision

The standard (very good) advice is to carry out all calculations with all the decimal places that you have and only round off the final result. This is usually followed by the rule that if your number ends in 5, round it to the nearest even digit, so that 3.235 goes up to 3.24 (not 3.23, which ends in an odd digit), while 3.265 goes down to 3.26. This second rule is usually important only if a large number of rounded numbers will be used later by someone else; the rule helps cancel roundoff errors.

The standard advice to use your experience in the field when deciding how many decimal places to throw away is often only annoying. Based only on the numbers themselves, you could use the first significant figure of $s/5$ as the last figure for $\bar{y}$. Thus, $\bar{y} = 1.4213$ is reported as 1.4 if $s = .71$, as 1.42 if $s = .071$, etc. This rule takes into account only the variation of the data within the sample, and therefore it applies strictly only to the sample itself. Unless you are working with a very carefully designed and calibrated test setup and have a very good reason to believe the answers you got, this rule may be a good one to follow anyway.

Summarizing Data Graphically

High-Low Plot. The high-low plot depicted in Figure 2.12 shows a very simple way to display data. Only the minimum, maximum, and mean or median are shown. The means or medians are usually connected by a line, as shown. Obviously, a lot of information is wasted, but it is useful when you want to eliminate excessive detail. You may prefer to replace the maximum and minimum with other limits, such as the mean and the standard deviation. In any case, you should indicate what the symbols mean. Notice that the vertical axis displays statistics of the data samples, while the horizontal axis displays only labels (lot numbers, etc.). This is the usual form of the graph, but others are certainly possible. The columns of data from Table 2.3 were arbitrarily used as the source of the data. Modern spreadsheets will make such a graph with almost no effort on the part of the user. Notice that the vertical axis dis-

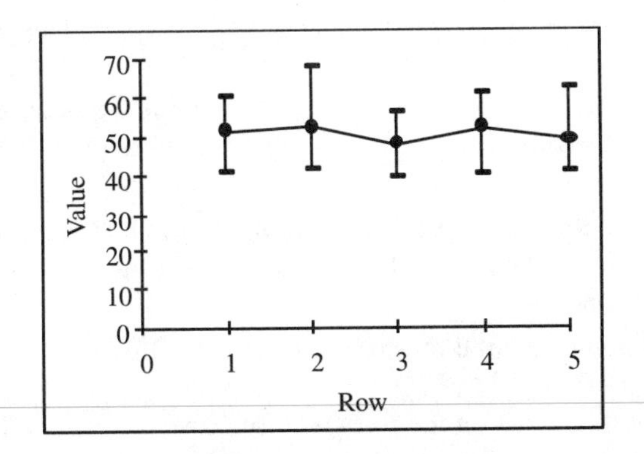

Figure 2.12 High-Low Plot

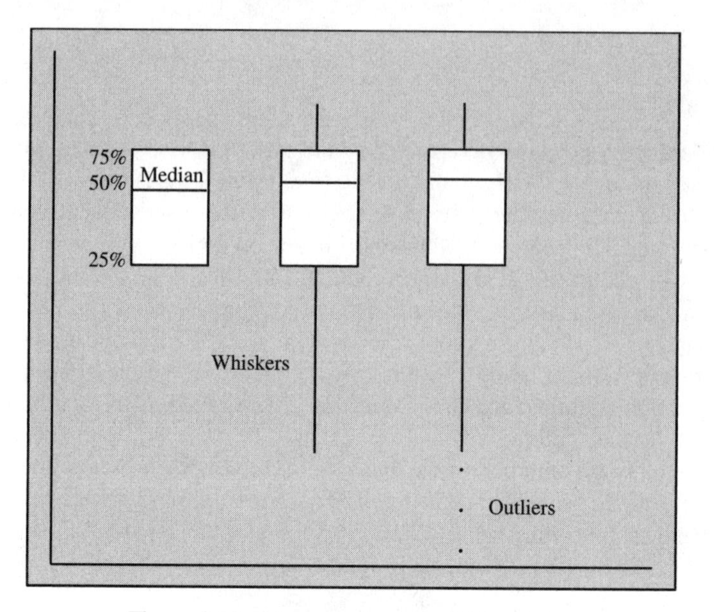

Figure 2.13 Making a Box-and-Whisker Plot

plays the values of the data, giving only a very crude idea of its distribution, and the horizontal axis displays labels.

Box Plot. Everybody seems to be in agreement on what a box plot, or box-and-whisker diagram, should look like, but there isn't a standard way of assigning the numbers to it! You should therefore be very cautious about what the numbers mean on a given box plot. The columns of data from Table 2.3 are again used for the graph. To make a box-and-whisker plot (see Figure 2.13), do the following:

Step 1. Find the median. This is the cross line on the box.

Step 2. Find the median of the data points above the median. This is the upper edge of the box.

Step 3. Find the median of the data points below the median. This is the lower edge of the box.

Step 4. From the top of the box, temporarily draw a line 1.5 times the length of the box, as shown. Starting from the top of the line, shorten the line until it hits the first actual data point.

Step 5. From the bottom of the box, temporarily draw a line 1.5 times the length of the box, as shown. Starting from the bottom of the line, shorten the line until it hits the first actual data point.

Step 6. Plot any points that fall beyond the ends of the lines.

The names for the points on the lines and for those beyond the lines are not the same in all texts and in all software, so check their definitions! The term interquartile

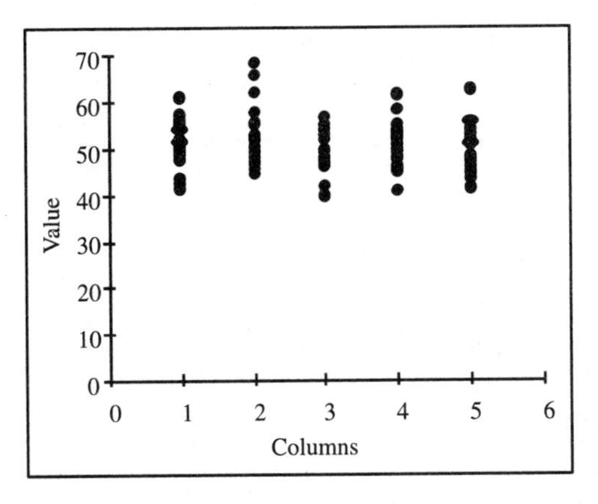

Figure 2.14 Dot Plot

range, or IQR, is sometimes used. It is simply the length of the box. The box plot gives somewhat more detail about how the data are distributed along the vertical axis than the high-low plot does, but the horizontal axis again consists of labels. If only one box is to be shown, it is not unusual to see the roles of the axes switched so the box is "on its side."

Dot Plot. A third option is the dot diagram, or dot plot, in which all of the data are exhibited. This is shown in Figure 2.14. The columns of data from Table 2.3 are again used. The dot plot is the only one of the three that allows stratification (splitting into groups) to be detected visually. Outliers can be displayed with a different symbol. If a very large amount of data are to be displayed, every n-th (e.g., every tenth) point could be shown. Again, the horizontal axis shows labels. The dot plot of Figure 2.14 showed a single column of dots and was turned on its side. There were so many dots representing identical values that they were stacked up.

Just by looking at the preceding example graphs, it becomes obvious that visual comparisons are possible, and interpretation techniques will be described later. These types of graphs are usually best for giving technical presentations.

Histograms. You can easily convert the dot diagram in Figure 2.14 to a histogram for administrative and other discussions where such detail is not necessary. To make a histogram, you can do the following:

Step 1. Decide how many bars you need. Table 2.7 from Jackson and Frigon [3] can be used as a guide.

Step 2. Determine the class interval. Find the maximum and minimum values of your data set and calculate the range = maximum value − minimum value. You should definitely check for outliers before making this calculation. The width of the class interval is simply the range divided by the number of bars:

$$\text{width of interval} = \text{range} / \# \text{ bars}$$

Data Points	Bars
< 50	5 – 7
50 – 100	6 – 10
100 – 250	7 – 12
> 250	6 – 20

Table 2.7 Suggested Number of Bars

Step 3. Label each bar by its *class mark*, which is the midpoint of its interval. A class mark can be found by averaging the *class boundaries* of its interval. For example, if the range of a data set is from 0 to 100 and you use 10 bars, then the width of each interval is 100/10 = 10, the class boundaries are 0, 10, 20, . . ., 100, and the class marks are 5, 10, . . ., 95.

Step 4. Count up the number of data points in each interval. These are the *counts*, which indicate how frequently a point fell in each interval. You now have a *bar graph.*

Step 5. To make a true histogram, you must divide the count for each bar by the total number of points to get a (relative) *frequency*. The sum of the frequencies is now equal to 1 (100%). To a statistician, frequency means "frequentness" or how often something happens, not the rate of variation. Figure 2.15 demonstrates a completed histogram.

An important type of histogram for comparing categorical (nominal) data may be formed by placing the categories in decreasing order. As mentioned in Chapter 1, this Pareto chart then identifies the key sources of error, profit, or whatever variable is being measured. The cumulative sum of the frequencies (heights of the bars) are often plotted using an axis to the right, as in Figure 2.16. Some authors will show "before and after" data by using the bars for one set of data and the lines for the other.

You will probably find yourself calculating with frequencies, so the following may be helpful. If you have the sum 3 + 3 + 4 + 4 + 5 + 5 + 5 + 5 + 5 + . . ., it

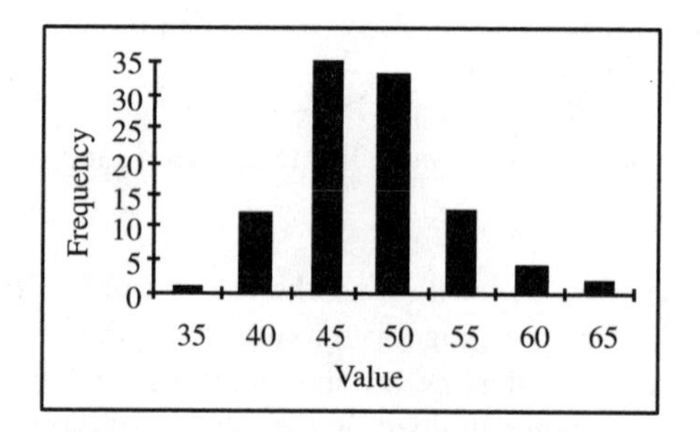

Figure 2.15 Completed Histogram

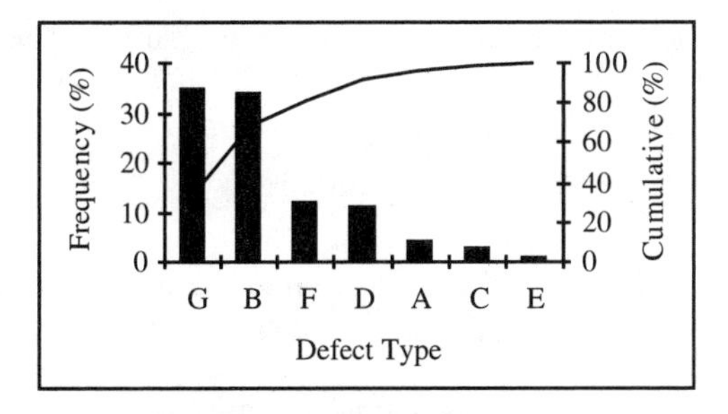

Figure 2.16 Pareto Chart

is simpler to write it as 2 × 3 + 2 × 4 + 5 × 5 + . . . If you call n_i the number of occurrences, or count, of the i-th number y_i (two 4's, five 5's, etc.), then many formulas can be simplified. The sum of the numbers is

$$\sum_i n_i y_i$$

The total count or total number of data points is

$$N = \sum_i n_i$$

Since the mean is equal to the sum of the numbers, divided by how many numbers there are, the mean is the frequency of a number y^i is obtained from the number of its occurrences by dividing by the total number of y's. Thus, $f_i = n_i / N$, and the sum of the frequencies is equal to 1. The mean simplifies to

$$\bar{y} = \sum_i f_i y_i$$

$$\bar{y} = \frac{\sum_i n_i y_i}{\sum_i n_i} = \frac{\sum_i n_i y_i}{N}$$

$$\tilde{y} = y_L + w(1/2 - F)/f_j$$

Using Existing Data to Calculate the Median. Many times data will be supplied to you in the form of a completed histogram. This data may come to you from a supplier, another company, or a different division of your company. In order to compare this data to data from other sources, it is necessary to calculate the median of this new information.

If you have data only in the form of a completed histogram and not the raw data, you can still estimate the median. The histogram shown in Figure 2.15 will be used as an example. Starting from the left in the histogram, sum up the bar frequencies (heights)

$$F = \sum_j f_j$$

until the total first exceeds 50%. You can use either percentages or decimals; just be consistent. The median will be in the bar in which:

$$F = 1 + 11 + 35 = 47\% \text{ is a little too small}$$

$$F = 1 + 11 + 35 + 34 = 81\% \text{ is the first bar above } 50\%$$

Let $F = 47\%$, and save it for later. The median is somewhere in bar $j = 4$, which has class mark $y = 50$ on the horizontal axis. The range of the bar is $y = 45$–55. Since the previous sum was $F = 47\%$, the median must be close to the left end of the $j = 4$ bar. Without knowing exactly how the data are distributed in the bar, you can only estimate the actual median.

To estimate the location of the median in bar $j = 4$, you estimate the proportion of counts in the bar that are to the left of the median. If the total number of counts is N, then bar j has Nf_j counts. The value of f_4 is $34\% = .34$. You don't need to know the actual value of N; it will cancel out of the final equation. The portion of Nf_j that lies to the left of the median is found by taking the ratio of areas in Figure 2.17. Since half of the N counts lie to the left of the median (by definition), and the sum of the counts up to the previous bar is NF, the unknown number of counts is also equal to $50N-NF$. Note: Use $.5N-NF$ if you have been using decimals instead of percentages. Since the two counts are equal,

$$(\tilde{y} - y_L)Nf_j/w = 50N - NF$$

where w is the range of bar 4, or $w = 55-45 = 10$. The left edge of the bar is $y_L = 45$. Solving for the median,

$$\tilde{y} = y_L + w(50 - F)/f_j$$

$$= 45 + 10(50 - 47)/34 = 45.88$$

with F and f_j in percentages, or

$$\tilde{y} = y_L + w(.5 - F)/f_j$$

$$= 45 + 10(.5 - .47)/.34 = 45.88$$

with F and f_j in decimal form. As a check, notice that $F = 47\%$ at $y_L = 45$ suggested that the median (50% point) wouldn't be much larger than 45. The estimated median at 45.88 therefore looks reasonable.

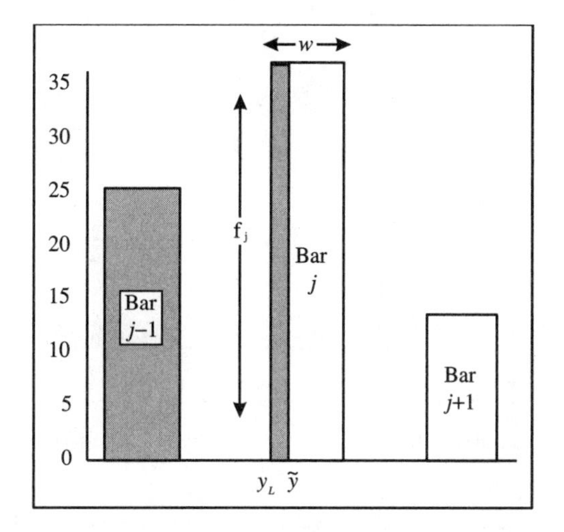

Figure 2.17 Finding the Median

DISTRIBUTIONS

A data distribution is a description of the relative percentage of the data falling in a given range of possible values of the variable measured. You have already seen that you can calculate statistics for your distribution and that repeated sampling produces a *sampling* distribution for the statistics themselves. Based on the type of statistic you want to compare (means, variances, etc.), you need a description of the relative likelihood of getting a given amount of difference in the samples, such as the percentage difference between the means of two samples that are supposedly from the same population.

You need one final step before you are ready to compare products and processes to determine which suppliers to select, procedural steps to change, environments to control, or new equipment to buy. You will need a way to decide how likely it is that the difference in samples that you observed is real or only due to chance variation in your data. This is done by setting up a test that assumes there is no difference in the samples, that is, that your data are homogeneous. The test then compares the amount of deviation actually measured with the amount predicted for purely random variation. Fortunately, these tables of the likelihood of these random variations have already been calculated for you and are available in Appendix A. This section covers the relevant distributions for performing designed experiments; topics include:

- Discrete Distributions
- Continuous Distributions

Discrete Distributions

The discrete distribution provides for only specific values of the variable measured. These are such values as the size of a drill bit, small, medium, or large or other dis-

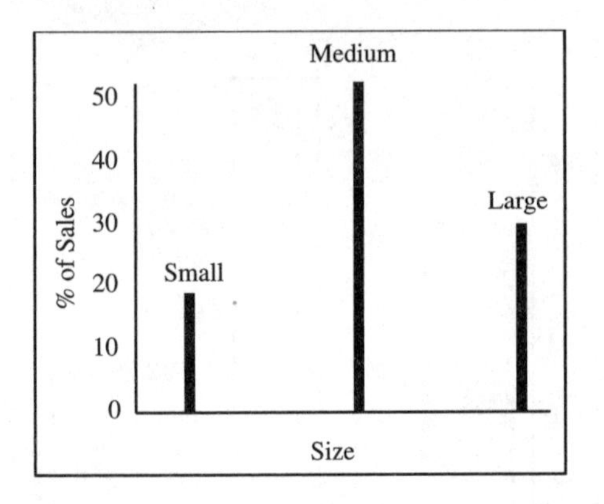

Figure 2.18 Discrete Probability Distribution

crete measures. A good example of its application in a service industry is the manager of a hamburger shop who knows from experience that, of the people who buy hamburgers at his shop, roughly 20% will buy a "small" hamburger, 50% a "medium" hamburger, and 30% a "large." These proportions also represent the distribution of the probabilities of future sales. He only sells three sizes, so it is a *discrete* distribution. The distribution is shown graphically in Figure 2.18.

The probability that a customer will want cheese on the hamburger may depend on the size of that hamburger. If so, it is a *conditional* probability. For example, say that one medium-sized hamburger in four (or 25%) is a cheeseburger. Then the probability of selling a medium-sized cheeseburger is found by multiplying:

$$P(\text{medium}) \times P(\text{cheese, given medium}) = (50\%) \times (25\%)$$

$$= (.5) \times (.25)$$

$$= .125 = 12.5\%$$

Notice that the probabilities *multiplied* because BOTH conditions (medium AND cheese-given-medium) were required. If one condition OR another is required, for example, small OR medium, then the probabilities *add*. If you are worrying about adding those proportions, think of the total number of possibilities as a pie and what went into its calculation: small (20%), medium (50%), large (30%). The total is 100%. The three options are mutually exclusive—a hamburger can't be both large and medium. A medium may or may not have cheese on it; the sum of the probabilities for medium with cheese (12.5%) and medium without cheese (50%×75% = 37.5%) must add up to the probability for medium (12.5% + 37.5% = 50%). You have cut up one of the slices. The place to be careful is in adding dissimilar conditional probabilities: A hamburger could have ketchup AND cheese, one or the other, or neither. Simply break these options up, making sure you have included every one that you really want to count. Drawing a pie chart will help if you are uncertain (see Figure 2.19).

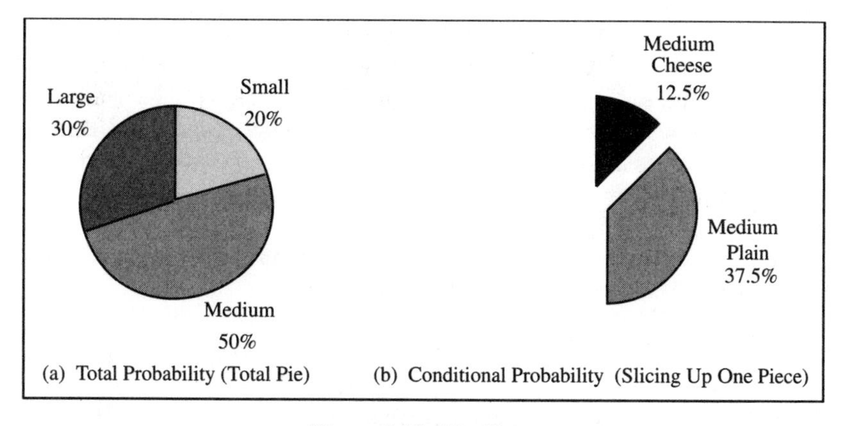

Figure 2.19 Pie Chart

Bernoulli Distribution. A test is called a *Bernoulli trial* if it can have only two possible outcomes (e.g., pass or fail), if its outcome is independent of the outcome of other trials, and if the probability of success of the test does not change if the test is repeated. Tossing a coin is an obvious example. The toss can only result in a head or a tail. The odds of a head are the same, regardless of whether the last toss was a head or a tail, and the odds of getting a head don't "drift" over from 50% to 45% as the tossing is repeated. The probability of "success" is usually given the letter p, and "failure" is $q = 1-p$. These designations are completely arbitrary and quite often reversed. The *Bernoulli distribution* is simply two spikes, one of height p and the other of height q, as demonstrated in Figure 2.20. The real interest in Bernoulli trials is that they form the basis for essentially all the discrete distributions you will encounter here.

Binomial Distribution. The *binomial distribution* provides the probable number of successes $p(x)$ in some number (n) Bernoulli trials:

$$p(x) = \binom{n}{x} p^x (1-p)^{n-x}$$

$$x = 0,\ 1,\ 2,\ \ldots,\ n$$

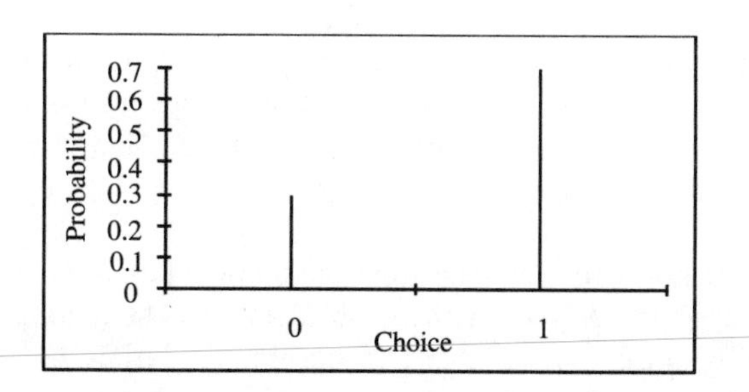

Figure 2.20 The Bernoulli Distribution

where $p^x(1-p)^{n-x}$ is the probability of x successes and $n-x$ failures out of n trials, and the combination $= n!/x!(n-x)!$ is the number of possible arrangements of those successes and failures. For example, there are five possible arrangements of four successes p and one failure q: $qpppp$, $pqppp$, $ppqpp$, $pppqp$, $ppppq$. Thus, the probability of two successes in five trials is $p(2)$. If you have sets of trials of differing sizes, you could modify the notation to $p(x; n)$.

Example: The probability of tossing a head once is simply p = .5, but the probability of tossing a head four times out of four tosses is

$$p(4) = (1)(.5)^4(1-.5)^{4-4} = 1/16$$

The probability of tossing a head on the fourth toss is .5, regardless of what you got on the first three tosses, and is still .5 on the eighth toss, even if you got a head on the first seven tosses.

Example: A die has six sides, so the probability of any side coming up is the same, 1/6. The probability of side number 3 coming up for the first time on the fourth roll is $qqqp = (5/6)(5/6)(5/6)(1/6) = .0965$, or about 10%. The probability of rolling a 3 on any of the first four rolls (i.e., one or more) is one minus the probability of not rolling a 3 on any of the first four rolls, or

$$1-p(0) = 1-(0)(1/6)^0(5/6)^{4-0} = .518$$

The probability of rolling a three twice in the first four rolls is

$$p(2) = (2)(1/6)^2(5/6)^{4-2} = (6)(.02778)(.6944) = .116$$

Notice that it never mattered what happened *after* the fourth roll in any of these cases. Also notice that the wording of any probability statement is very crucial.

Since you have a distribution, you can define a mean and a variance (and then, of course, a standard deviation) of a sample from it. If you like to work with series, you can calculate the mean of the sample. It is np. Similarly, the variance of the sample is $npq = np(1-p)$.

You have three distinctly different uses of the symbol p—as a marker for a success in a string of successes and failures, as the best estimate of the mean value of the proportion of successes for the population, and as the probability of successes in n trials. Unfortunately, this is standard notation. To avoid confusion, try using (1,0) or (Y,N) instead of (p,q) for successes and failures in pass/fail data, and write out p as $p(x)$ when it is used as a discrete probability function.

Proportions. While not really separate from the binomial distribution, proportions are significant enough to deserve separate consideration. The binomial distribution gives the probability of a certain number x of successes in n trials. If a sample of size n is taken, then, the *proportion* of successes in that sample is the ratio: x successes

YYYYYNYYYYNYYYYNNYYYYYYN

Figure 2.21 Inspection Results

actually measured divided by the total number of possible successes n. In symbols, $p = x/n$. Since the term proportion is such a part of everyday language, it has already been used freely up to this time. In statistical analysis, a specific technical meaning is attached to the term. Since a proportion is a number that is calculated from the sample data, it is a statistic. As just noted, this is also the best estimate of the mean value of the proportion of successes for the entire population. The mean and variance for proportions, corresponding to np and npq, are p and pq/n, respectively.

To illustrate, consider the string of passes (Y) and failures (N) in an inspection, as shown in Figure 2.21.

Thus, $n = 25$, $x = 19$, $n-x = 6$. Then $p = x/n = 19/25 = .76$, $q = (n-x)/n = 1-p = .24$, $np = 19$, $npq = 4.56$, $\sqrt{npq} = 2.14$, $pq/n = .0073$, and $\sqrt{pq/n}$=.0854. Although this is hardly enough data to make realistic estimates of what the population is really doing, it is sufficient for demonstrating the calculations.

Hypergeometric Distribution. The only thing hard about the hypergeometric distribution is its name. It tells you how likely you are to measure a certain proportion of defects, given that a sample had a certain proportion. As shown in Figure 2.22, the total population consists of N items, D of which are defects, and $N-D$ of which are not. Ideally, if you take a sample of size n, it should contain d defects and $n-d$ "good" items, provided that you do not put the test item back after you select it.

$$D / N = d / n$$

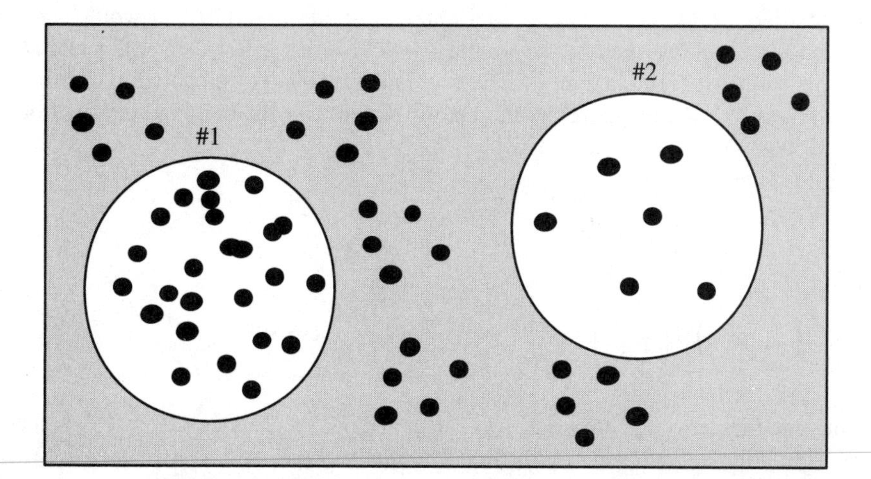

Figure 2.22 Sampling

Obviously, there must be some roundoff; you can't have exactly $d = 13.754$ defects in your sample. The agreement should improve as N and n increase. In practice, some samples will have a slightly higher proportion than d/n, as in Sample #1, while others will have a slightly lower proportion, as in Sample #2.

The probability of drawing a sample of size n containing x defects, where x may or may not equal d, is given by:

$$p(x) = \frac{\begin{pmatrix}\text{Ways to get}\\ x \text{ of the}\\ D \text{ defects}\end{pmatrix}\begin{pmatrix}\text{Ways to}\\ \text{miss } D\\ \text{entirely}\end{pmatrix}}{\begin{pmatrix}\text{Ways to get}\\ \text{sample } n\end{pmatrix}} = \frac{\binom{D}{x}\binom{N-D}{n-x}}{\binom{N}{n}}$$

where x may be 0, 1, 2, up to the lesser of n and D.

Clearly, the number of defects x in the sample can't exceed the total size of the sample n, and it can't exceed the total number of defects D in the population. If N is large, the factors in your answer can get large very quickly and make computation difficult. If $n << N$, then use the binomial distribution with $p = D / N$. If N is very large, you may want to use Stirling's approximation for factorials:

$$n! \approx \sqrt{2\pi} n^{n+1/2} e^{-n} = \sqrt{2\pi n}\left(\frac{n}{e}\right)^n$$

For $n - 50$, $n! = 3.041 \times 10^{64}$, and the approximation gives 3.036×10^{64}, or about 1/6 of 1% error. If more accuracy is needed, replace e^{-n} by $e^{-n-1/2n}$ or $(n/e)^n$ by $(n/e)^{ne-1/12n}$.

As an example, suppose that you have been making an item for years, and typically there are 50 defects in a run of 10,000. You pull out a sample of 100 from the latest run, and 15 are defective. What is the probability of getting 15 defects, given your previous experience? Here, $N = 10{,}000$, $D = 50$, $n = 100$. You would have expected $d = .5$, or less than 1, but you got $x = 15$. Since x is less than both n and D, the hypergeometric formula applies. The population is so large compared to the sample size ($n << N$) that an approximation is warranted. Use the binomial distribution with p $D/N = 50/10000 = .005$. Then

$$p(x) \approx \binom{N}{x} p^x (1-p)^{N-x}$$

$$p(15) \approx \binom{1000}{15} .005^{15} (.995)^{9985} = \binom{1000}{15} \left(5.598 \times 10^{-57}\right)$$

Only the first term was not easily evaluated on a pocket calculator. To evaluate the combination in the first term, use Stirling's approximation:

$$\binom{N}{x} \approx \frac{1}{\sqrt{2\pi}} \left(\frac{N}{x}\right)^{x} \left(1 - \frac{x}{n}\right)^{-N-x+\frac{1}{2}}$$

$$\binom{1000}{15} \approx .3989\left(2.284 \times 10^{42}\right)(3.235 \times 10) = 2.95 \times 10^{48}$$

Substituting back, $p(15) = 1.65 \times 10^{-8}$, a negligible chance that you could have gotten that many defects in a sample from a "good" run. Since sample size n is large in this example, you could also use the hypothesis test for proportions described later. In general, small sample tests involving proportions are a problem, and the hypergeometric analysis shown previously is the only recourse. To visualize the problem, imagine having one defect more or less in a sample size of 100. Compare this to having one defect more or less in a sample size of 10. Clearly, the smaller the sample size, the greater the impact of chance variation or experimental error.

Continuous Distributions

The continuous distribution provides a description of all values measured on a continuous scale. These may include voltage, pressure, time to failure, and yield from a process. The frequency of occurrence of values of a continuous variable is called a *probability density function.* It provides a graphical description of the likelihood of occurrence of values of the variable, so you can see what values are most likely to occur. Because of the way it is set up, the area "under" the curve (between the horizontal axis and the curve) in some interval along the horizontal axis represents the probability of measuring a value of the variable in that interval. Thus, the total area under the curve must equal 100% (a probability of 1) that you will get *some* value when you make a measurement. When discrete distributions have many, closely spaced values, they are often approximated by a continuous distribution.

Normal Distribution. Usually, random variations (noise) in a measurement have a "bell-shaped" distribution called the *normal distribution.* Knowledge of how to work with this distribution will allow you to make comparisons of test results in the presence of noise. In order to have a universal distribution for making comparisons, you code your variable, just as you did with your raw data:

$$z = \frac{y - \mu}{\sigma}$$

This ratio is called the z score. The frequency of occurrence of small random variations and of all random variations involving a very large sample size has been found to be well represented by $e^{-(y-\mu)^2/2\sigma^2} = e^{-z^2/2}$, as shown in Figure 2.23. This

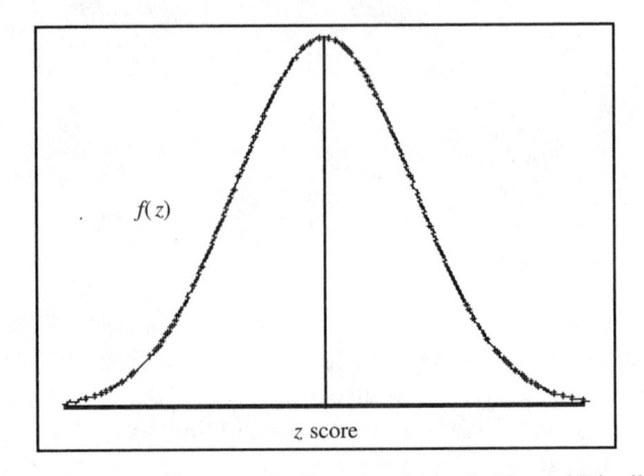

Figure 2.23 Probability Density Function $\phi(z)$ of a Normal Distribution

frequency distribution (or probability density function) is called the "normal" distribution, because it occurs so often in nature. In order for the total area under the $e^{-z^2/2}$ to equal one (100%), a coefficient of $1/\sqrt{2\pi}$ must multiply the $e^{-z^2/2}$.

The area $\Phi(z)$ to the left of z under the curve $\phi(z)$ is the probability of a measurement being less than z. Since the area is hard to read from a graph, even when area is plotted directly against z as in Figure 2.24 or 2.25 (which is $\alpha = 1-\text{area}$), a table is given in Appendix A. The population mean and standard deviation μ and σ are called population parameters, since they are constants in a known normal distribution, whereas $\bar{y}$ and s are called sample statistics, because they are derived from a sample, which is a finite number of discrete data points from the population. In the real world, μ and σ are usually unavailable or meaningless, and you must replace them with their approximations, $\bar{y}$ and s.

Properties of the Normal Distribution. The normal distribution has certain properties that make it unique among the common distributions.

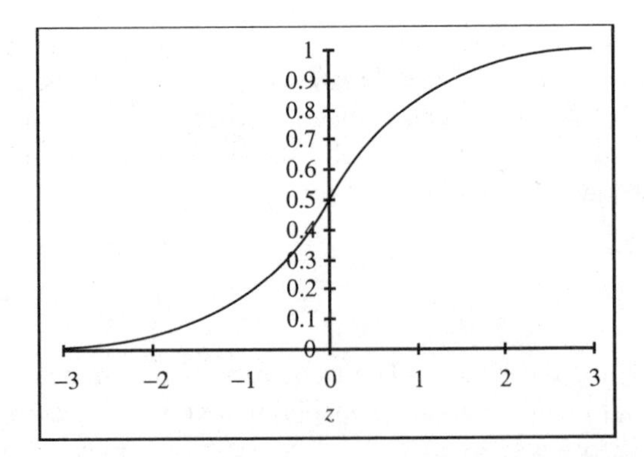

Figure 2.24 Cumulative Distribution Function $\Phi(z)$ of a Normal Distribution

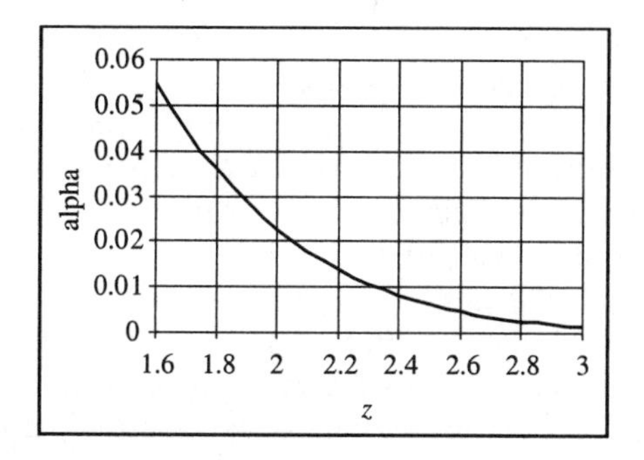

Figure 2.25 Alpha ($\alpha = 1 - \Phi(z)$), the Tail Area

- The population parameters μ and σ of a normal distribution may be changed independently of each other. Notice that the mean is set by the population parameter m (since they are equal), not the other way around. The same is true for s and the standard deviation. In all the other common distributions, changing a single population parameter changes *both* the mean and the standard deviation.
- If n independent, normal random variables are added, the mean of the sum is the sum of the individual means, and the variance of the sum is the sum of the individual variances.

$$\mu = \sum_{i=1}^{n} \mu_i$$

$$\sigma^2 = \sum_{i=1}^{n} \sigma_i^2$$

- The sum of a large number of distributions approaches a normal distribution.
- The means of identical distributions tend to follow a normal distribution.
- Related distributions that have sample size n as a parameter tend to approach a normal distribution in shape as n gets large.

The median is the point at which the probability is 50% for a number being on either side of it (the "50:50" point). For a symmetric distribution, it is therefore the center point. For the normal distribution, the mean, median, and mode all fall at the same place, the location of the peak. The population median is often labeled $\tilde{\theta}$, or just θ, to distinguish it from the sample median, $\tilde{x}$.

WHAT HAVE YOU LEARNED?

Although it may not seem so, this has been a very short introduction to statistics. In fact, it covers only those aspects directly related to design of experiments. You probably have noticed that a great deal of time was spent in the planning aspects of the tests, trying to figure out what to look for, how to do it, and so on. The three parts of a DOE are planning, testing, and statistical analysis.

The testing is something you already know about. It requires the special expertise of your field. The statistical tests themselves are actually relatively simple, once you have selected the right one. The planning stage drives the type of process testing you will do and the type of statistical testing as well.

All of the tests described in this book involve comparison in some fashion. You are not simply gathering data, you are trying to improve something, either by selecting a better process or a better input, or you are trying to improve something by trying to fix it. It is not enough to compare two things—you must also decide how important that result is. It is because you must decide importance in a world of variation that you must have at least a speaking acquaintance with statistics. As you continue reading, try to make sure you understand which test to use and its limits of validity.

Graphical methods are extremely important in understanding and comparing your data. You will be seeing much more of them in the coming chapters. Usually, however, they do not provide a quantitative measure of the statistical significance of an observed difference between two or more data sets. A bit more statistical background is needed before you are ready to make such tests. That comes next.

REFERENCES

1. R. A. Fisher, *Statistical Methods for Research Workers, 13th Edition*, Edinburgh: Oliver and Boyd, 1958.
2. W. Cheney and D. Kinkaid, *Numerical Mathematics and Computing, 2nd Edition*, Pacific Grove, CA: Brooks/Cole, 1985.
3. H. Jackson and N. Frigon, *Achieving the Competitive Edge*, New York: John Wiley & Sons, 1996.

3

STATISTICAL TESTS

This chapter covers statistical tests, including hypothesis tests and contingency tables. We take full advantage of the fact that sample means from the same parent distribution tend to follow a normal distribution, even though the parent distribution itself may not be normal. You will see that variances from the same distribution also have their own useful properties which you can use to advantage.

Some comparisons are made which require answers like, "We are 95% confident that the worldwide availability of cobalt has decreased by $x\%$." Others simply have answers like, "Japanese buyers have a definite preference for red cars." The more quantitative an answer must be, the easier it is to estimate a required sample size, but the number is not necessarily larger. This section covers the more quantitative data; tests for preferences and other variables that are less quantitative are covered in a subsequent section.

Comparison tests are set up on the basis of a *null hypothesis* that there is no difference between two populations (e.g., no effect of a treatment), a sample is taken from each population, and their statistics are compared to determine how likely it is that the two samples came from the same population. The important point is that we compare populations by comparing samples. Tests can also be set up to compare more than two populations, which is the primary interest of this book. Unlike the American legal system of "innocent until proven guilty," a test is set up along the line of the British system of showing that the charge is "not proved." Since our ability to demonstrate a difference depends strongly on sample size, this is a very reasonable procedure.

The three main types of population parameters that are compared by hypothesis tests are means, variances, and proportions. The tests are usually for equality (called two-sided, because the second parameter can fail the test by falling on either side of the first) or for one parameter to be larger than the other (called one-sided, because the second parameter can fail only by falling on one side of the first). Other tests, such as comparisons of distributions, certainly exist, but these three are of primary importance here. There are two general types of tests: parametric and nonparametric. *Parametric* tests for means and variances assume that the populations follow normal distributions and rely on the special properties of the normal distribution in carrying out the test. Often, the populations must have the same variance, which means that it must be possible just to shift one population to get the other, in other words, that the means only differ by a constant. Many common tests demand that the data be very close to a true normal distribution for validity. These tests helped to pave the way for nonparametric tests. *Nonparametric* tests do not require the comparison of parameters (mean, standard deviation, etc.) in comparing distributions. *Distribution-free* testing requires no assumption on the form of the distribution. The term nonparametric is often used interchangebly with distribution-free, since it is sometimes difficult to decide into which category a test falls, and many tests fall into both.

SAMPLING DISTRIBUTION

Most comparison tests of interest in designed experiments fall into two categories: those based on comparing means, and those based on comparing variances. Since sampling distributions are approximately normal, even when the underlying distribution is not, they are a good choice for making comparisons.

The distribution of sample means, or *sampling distribution*, follows the normal distribution with y replaced by $\bar{y}$:

$$z = (\bar{y} - \mu)/\sigma_{\bar{y}} = (\bar{y} - \mu)/\left(\sigma/\sqrt{k}\right) \approx \left(\bar{y} - \bar{\bar{y}}\right)/s_{\bar{y}}$$

where

$\sigma_{\bar{y}}$ = standard deviation of the sampling distribution

k = number of sample means (sample size, but the sample consists of means)

$s_{\bar{y}}$ = estimated standard deviation of the sampling distribution

The sampling distribution is the key to why the normal distribution is so important. For a reasonable number of samples (k = at least 30), the sampling distribution can be approximated as a normal distribution, and $\bar{\bar{y}}$ approaches μ. If the samples themselves are approximately normal, k can be much less than 30. The mean of the distribution of all nk points in k samples of size n will be exactly the same as the grand mean $\bar{\bar{y}}$ of the k sample means. As k increases, $\sigma_{\bar{y}}$ decreases, because σ is a population parameter and is fixed; thus, the variance of the sampling distribution will be smaller than the variance of the population distribution. If all nk points are taken as a single sample, its variance will lie somewhere between σ^2 and $\sigma^2_{\bar{y}}$. If many measurements of $s_{\bar{y}}$ are made, they may be "pooled" (explained later) to estimate $\sigma_{\bar{y}}$.

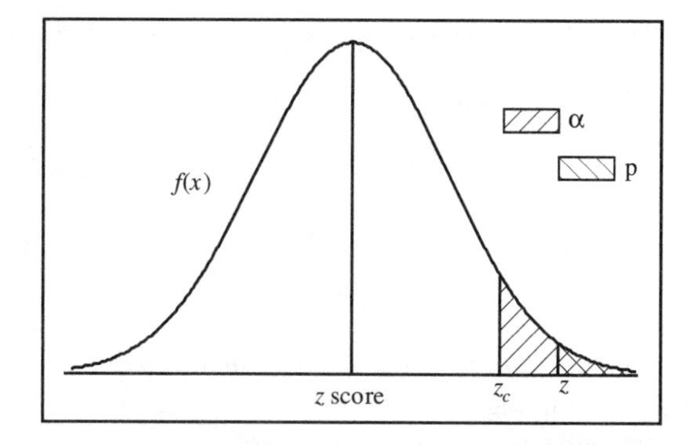

Figure 3.1 Definition of Symbols

For an example of how to use this information, suppose that you know very good estimates of population mean and variance from samples taken over a long period of time, and you want to know how likely it is that a new sample with a much larger mean than μ is actually from the same population. To do this, you must see how likely it is that $\bar{y}$ could be this large from random variation. Figure 3.1 shows the probability density function $\phi(z)$, illustrating the meaning of α, the area of the tail where there is $\alpha\%$ probability of z lying beyond the critical z_c. You will be determining how likely it is that $z_{\bar{y}} = (\bar{y} - \mu)/\sigma$ falls beyond the critical z_c just because of random variation. This chance that the answer was unusually large but "just happened" as a rare occurrence is the key issue here. By setting the amount of risk you are willing to accept, you also set a critical $\bar{y}$. The figure also shows the p-value, the area in the tail beyond $z_{\bar{y}}$. If the p-value is less than α (i.e., that $z_{\bar{y}} > z_c$), then reject the hypothesis that $\bar{y}$ came from the usual population. For $\alpha\% = 5\%$ probability that $z_{\bar{y}}$ is random, or $100(1 - \alpha)\% = 95\%$ chance that the new sample doe not fit the usual population, $z_c = 1.645$.

Since α represents the probability of getting a z beyond z_c, the way to test that the *magnitude* of z does not lie beyond some range, that is, that $|z| < z_c$, is to split α into $\alpha/2$ in each tail. This changes z_c. For a 95% chance that a large z is not random, the new $z_c = 1.96$. The population mean and standard deviation for the standard data set in Table 2.3 are $\mu = 50$ and $\sigma = 5$ The grand mean for the entire sample of size 100 is 50.6755.

To do the simple "one-sided" test that $z_{\bar{y}} > z_c$ for the entire sample of size 100:

Step 1. Calculate the sample mean $\bar{y}$.

$$\bar{y} = 50.6755$$

Step 2. Calculate the z corresponding to $\bar{y}$, using the known population mean and standard deviation.

$$z = (\bar{y} - \mu)/\sigma = (50.6755 - 50)/5 = .1351$$

Step 3. Look up the critical z_c corresponding to the accuracy you want.

$$\text{For } \alpha = .05, z_c = 1.96$$

(For the two-sided test, simply use the z_c for $\alpha/2$.)

Step 4. Test $|z|$ versus z_c.

Since .1351 is much less than 1.96, you can be 95% confident that the sample is from the parent distribution.

In the real world, the number of samples k is usually too small for the approximation to apply, and the sampling distribution does not look "normal" enough. In other words, you can't prove that the underlying assumptions and approximations are valid, even though you may know that they really are. In this (usual) case, you have to make some adjustments. Your best estimate of μ is still $\bar{\bar{y}}$, but now a "pooled" estimate of the sample variance is used for σ^2. Details are given later. You also use the t distribution in place of the normal distribution.

$$t = (\bar{y} - \bar{\bar{y}})/s_p$$

If the actual population is only "close" to a normal distribution, this is also the way to go, because the sampling distribution will be closer to normal than the population distribution. In the preceding steps, simply replace the letter z by the letter t, and use the t distribution instead of the normal distribution. The only added step is to look in the row of the table corresponding to the sample size minus 1.

Regardless of whether you use z or t, you can take advantage of the fact that $\bar{y}_1$- $\bar{y}_2$, $\bar{y}$ − const, $\bar{y}/n$, and other differences and ratios of sampling distributions are themselves also sampling distributions. The trick is in getting the variance right. Here are some rules:

- Taken as a single number (not a set of n values), $\bar{y}$ is a random variable with $df = 1$, and it has a variance.
- A known constant, such as a specification ($y_{max} = 14$), the μ for a population, or a difference $\mu 1 - \mu 2$ is not a random variable. It has $df = 0$ and no variance.
- Proportions may be compared by letting $\mu = np$ and $\sigma^2 = npq$, where n = sample size, p = proportion of successes, and $q = 1 - p$ = proportion of failures. Let P = true proportion of successes for the population, even if population size N is undefined. Then p is the best estimate of P, based on the sample of size n.
- The variances of differences *add*. Thus,

$$\bar{y}_{1,2} = \bar{y}_1 - \bar{y}_2 \quad \text{but} \quad s_{1,2}^2 = s_1^2 + s_2^2$$

- When pooling variances, however, you are looking for the best estimate of the variance of the sampling distribution, the mean square error s_p^2 = MSE. Thus,

$$S_p^2 = \frac{s_1^2}{n_1} + \frac{s_2^2}{n_2}$$

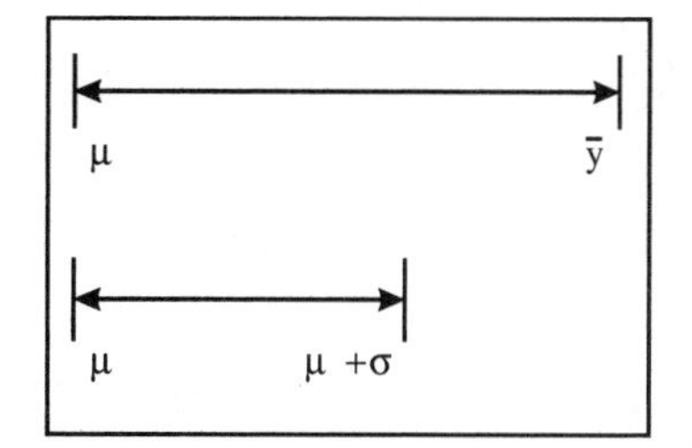

Figure 3.2 Examining Measurement Error

CHI-SQUARE DISTRIBUTION

The chi-square (χ^2) distribution is important, because it is used in determining whether the measured difference, or "distance," of a random variable from a constant value, for example, $(\bar{y}-\mu)$, is significantly larger than some expected value, for example, σ. See Figure 3.2. As an example, if measurements of the mean density of some material have always been 94.3 ± 0.8 in the past, but a sample of size $n = 25$ from the most recent lot has a mean of 96.1, a chi-square test can answer the question, "Is there less than a 5% chance that the sample mean $\bar{y}$ would be so far from the population mean μ?"

The only random variable involved here is the statistic $\bar{y}$; constants μ and σ are population parameters. Measured values of $\bar{y}$ can be used to define a new random variable

$$\chi^2 = \frac{(n-1)(\bar{y}-\mu)^2}{\sigma^2}$$

where n is the size of the sample used to calculate $\bar{y}$. Choosing this form for the new random variable causes it to have a universal distribution for a given n. It is said to be distributed as χ^2_{n-1}. This distribution is shown in Figure 3.3.

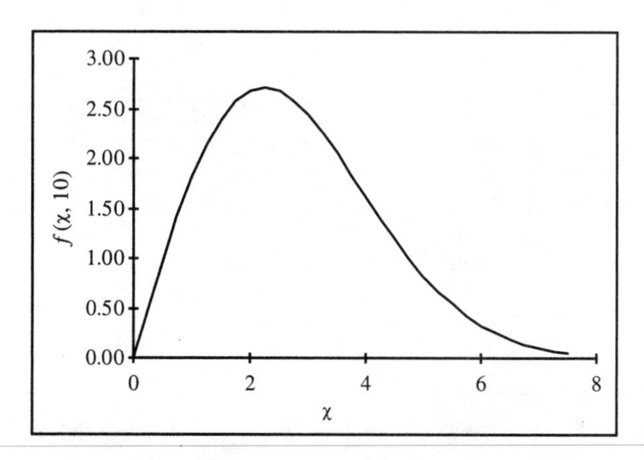

Figure 3.3 Chi-square Probability Density Function

The sum of chi-square variables is still a chi-square variable. Thus, sample variances "transformed" in this way may be summed to form a chi-square variable. The chi-squared variable for total sum of squares *SST* for *k* samples is given in the following equation:

$$\frac{SST}{\sigma^2} = \sum_{i=1}^{k} (n_i - 1)\frac{s_i^2}{\sigma^2}$$

The sampling distribution of $\chi^2 = (n-1)s^2/\sigma^2$ is chi-squared with $df = n - 1$ degrees of freedom. Then the ratio of the sample variance to the population variance will lie in the confidence interval

$$\chi^2_{1-\alpha/2,df} / df \le s^2/\sigma^2 \le \chi^2_{\alpha/2,df} / df$$

For applications, you will probably just want s/σ. Table 3.1 shows a short table of s/σ ranges for for various values of n, where $df = n - 1$.

You can look at this in two different ways: You can decide that you will need to take an enormous sample to pin down the population variance, or you can decide that the distribution is relatively ìforgivingî and generous about how bad a variance has to be before it is considered unusual.

There is a way to compare means using chi-square: Use a test to see whether the relative variance of the means is greater than a critical value from a chi-square table. This relies on

$$\frac{SST}{\sigma^2} = \sum_{i=1}^{k} (n_1 - 1)\frac{s_i^2}{\sigma^2}$$

where you have k means, each with sample size n_i. You don't really need to know the population variance in the denominator, since chi-square is looking at the *relative* variance.

95% Confidence	Variance		Standard Deviation	
n	min	max	min	max
5	0.157	6.388	0.396	2.527
6	0.198	5.05	0.445	2.247
8	0.264	3.787	0.514	1.946
10	0.315	3.179	0.561	1.783
15	0.403	2.484	0.635	1.576
20	0.461	2.168	0.679	1.472
25	0.504	1.984	0.71	1.408
30	0.537	1.861	0.733	1.364
40	0.587	1.704	0.766	1.306
50	0.622	1.607	0.789	1.268

Table 3.1 Confidence Intervals for Variances and Standard Deviations

The population variance is built into the table and never appears in the test, which only uses *SS*, the sum of the squares (numerator), for each *i*. You simply compare the sum of the SS_i, that is, *SST* with the table value.

t DISTRIBUTION

Although the *t* distribution may be used in place of the normal distribution when you have small sample sizes, it is used primarily for comparing means to see whether they come from the same sampling distribution. This is because $(\bar{y}-\mu)/(s/\sqrt{n})$ follows a *t* distribution with $n-1$ degrees of freedom. Notice that this is essentially a *z* score with σ replaced by $s/\sqrt{n}$, the standard deviation for a sampling distribution of a set of means $\bar{y}_i$. Figure 3.4 shows the *t* distribution for 10 degrees of freedom. It would therefore be used for doing a test on a sample size of $n = 11$. The width of the curve depends on the sample size *n* to be used in making a test. The *t* distribution (Figure 3.4) has been described as being "like the normal distribution, but with fatter tails."

In doing a *t* test, we replace *z* by

$$t = \frac{\bar{x} - \mu}{s/\sqrt{n}}$$

where *n* is the sample size, and use a table of the *t* distribution in place of a table of the normal distribution. Be especially careful of table values of the *t* distribution. Some authors use α where others would use $\alpha/2$. Graphically, some are using α for the area to the right of $+t$, while others add in the area to the left of $-t$ as part of α. For them, the area to the right of $+t$ is then only $\alpha/2$. Since the *t* distribution is symmetric, you just need to know whether the column heading is α or $\alpha/2$.

To see how your *t* table is set up, look to the right of the 5, under the $\alpha = .05$ heading, and find the critical value of *t*. If it is 2.015, then all the area is assumed to lie in the tail to the right of t_c as in Figure 3.5. You can use this number directly as the critical value of *t* in a one-sided test at the $\alpha = .05$ level. To do a two-sided test at the $\alpha = .05$ level, simply use the .025 column in your table and get 2.571. On the other hand, if you look to the right of the 5, under the $\alpha = .05$ heading, and you find 2.571,

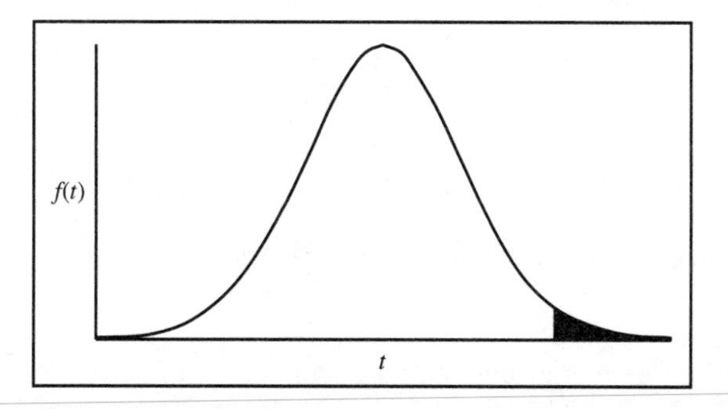

Figure 3.4 Probability Density Function for the *t* Distribution

df	......	0.05	......	0.025
.		.		.
.		.		.
.		.		.
.		.		.
5	......	2.015	......	2.571
.				
.				
.				
.				

Figure 3.5 Reading a *t* Table

then your table is already set up for a two-sided test. You can look under the column for $\alpha = .1$ setting in this kind of table to get the critical value for the one-sided test at the $\alpha = .05$ level. This trick is based on the symmetry of the *t* distribution.

For an example of how to use the *t* test for comparing the means of two samples, consider the data in Table 3.2 for the daily sales of an item in two different convenience stores in the same chain. The data were taken over the same week for each store. A check of the data indicates that there are no obvious outliers (e.g., 24 were sold on the first day by store A; if this had been 81, this data point would be removed using Dixon's test).

To perform the *t* test, do the following:

Step 1. Calculate the variances for *A* and *B*. These are $s_A^2 = 9.62$ and $s_B^2 = 9.81$. The variances are within 1% of the pooled variance (well within the maximum recommended ratio of 3), so equal variances may be assumed. The case of unequal variances will be discussed later.

Step 2. Calculate the sample means $\bar{y}_A = 26.43$ and $\bar{y}_B = 27.86$.

Step 3. Calculate t_0. Since the quantity of interest is the difference in population means, the general formula is

$$t_o = \frac{(\bar{y}_A - y_B) - (\mu_A - \mu_B)}{\sqrt{\dfrac{s_A^2}{n_A} + \dfrac{s_B^2}{n_B}}}$$

where the root mean square of the sampling variances is used as the pooled sampling variance.

Step 4. Set the difference in population means equal to zero, which is the null hypothesis, and calculate t_0. The result is $-.857$. Ignore the sign, since this is a two-sided test.

A	B
24	25
31	30
28	33
25	26
29	28
26	29
22	24

Table 3.2 Input Data for *t* Test

Step 5. Look up the critical value for t, based on $n - 1$ degrees of freedom for each of the two data sets, or $df = (7 - 1) + (7 - 1) = 12$. For $\alpha = 12$, the critical value is 2.179.

Step 6. Compare $t_0 = .857$ to 2.179, showing that the difference in means is not significant at the .05 level. If you do this with software, you will be shown a p-value of .408; it would be smaller than .05 if the difference in means were significant at the 5% (.05) level.

Relative sizes of the tail area, the p-value, are shown in Figure 3.6 for three different degrees of freedom, $df = 10, 20, 30$, and a range of t. For a particular, critical t, the corresponding p-value is called α.

Suppose that columns 2 and 3 in the standard data set of Table 2.31 were random samples from two lots A and B, made by the same process, but the suppliers of raw materials differ. Is A different from B? It sure looks like it. Why? Probably because the difference in means is at least comparable to half the range of the data. You need a way to quantify the distinction between those two cases. There is also a hidden factor here. What would you get if you repeated both sets of measurements k times? Would the means of A and B move up and down very much? Would the range of each vary wildly? You need to understand what repeated measurement "buys" you.

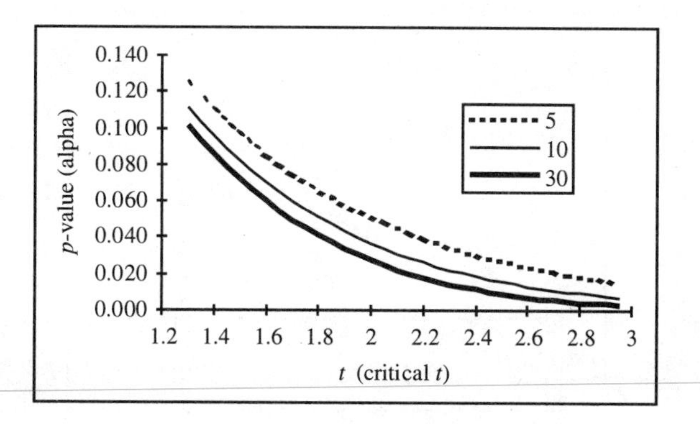

Figure 3.6 Tails of the *t* Distribution

	Col 1	Col 2	Col 3	Col 4
Col 2	-0.5921			
Col 3	1.917	2.1975*		
Col 4	0.0702	0.6597	-1.8899	
Col 5	1.3182	1.7024	-0.4902	1.2763

Table 3.3 Results of Comparing Columns in the Standard Data Set

If you perform a *t* test on pairs of columns in the standard data set (Table 2.3), you get the result shown in Table 3.3, where ** stands for $p < .01$ and * stands for $p < .05$. This is a standard notation. Since the result of pairing column 2 with column 3 is the same as the result of pairing column 3 with column 2, only one of the results is listed. The *p*-value for a significant difference between the means of columns 2 and 3 is $.01 < p < .05$.

But wait; something appears to be wrong here. These five samples came from the same normal distribution: How is it that a *t* test indicates a 95% probability of a significant difference in columns 2 and 3? Why do columns 2 and 3 each "pass" the *t* test with 1, 4, and 5 but not with each other?

First, let's look at some behind-the-scenes information. The sample data set was created with a random number generator. The dashed line in Figure 3.7 is the theoretical curve $N(\mu,\sigma^2) = N(50, 25)$. Data points above the line indicate skewness to the right. The skewness is .562. Actually, only about 6 points are more than 3% high. The mean and variance of the 100-point sample are 50.68 and 28.58. More points would have pushed the data curve toward the theoretical curve.

The 100 points were broken into 5 groups of 20. As expected (because the grand mean was high and the distribution was skewed to the high side), the group means were all high. These are shown in Figure 3.8.

Now let's turn this all the way around and suppose that you are taking samples of size 20, and the first two you take happen to be the two end points (columns 2 and 3) on the graph. A *t* test with unequal variances (47.19 and 16.65) shows that the difference in means is "significant" at the .05 level (2.20 measured vs. 2.04 table). This

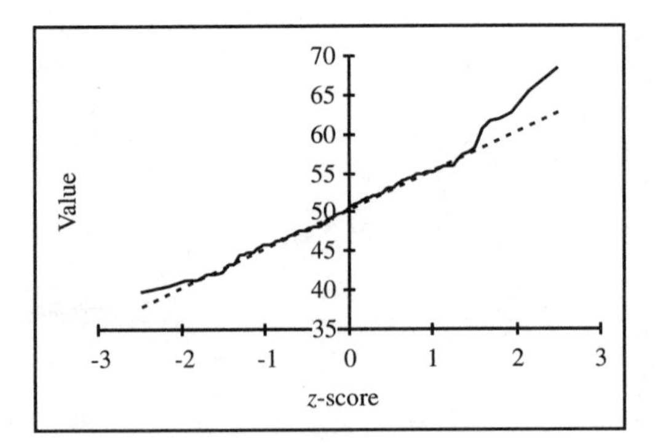

Figure 3.7 Sample of 100 Values

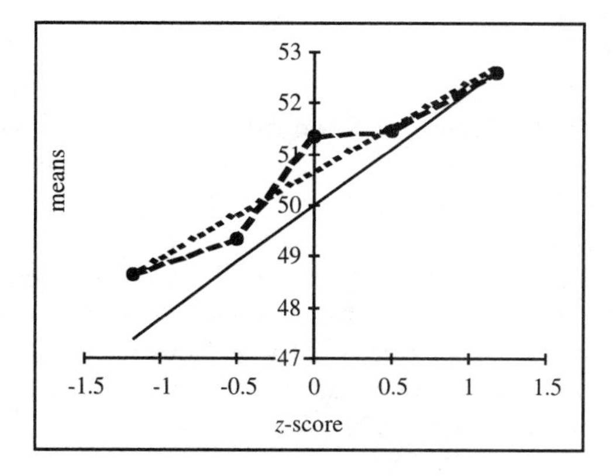

Figure 3.8 Means of 5 Groups of 20 from Sample

says that it is unsafe to assume that the two samples came from the same population. There is only one chance in 20 (the .05 level) that you could get this pair of means, choosing two samples at random. If you take three more samples *under the same test conditions* and look at all five means, then you see that the first two means are not so abnormal after all. What this shows is that there was a larger than normal risk in assuming that those first two samples came from the same parent population. The test advised you not to take that risk. It turned out that it was unnecessarily cautious—this time. Chapter 4 shows you how to test more than two samples at a time. If you test the data with that technique, you'll find that the added information indicates it really is safe to assume that the five samples come from the same population.

F DISTRIBUTION

The *F* distribution was named by Snedecor in honor of Fisher, since Fisher did the early work that Snedecor then extended. For that reason, you may find *F* called the Snedecor's *F* distribution, Fisher's *F* distribution, or simply the *F* distribution. The *F* distribution may be used to compare two population variances by comparing the ratio of the variances of a sample from each to a value from a table. The samples need not be of equal size.

The following three chapters go into great detail about using the *F* test to compare variances, so only the following remarks will be made. As used here, the *F* test compares ratios of mean square errors. You will be getting a sum of squares *SS* for each of two distributions, which need not be of the same size, and you will be testing to see whether the ratio of the sampling variances MSE could be equal to 1. Critical values of the function *F* are tabulated as $F(\alpha, df_1, df_2)$, where df_1 is the number of degrees of freedom for the MSE in the numerator and df_2 is the corresponding number for the MSE in the denominator. In typical applications shown later, the numerator will relate to the sampling distribution of the difference between individual sample means and the grand mean for their sampling distribution (as estimated using the mean of the sample means). The denominator will be an estimate of the MSE of the sampling distribution, given a sample of that many degrees of freedom.

CONFIDENCE INTERVALS

You already saw the importance of the confidence interval of χ^2 in a previous section. A similar confidence interval can be built up from other hypothesis tests (or vice-versa). for example, a $\bar{y}$ to be measured will lie within an interval of width $d = \bar{y}_{max} - \mu = z_\alpha \sigma$ above the true population mean μ. On the other hand, if you want an interval around (straddling) the mean, simply replace z_α by $z_{\alpha/2}$. Now you have a new critical z, since $y_{\max}$ can only be $d/2$ away from μ. For 95% confidence, $\alpha/2 = .025$. As just noted, since these confidence intervals are based on the tests described earlier, each type of test defines its own confidence interval. However, for proportions, much simpler and better equations are available for $p_{\max}$ and $p_{\min}$. The $100(1-\alpha)\%$ confidence bounds for proportions are given below.

Upper bound on p:

$$p_{\max} = \frac{1}{1+\dfrac{N-x}{x+1}F_{1-\alpha/2;2(N-x);2(x+1)}}$$

Lower bound on p:

$$p_{\min} = \frac{x}{x+(N-x+1)F_{\alpha/2;2(N-x+1);2x}}$$

where F is another distribution, to be covered later. For the moment, you can simply look up the table values for F without worrying too much about their origin. To read a table like the one in Appendix A, pick the table corresponding to the alpha you want. There are two degrees of freedom: Go across the top to df_1, the degrees of freedom in the numerator, then go down that column to the number corresponding to df_2 down the side, the degrees of freedom in the denominator. The number where column df_1 and row df_2 meet is the number to insert into the formula as F. Now you can calculate $p_{\max} - p$ and $p - p_{\min}$. In general, they will not be the same.

DIXON'S TEST FOR OUTLIERS

Dixon's test provides an easy way to check for outliers. It was developed for comparing means, which tend to follow a normal distribution. If your data distribution is reasonably normal, you can use it for individual data points, especially as a screen for deciding whether to go back and check certain points. It uses Table A.32 of Appendix A. From left to right, the table shows: a formula for a ratio r corresponding to your sample size n, the value of n, and the minimum value of r for 90%, 95%, and 99% confidence that you have an outlier. Be sure to note that each range of n has its own equation. As an example, consider the data set {14, 18, 13, 31, 15}. Rank the data in increasing order: {13, 14, 15, 18, 31} = $\{x_1, x_2, x_3, x_4, x_5\}$. There are 5 data points, so look under $n = 5$ in the table. The ratio r for $n = 5$ is

$$r = (x_2 - x_1)/(x_n - x_1) = (15 - 14)/(31 - 14) = .0588$$

It is much less than $r = .642$, the minimum value for 95% confidence that x_1, the left-most point, is an outlier. You must keep x_1. To look at the other end of the distribution, reverse the order of numbering. Now $x_1 = 31$, $x_2 = 18$, and $r = (31 - 18)/(31 - 14) = .765$. This is bigger than the minimum value for 95% confidence but not bigger than the .780 needed for 99% confidence. For ordinary work, you only need 95% confidence and would throw out the $x_5 = 31$. For a very crucial result (for which you probably should have taken a larger sample size), you ought to keep the 31. It is easy to speculate that the 31 should have been a 13. If you have more than one "outlier," then you probably should not repeat the test on the second one. Rather, you should go back and review your test setup and procedures. If you watch for outliers as you are taking your data, you can make corrections and do retests under essentially the same conditions as the original test. If not, you are left with speculations and unpleasant decisions.

USING A PROGRAMMABLE POCKET CALCULATOR

There are times when all you have to work with is a pocket calculator: at a remote test site, on a production floor, on a survey team, or in a start-up business that so far has limited capital for computers. Even a laptop computer can be too big and bulky to take some places (especially meetings), it is too expensive to risk damaging it, and not everyone can afford one. Quite often, you want to do "sanity checks" on intermediate results during a test to make sure nothing is going badly wrong. Even a simple pocket calculator will allow you to calculate the mean and variance "by hand" if the sample size is not too large. The following ideas may help you to create your own shortcuts.

Simple Tests

Many programmable pocket calculators that do regression will show the intermediate sums, plus the mean and standard deviation of each variable. This added information can be used to do two-variable tests: correlations, *t* tests, and *F* tests. They may also be used to calculate r^2 and other data associated with regression. Then you can use the formulas in the next section to calculate the critical values instead of looking them up in a table.

If you have the intermediate results, the shortcut to calculating r, which can then be squared to get r^2 if you wish, is to use the summation formulas:

$$S_{xx} = \sum (x - \bar{x})^2 = \sum x^2 - n\bar{x}^2$$

$$S_{yy} = \sum (y - \bar{y})^2 = \sum y^2 - n\bar{y}^2$$

$$S_{xy} = \sum (x - \bar{x})(y - \bar{y}) = \sum (x - \bar{x})y = \sum x(y - \bar{y}) = \sum xy - n\bar{x}\bar{y}$$

Then

$$r = \frac{S_{xy}}{\sqrt{S_{xx}S_{yy}}}$$

You only need the mean and standard deviation of each data set to do the t and F tests. Since you should be doing these on equal sample sizes $n_x = n_y = n$ for x and y anyway (for best results), the shortcut here is that you can do both sets at once, including the standard deviations. With equal sample sizes, t_o reduces to

$$t_o = \frac{\sqrt{n}(\bar{x} - \bar{y})}{\sqrt{s_x^2 + s_y^2}}$$

If you assume that $\sigma_x = \sigma_y$, then the degrees of freedom are $df = 2(n - 1)$. If you don't, then $df = n - 1$. Since a smaller df means a bigger critical t_c, the unequal variance test is "tougher" than the equal variance test. The same data may be used for the F test to compare the variances.

Obtaining Tabulated Values

The following formulas[1] are simple enough that you can program them into a pocket calculator or just use the calculator to do the arithmetic. If you have a programmable calculator, you should put the constants in storage registers and call them into the program. Some calculators will allow you to evaluate the integral of a function, and you can find tabulated values this way. The following approximations are probably as accurate as you need them to be, and the calculation time is significantly less.

Normal Distribution

Alpha. For the normal distribution, to find alpha (or the point estimate), given nonnegative z:

$$\alpha = \frac{1}{2\left(1 + c_1 z + c_2 z^2 + c_3 z^3 + c_4 z^4\right)^4}$$

where

$$c_1 = .196854 \qquad c_3 = .000344$$

$$c_2 = .115194 \qquad c_4 = .019527$$

The magnitude of the error is less than 2.5×10^{-4}, which is usually better than you need in most cases. Table 3.3 compares the tabulated values to the approximate values. Note: On your programmable pocket calculator, you would make the substitution (Hörner's method)

$$1 = c_1 z + \bullet\bullet\bullet = 1 + z\left(c_1 + z\left(c_2 + z\left(c_3 + z c_4\right)\right)\right)$$

to save storage. Table 3.4 shows the check of results.

z	alpha(z)	poly(z)
1.6	0.0548	0.0546
1.8	0.0359	0.0357
2	0.0228	0.0226
2.2	0.0139	0.0138
2.4	0.0082	0.0083
2.6	0.0047	0.0048
2.8	0.0026	0.0028
3	0.0013	0.0016

Table 3.4 Accuracy Check on Approximated Alpha

c_0 = 2.51552	d_1 = 1.43279
c_1 = 0.80285	d_2 = 0.18927
c_2 = 0.01033	d_3 = 0.00131

Table 3.5 Coefficients

For the normal distribution, to find the critical z, given α:

Let $x = \sqrt{ln\left(\frac{1}{\alpha^2}\right)}$

$$z = x - \frac{c_0 + c_1 x + c_2 x^2}{1 + d_1 x + d_2 x^2 + d_3 x^3}$$

The error is less than 4.5×10^{-4}. Table 3.6 shows the check of results.

Chi-Squared Distribution

If your calculator is not programmable or has limited space, use the approximation

$$\chi^2\,(.95, n) = .753876\,n - 4.0155 \approx .7539\,n - 4.016 \text{ for } 10 < n < 40$$

alpha	z	z-fit
0.5	0	0
0.25	0.6745	0.6742
0.1	1.2816	1.2817
0.05	1.6449	1.6452
0.01	2.3263	2.3268
0.005	2.5758	2.5762
0.001	3.0902	3.0905

Table 3.6 Check of Predicted z

	0.1		0.05		0.01	
n	table	fit	table	fit	table	fit
1	3.84	3.75	5.02	4.93	7.88	7.9
2	5.99	5.94	7.38	7.34	10.6	10.67
3	7.81	7.78	9.35	9.32	12.84	12.92
4	9.49	9.46	11.14	11.13	14.86	14.94
5	11.07	11.04	12.83	12.82	16.75	16.83
10	18.31	18.29	20.48	20.48	25.19	25.25
15	25	24.98	27.49	27.49	32.8	32.86
20	31.41	31.4	34.17	34.17	40	40.05
30	43.77	43.77	46.98	46.98	53.67	53.71
50	67.5	67.5	71.42	71.42	79.49	79.52
75	96.22	96.21	100.84	100.84	110.29	110.31
100	124.34	124.34	129.56	129.56	140.17	140.19

Table 3. 7 Test of Chi-square Formula

This provides answers that are accurate to within 3%. For better accuracy (error less than 1%) and n greater than 2 or 3:

$$\chi^2_{\alpha,\nu} = \nu\left(1 - \frac{2}{9\nu} + z_\alpha\sqrt{\frac{2}{9\nu}}\right)^3$$

Table 3.7 shows the check of results.

t Distribution

For one-sided t tests with $n = 5$ or greater:

$$t(.05, n) = n / (.607793\, n - .55934) \approx n / (.6078\, n - .5593)$$

$$t(.1, n) = n / (.780241\, n - .51419) \approx n / (.7802\, n - .5142)$$

These are correct to three places beyond the decimal. When you extrapolate in a t table, you should use $1/t$ and $1/n$, since they are linear, as these equations show.

Total area in both tails

$$\alpha_t = 2\alpha_\Phi\left(\frac{t\left(1 - \frac{1}{4\nu}\right)}{\sqrt{1 + \frac{t^2}{2\nu}}}\right)$$

with less than 2% error for $\nu > 13$ and less than 1% error for $\nu > 18$. Note: $\alpha_\Phi(x) = 1 - \Phi(x)$ = one-sided tail area under normal distribution.

Table 3.8 shows the check of results.

alpha(n, t) = .05		
n	t	alpha
5	2.571	0.0581
10	2.228	0.0519
15	2.131	0.0508
20	2.086	0.0504
30	2.04227	0.0502
40	2.02108	0.0501
60	2.0003	0.05
120	1.97993	0.05

(a)

alpha(n, t) = .01		
n	t	alpha
5	4.032	0.0181
10	3.169	0.0117
15	2.947	0.0107
20	2.845	0.0104
30	2.75	0.0102
40	2.704	0.0101
60	2.66	0.01
120	2.617	0.01

(b)

Table 3.8 Predicted t

With a stated amount of confidence, you can perform comparison tests to determine whether your sample mean is equal to the population mean or some other specific constant or whether two populations have the same mean, based on a sample from each. You can perform similar tests to see whether the sample mean is greater than (or less than) the population mean. You can also perform similar tests on variances and proportions if the sample sizes are large, but small samples are more complicated.

WHAT HAVE YOU LEARNED?

- Most of the important work in making a comparison is done before you get to the calculations, and most of the important work in doing the calculations is done before you look at the numbers.
- There is a big difference between data looking different and being significantly different statistically. There is also a big difference between data being statistically different and practically different.
- The more qualitative, categorical, and subjective your data are, the more you should consider a nonparametric/distribution-free test (Chapter 7). The more quantitative and precise your data are, especially if they arise from a continuous variable, the more you should consider a classical technique.
- Working with distributions of sampling statistics makes meaningful comparisons possible.
- Confidence intervals may be used to express the spread of your data. They may be solved for sample size to provide an estimate of the *minimum* sample size when you plan your test.
- Since most of your analysis is likely to be on-the-job rather than for a journal publication, some formulas are provided for use with your pocket calculator. You'll probably find, however, that the answers you get are as good as a table provides when the sample size is above 3 or 4.

REFERENCES

1. C. Hastings, Jr., Approximations for Digital Computers. Princeton Univ. Press, 1955.

4

ANALYSIS OF VARIANCE

Analysis of Variance (ANOVA) is an effective analysis tool that allows for the simultaneous comparison of populations to determine if they are identical or if they are significantly different. ANOVA is an important resource for the evaluation of existing data and as a tool to be used in conjunction with a Designed Experiment. ANOVA tells us if the variance we see in our data is significant and measures that significance. ANOVA accomplishes this by measuring the variance of different treatments and treatment levels, such as time, temperature, cost, manufacturer, or process. These treatments all have specific values and are also called input variables or independent variables. These are the variables that we can control as input to our evaluation. The variable(s) whose values are affected by these treatments are often called response variables, outcome variables, or dependent variables.

ANOVA determines whether the means for several treatments are equal by examining population variances using Fisher's F Statistic. The F statistic is based on the evaluation of the variance (s^2) of the data. ANOVA compares two estimates of this variance, one estimate attributable to the variance within treatments (S^2_{Within}) which is also called error, and one estimate from between treatment means ($S_{2Treatments}$).

We calculate the first estimate from the variance within all the data from several distinct treatments or different levels of one treatment. This within/error-treatment estimate is the unbiased estimate of variance that remains the same, whether the means of the treatments are the same or different. It is the average, or mean, of the variances found within all the data.

The second estimate of the population variance is calculated from the variance between the individual treatment means ($S^2_{Treatments}$). This estimate is a true representation of the variance only if there is no significant difference between it and the variance within sample means (S^2_{Within}). Fisher's F statistic measures that difference in means based on the ratio of the variance between and the variance within treatments, and we compare that ratio to the critical value of F ($F_{Critical}$). In this overview we will discuss two types of ANOVA, One-Way ANOVA and Two-Way ANOVA.

- One-Way ANOVA
- Two-Way ANOVA
- Linear Contrasts
- Multivariant ANOVA

ONE-WAY ANOVA

One-Way ANOVA provides for the analysis of two populations with a single treatment. This will assist the analyst in determining if there are differences in such things as the quality of materials coming from two suppliers, the warranty returns from two different areas, the differences between two processes producing the same product, and any other combination of inputs to a single treatment.

The computation for ANOVA can be complex and tedious. In most cases, ANOVA is accomplished using a software program. We strongly recommend this approach. For the purpose of gaining an understanding of variance and to have the ability to execute some ANOVA evaluations, we have also simplified a manual approach to ANOVA by developing a generic computation and decision table, Table 4.1, which demonstrates all the intermediate steps in ANOVA.

This table also establishes the sequence of steps to perform the ANOVA as we implement it for the evaluation of process data and for use in designed experiments. There are six steps in our approach to the ANOVA:

1. Calculate the sum of the squares.
2. Determine the degrees of freedom.
3. Calculate the mean squares.
4. Calculate the F ratio.
5. Look up the critical value of F.
6. Calculate the % contribution.

To demonstrate these steps, we will use the data from Table 4.2. These data represent the acceptance yield of six lots of 100 received from two different suppliers.

Source of Variation	Sum of the Squares	Degrees of Freedom	Mean Squares	F Ratio	F' Critical	% Contribution
Treatment	$SS_{Treatment}$	$df_{Treatment}$	$MS_{Treatment}$	$F_{Treatment}$	$F'_{Treatment}$	$\%_{Treatment}$
Within	SS_{Within}	df_{Within}	MS_{Within}			$\%_{Within}$
Total	SS_{Total}	df_{Total}				

Table 4.1 One-Way ANOVA Computation and Decision Table

Supplier A	
A1	A2
98	89
94	99
97	94
98	99
97	92
100	96

Table 4.2 One-Way ANOVA Data

These data, the ANOVA equations and calculations, and the evaluation of linear contrasts have all been specifically formulated to use simple spreadsheets during your analysis.

Step 1. Sum of the Squares

The sum of the squares is the first calculation on our ANOVA table. We must calculate the sum of the squares for total variation (SS_{Total}) for the variation attributed to the treatment ($SS_{Treatment}$) and the variation attributed to error (within) (SS_{Within}). We have significantly simplified the approach to calculating the sum of the squares and designed statistical notation that is comprehensible and easily translated to a spreadsheet.

Sum of the Squares Total (SS_{Total}) The sum of the squares for the total variation SS_{Total} is calculated by subtracting the squared value of the summation of y $(\Sigma y)^2$ divided by the total number of data points (N) from the summation of the squared individual values of y Σy^2, as indicated in Eq. (4.1).

$$SS_{Total} = \sum y^2 - \frac{\left(\sum y\right)^2}{N} \tag{4.1}$$

This equation is easily translated into a simple spreadsheet as indicated in Table 4.3. The spreadsheet is used to calculate the SS_{total}.

$$SS_{Total} = \sum y^2 - \left(\sum y\right)^2 / N$$

$$SS_{Total} = 110901 - 110784.08 = 116.92$$

Sum of the Squares Treatment ($SS_{Treatment}$) To calculate the sum of the squares for the treatment variation ($SS_{Treatment}$), first summarize the squared values of each level

n	y	y^2	n	y	y^2
1	98	9604	7	89	7921
2	94	8836	8	99	9801
3	97	9409	9	94	8836
4	98	9604	10	99	9801
5	97	9409	11	92	8464
6	100	10000	12	96	9216
$SS_{Total} = \sum y^2 - (\sum y)^2 / N$				$\sum y$ 1153	$\sum y^2$ 110901
				$(\sum y)^2$ 1329409	
$SS_{Total} = 110901 - 110784.08 = 116.92$				$(\sum y)^2 / N$ 110784.08	

Table 4.3 SS_{Total} Calculation Spreadsheet

of our data for each treatment ($A1$, $A2$, etc.) divided by the number of data points in each level (n) and subtract the squared value of the summation of y ($(\Sigma y)^2$) divided by the total number of data points (N), as indicated in Eq. (4.2).

$$SS_{Treatment} = \left(\frac{(\sum Y_{A2})^2}{n} + \frac{(\sum Y_{A2})^2}{n} \right) - \frac{(\sum y)^2}{N} \tag{4.2}$$

Applying this equation to our data from the data table, we can use a spreadsheet to calculate the values for the Treatment Sum of the Squares as indicated in Table 4.4.

$$SS_{Treatment} = \left(\frac{(\sum Y_{A2})^2}{n} + \frac{(\sum Y_{A2})^2}{n} \right) - \frac{(\sum y)^2}{N}$$

$$SS_{Treatment} = (56842.67 + 53960.17) - 110784.08 = 18.75$$

Sum of the Squares Within (SS_{Within}) The sum of the squares for within is the total sum of the squares minus the sum of the squares for the treatment as indicated in Eq. (4.3).

$$SS_{Within} = SS_{Total} - SS_{Treatment} \tag{4.3}$$

Using this equation with the data we used in developing the calculation of the sum of the squares element of the ANOVA table, we can determine the SS_{Within} as follows:

n	y_{A1}	n	y_{A2}
1	98	1	89
2	94	2	99
3	97	3	94
4	98	4	99
5	97	5	92
6	100	6	96
$\sum Y_{A1}$	584	$\sum Y_{A2}$	569
$\left(\sum Y_{A1}\right)^2$	341056	$\left(\sum Y_{A2}\right)^2$	323761
$\left(\sum Y_{A1}\right)^2/n$	56842.67	$\left(\sum Y_{A2}\right)^2/n$	53960.17

$$SS_{Treatment} = \left(\left(\sum Y_{A2}\right)^2 \Big/ n + \left(\sum Y_{A2}\right)^2 \Big/ n\right) - \left(\sum y\right)^2 \Big/ N$$

$$SS_{Treatment} = (56842.67 + 53960.17) - 110784.08 = 18.75$$

Table 4.4 Spreadsheet Calculations for $SS_{Treatment}$

$$SS_{Within} = SS_{Total} - SS_{Treatment}$$

$$SS_{Within} = 116.92 - 18.75 = 98.17$$

The calculations for the sum of the squares can now be added to Table 4.5.

Step 2. Degrees of Freedom

Degrees of Freedom (*df*) are the number of independent comparisons available to evaluate the data. It is necessary to determine the degrees of freedom for the Treatment, Within, and Total. The degrees of freedom for the Treatment is the number of treatments minus one, $T - 1$; in this case we have two treatments ($A1$ and $A2$). The total degrees of freedom are the total number of data elements in the analysis, minus one, $N - 1$; in this case we have 12 data elements. The degrees of freedom for within

Source of Variation	Sum of the Squares	Degrees of Freedom	Mean Squares	F Ratio	F' Critical	% Contribution
Treatment	18.75	$df_{Treatment}$	$MS_{Treatment}$	$F_{Treatment}$	$F'_{Treatment}$	$\%_{Treatment}$
Within	98.17	df_{Within}	MS_{Within}			$\%_{Within}$
Total	116.92	df_{Total}				

Table 4.5 ANOVA Table with Sum of the Squares Applied

Source of Variation	Sum of the Squares	Degrees of Freedom	Mean Squares	F Ratio	F' Critical	% Contribution
Treatment	18.75	1	$MS_{Treatment}$	$F_{Treatment}$	$F'_{Treatment}$	$\%_{Treatment}$
Within	98.17	10	MS_{Within}			$\%_{Within}$
Total	116.92	11				

Table 4.6 ANOVA Table with Degrees of Freedom Applied

are calculated by subtracting the treatment degrees of freedom from the total degrees of freedom. This information can then be applied to Table 4.6.

$$df_{Treatment} = T - 1 = 2 - 1 = 1$$

$$df_{Total} = N - 1 = 12 - 1 = 11$$

$$df_{Within} = df_{Total} - df_{Treatment} = 11 - 1 = 10$$

Step 3. Mean Squares

The Mean Squares (MS) element of the ANOVA table is the quotient of the Sum of the Squares of the treatment and within, and the degrees of freedom (*df*) as indicated in Eqs. (4.4) and (4.5):

$$MS_{Treatment} = \frac{SS_{Treatment}}{df_{Treatment}} \tag{4.4}$$

$$MS_{Within} = \frac{SS_{Within}}{df_{Within}} \tag{4.5}$$

Applying this equation to our data from the sum of the squares, we can calculate the values for the Mean Squares as indicated in the following equations. This information can then be applied to Table 4.7.

$$MS_{Treatment} = \frac{SS_{Treatment}}{df_{Treatment}} = \frac{18.75}{1} = 18.75$$

$$MS_{Within} = \frac{SS_{Within}}{df_{Within}} = \frac{98.17}{10} = 9.82$$

Source of Variation	Sum of the Squares	Degrees of Freedom	Mean Squares	F Ratio	F' Critical	% Contribution
Treatment	18.75	1	18.75	$F_{Treatment}$	$F'_{Treatment}$	$\%_{Treatment}$
Within	98.17	10	9.82			$\%_{Within}$
Total	116.92	11				

Table 4.7 ANOVA Table with Mean Squares Applied

Source of Variation	Sum of the Squares	Degrees of Freedom	Mean Squares	F Ratio	F' Critical	% Contribution
Treatment	18.75	1	18.75	1.91	$F'_{Treatment}$	$\%_{Treatment}$
Within	98.17	10	9.82			$\%_{Within}$
Total	116.92	11				

Table 4.8 ANOVA Table with the *F* Ratio Applied

Step 4. *F* Ratio

The *F Ratio* is the quotient of the $MS_{Treatment}$ and MS_{Within}. It is calculated using Eq. (4.6):

$$F_{Ratio} = \frac{MS_{Treatments}}{MS_{Within}} \tag{4.6}$$

Applying this equation to the Mean Squares data, we can calculate the value for the F_{Ratio} as indicated in the following equation. This information can then be applied to Table 4.8.

$$F_{Ratio} = \frac{MS_{Treatments}}{MS_{Within}} = \frac{18.75}{9.82} = 1.91$$

Step 5. Look Up the Critical Value of F'

We compare the F_{Ratio} to the critical value of F' ($F'_{Critical}$) to determine if the variance demonstrated is significant. The critical value of F' is determined by referring to the *F* table (Table A-2 in Appendix A) for the applicable degrees of freedom and significance level selected for your evaluation.

The level of significance applied to the analysis can be a very subjective choice where no specific standards exist. It is important that some standard for selection of the significance level be implemented and applied to analysis uniformly throughout. The selection of the level of significance often reflects the consequences of the decision that will result from the analysis. A typical decision table for level of significance is provided in Table 4.9.

$$F': df_{Numerator}, df_{Denominator}, \alpha$$

We will select a significance level (a) of .05. Therefore, our $F'_{Critical}$ will be:

$$F': df_{1}, df_{11}, .05 = 4.96$$

$$F_{Ratio} > F_{Critical} = \textit{Significant}$$

$$F_{Ratio} < F_{Critical} = \textit{Not Significant}$$

1.91 < 4.96 Not Significant

Decision	Level
Critical Systems Parameters System Reliability System Safety Requirements System Performance Systems Competitive Capability	.01 >
Process Efficiency or Effectiveness Process Improvement Options Process Selection Process Differentiation Equipment Selection	.05 >
Administrative/Business Decisions Payment of Bonuses Return on Investment Marketing Decisions	.10 >

Table 4.9 Significance Decision Table

This information can then be applied to ANOVA Table 4.10.

Step 6. Calculate the % Contribution

The % Contribution element of the ANOVA table is the quotient of the Sum of the Squares for the treatment and within, and the Sum of the Squares for Total as indicated in Eqs. (4.7) and (4.8):

$$\%Contribution\ Treatment = \frac{SS_{Treatment}}{SS_{Total}} \quad (4.7)$$

$$\%Contribution\ Within = \frac{SS_{Within}}{SS_{Total}} \quad (4.8)$$

Using this equation, we can calculate the % Contribution using the information from our calculations of the sum of the squares.

Source of Variation	Sum of the Squares	Degrees of Freedom	Mean Squares	F Ratio	F' Critical	% Contribution
Treatment	18.75	1	18.75	1.91	4.96	$\%_{Treatment}$
Within	98.17	10	9.82			$\%_{Within}$
Total	116.92	11				

Table 4.10 ANOVA Table with the Critical Value of F′ Applied

Source of Variation	Sum of the Squares	Degrees of Freedom	Mean Squares	F Ratio	F' Critical	% Contribution
Treatment	18.75	1	18.75	1.91	4.96	16.00
Within	98.17	10	9.82			84.00
Total	116.92	11				

Table 4.11 Completed ANOVA Computation and Decision Table

$$\%Contribution\ Treatment = \frac{SS_{Treatment}}{SS_{Total}} = \frac{18.75}{116.92} = .16$$

$$\%Contribution\ Within = \frac{SS_{Within}}{SS_{Total}} = \frac{98.17}{116.92} = .84$$

The final step can now be completed by applying the data to the ANOVA Calculation and Decision Table as indicated in Table 4.11:

Evaluating the Results of ANOVA

We can now use the completed table in Table 4.11 as a fact-based decision-making tool. Evaluating the data in this form becomes almost intuitive. A few of the important facts that we can extract from the table are listed here. These *facts* form the basis for informed business and engineering decisions concerning such Critical Business Success Factors as design, materials selection, procurement, training, variability reduction programs, process selection, and the management of further designed experiments. In this specific case:

- The Product variance caused by the treatment (Supplier) was not significant.
- The supplier treatment is contributing 16% to the overall product variability.
- Eighty-four percent of the product variability is not accounted for in this analysis.

TWO-WAY ANOVA

One-way ANOVA deals only with one treatment. One hypothesis is tested, namely, that all means are equal. It is often necessary to determine if two different treatments are effecting a process or product and if their effect is significant. Two-way ANOVA provides a tool to make that assessment of two treatments. To demonstrate this, the data table used for one-way ANOVA has been further divided into two treatments, Treatment *A* (Supplier) and Treatment *B* (Test set). This division is often accomplished as a result of the cause-and-effect analysis performed before every DOE or through the evaluation of existing data and using engineering assessments. These data now represent the acceptance yield of six lots of 100 received from two

Supplier A			Test Set B
A1	A2		
98	89	B1	
94	99		
97	94		
98	99	B2	
97	92		
100	96		

Table 4.12 Two-Way ANOVA Data

different suppliers and accepted on two different acceptance test sets, as indicated in Table 4.12.

Two-way ANOVA can be accomplished in the same six steps as one-way ANOVA, by adding the second treatment to the ANOVA Computation and Decision Table. This is presented in Table 4.13.

1. Calculate the sum of the squares.
2. Determine the degrees of freedom.
3. Calculate the mean squares.
4. Calculate the F ratio.
5. Look up the critical value of F.
6. Calculate the % contribution.

Step 1. Sum of the Squares

The sum of the squares is the first calculation on all ANOVA tables. We must calculate the Sum of the Squares for total variation (SS_{Total}) for the variation attributed to the treatments (SS_A and SS_B) and the variation attributed to error (SS_{Within}).

Source of Variation	Sum of the Squares	Degrees of Freedom	Mean Squares	F Ratio	F' Critical	% Contribution
A	SS_A	df_A	MS_A	F_A	F'_A	% A
B	SS_B	df_B	MS_B	F_B	F'_B	% B
Within	SS_{Within}	df_{Within}	MS_{Within}			% Within
Total	SS_{Total}	df_{Total}				

Table 4.13 ANOVA Analysis and Decision Table

n	y	y^2	n	y	y^2
1	98	9604	7	89	7921
2	94	8836	8	99	9801
3	97	9409	9	94	8836
4	98	9604	10	99	9801
5	97	9409	11	92	8464
6	100	10000	12	96	9216
$SS_{Total} = \sum y^2 - (\sum y)^2 / N$				$\sum y$ 1153	$\sum y^2$ 110901
				$(\sum y)^2$ 1329409	
$SS_{Total} = 110901 - 110784.08 = 116.92$				$(\sum y)^2 / N$ 110784.08	

Table 4.14 SS_{Total} Calculation Spreadsheet

Sum of the Squares Total (SS_{Total}) Calculating the sum of the squares for Total is accomplished in exactly the same way for all ANOVA calculations (One-way, Two-way, and Multivariant). The process is repeated here only for continuity. The sum of the squares for the total variation SS_{Total} is calculated by subtracting the squared value of the summation of y $(Sy)^2$ divided by the total number of data points (N) from the summation of the squared individual values of y Sy^2, as indicated in Eq. (4.9).

$$SS_{Total} = \sum y^2 - \frac{(\sum y)^2}{N} \tag{4.9}$$

This equation is easily translated into a simple spreadsheet as indicated in Table 4.14. The spreadsheet is used to calculate the SS_{Total}

$$SS_{Total} = \sum y^2 - (\sum y)^2 / N$$

$$SS_{Total} = 110901 - 110784.08 = 116.92$$

Sum of the Squares Treatments ($SS_{Treatments}$) To calculate the sum of the squares for the variation of two treatments ($SS_{Treatments}$), first summarize the squared values of each level of one treatment, then perform the same function for the other treatment, of the data for each treatment ($A1$, $A2$, $B1$, $B2$, etc.) divided by the number of data points in each level (n) and subtract the squared value of the summation of y $((\Sigma y)^2)$ divided by the total number of data points (N), as indicated in Eq. (4.10).

$$SS_A = \left(\left(\sum Y_{A1}\right)^2 \Big/ n + \left(\sum Y_{A2}\right)^2 \Big/ n \right) - \left(\sum y\right)^2 \Big/ N$$

$$SS_B = \left(\left(\sum Y_{B1}\right)^2 \Big/ n + \left(\sum Y_{B2}\right)^2 \Big/ n \right) - \left(\sum y\right)^2 \Big/ N \tag{4.10}$$

Applying this equation to our data from the data table, we can use a spreadsheet to calculate the values for the Treatments Sum of the Squares as indicated in Table 4.15. Since all treatments for sums of the squares are calculated identically and the spreadsheets are constructed in the same way, only the spreadsheet for SS_B is presented.

$$SS_A = \left(\left(\sum Y_{A2}\right)^2 \Big/ n + \left(\sum Y_{A2}\right)^2 \Big/ n \right) - \left(\sum y\right)^2 \Big/ N$$

$$SS_A = (56842 + 53960.17) - 110784.08 = 18.75$$

$$SS_B = \left(\left(\sum Y_{B2}\right)^2 \Big/ n + \left(\sum Y_{B2}\right)^2 \Big/ n \right) - \left(\sum y\right)^2 \Big/ N$$

$$SS_B = (52828.17 - 58016.67) - 110784.08 = 60.75$$

n	y_{B1}	n	y_{B2}
1	97	1	98
2	94	2	98
3	97	3	100
4	89	4	99
5	92	5	99
6	94	6	96
$\sum y_{B1}$	563	$\sum y_{B2}$	590
$\left(\sum y_{B1}\right)^2$	316969	$\left(\sum y_{B2}\right)^2$	348100
$\left(\sum y_{B1}\right)^2 \Big/ n$	52828.17	$\left(\sum y_{B2}\right)^2 \Big/ n$	58016.67

$$SS_A = \left(\left(\sum Y_{B1}\right)^2 \Big/ n + \left(\sum Y_{B1}\right)^2 \Big/ n \right) - \left(\sum y\right)^2 \Big/ N$$

$$SS_A = (52828.17 - 58016.67) - 110784.08 = 60.75$$

Table 4.15 SS_B Calculation Spreadsheet

Source of Variation	Sum of the Squares	Degrees of Freedom	Mean Squares	F Ratio	F' Critical	% Contribution
A	18.75	df_A	MS_A	F_A	F'_A	$\% A$
B	60.75	df_B	MS_B	F_B	F'_B	$\% B$
Within	37.420	df_{Within}	MS_{Within}			$\%_{Within}$
Total	116.920	df_{Total}				

Table 4.16 ANOVA Table with Sum of the Squares Applied

Sum of the Squares Within (SS_{Within}) The sums of the squares for the treatments are cumulative; therefore, we can calculate the sum of the squares for within as the total sum of the squares minus the sum of the squares for the treatments, as indicated in Eq. (4.11).

$$SS_{Within} = SS_{Total} - SS_{Treatments} \tag{4.11}$$

Using this equation with the data we have used in developing the calculation of the sum of the squares element of the ANOVA table, we can determine the SS_{Within} as follows:

$$SS_{Within} = SS_{Total} - SS_{Treatments}$$

$$SS_{Within} = 116.92 - (18.75 + 60.75) = 37.42$$

The calculations for the sum of the squares can now be added to Table 4.16.

Step 2. Degrees of Freedom

Degrees of Freedom (*df*) are the number of independent comparisons available to evaluate the data. It is necessary to determine the degrees of freedom for the Treatment, Within, and Total. The degrees of freedom for the Treatment is the number of treatments minus one, $T - 1$; in this case we have two treatments ($A1$ and $A2$). The total treatments are the total number of data elements in the analysis minus one, $N - 1$; in this case we have 12 data elements. The degrees of freedom for within are calculated by subtracting the treatment degrees of freedom from the total degrees of freedom. This information can then be applied to Table 4.17.

$$df_A = T_A - 1 = 2 - 1 = 1$$

$$df_B = T_B - 1 = 2 - 1 = 1$$

$$df_{Total} = N - 1 = 12 - 1 = 11$$

$$df_{Within} = df_{Total} - df_{Treatments} = 11 - 2 = 9$$

Source of Variation	Sum of the Squares	Degrees of Freedom	Mean Squares	F Ratio	F' Critical	% Contribution
A	18.75	1	MS_A	F_A	F'_A	$\% A$
B	60.75	1	MS_B	F_B	F'_B	$\% B$
Within	37.420	9	MS_{Within}			$\%_{Within}$
Total	116.920	11				

Table 4.17 ANOVA Table with Degrees of Freedom Applied

Step 3. Mean Squares

The Mean Squares (MS) element of the ANOVA table is the quotient of the sum of the squares for each treatment and the sum of the squares for within, and the degrees of freedom (*df*) as indicated in Eqs. (4.12) and (4.13):

$$MS_A = \frac{SS_A}{df_A} \quad MS_B = \frac{SS_B}{df_B} \tag{4.12}$$

$$MS_{Within} = \frac{SS_{Within}}{df_{Within}} \tag{4.13}$$

Applying this equation to our data from the sum of the squares, we can calculate the values for the Mean Squares as indicated in the following equations. This information can then be applied to Table 4.18.

$$MS_A = \frac{SS_A}{df_A} = \frac{18.75}{1} = 18.75$$

$$MS_B = \frac{SS_B}{df_B} = \frac{60.75}{1} = 60.75$$

$$MS_{Within} = \frac{SS_{Within}}{df_{Within}} = \frac{37.42}{9} = 4.16$$

Source of Variation	Sum of the Squares	Degrees of Freedom	Mean Squares	F Ratio	F' Critical	% Contribution
A	18.75	1	18.75	F_A	F'_A	$\% A$
B	60.75	1	60.75	F_B	F'_B	$\% B$
Within	37.420	9	4.16			$\%_{Within}$
Total	116.920	11				

Table 4.18 ANOVA Table with Mean Squares Applied

Source of Variation	Sum of the Squares	Degrees of Freedom	Mean Squares	F Ratio	F' Critical	% Contribution
A	18.75	1	18.75	4.50	F'_A	% A
B	60.75	1	60.75	14.60	F'_B	% B
Within	37.420	9	4.16			% Within
Total	116.920	11				

Table 4.19 ANOVA Table with the F_{Ratio} Applied

Step 4. *F* Ratio

The *F Ratio* is the quotient of the $MS_{Treatments}$ for each treatment and MS_{Within}. It is calculated using Eq. (4.14):

$$F_{Ratio} = \frac{MS_{Treatments}}{MS_{Within}} \tag{4.14}$$

Applying this equation to the Mean Squares data, we can calculate the value for the *F* Ratio as indicated in the following equations. This information can then be applied to Table 4.19.

$$F_A = \frac{MS_A}{MS_{Within}} = \frac{18.75}{4.16} = 4.50$$

$$F_B = \frac{MS_B}{MS_{Within}} = \frac{60.75}{4.16} = 14.60$$

Step 5. Look Up the Critical Value of *F'*

We compare the F_{Ratio} to the critical value of F' ($F_{Critical}$) to determine if the variance demonstrated is significant. The critical value of F' is determined by referring to the *F* table (Table A-2 in Appendix A) for the applicable degrees of freedom and significance level selected for your evaluation.

$$F': df_{Numerator}, df_{Denominator}, a$$

$$F': df_1, df_9, .05 = 5.12$$

$$F_{Ratio} > F_{Critical} = Significant$$

$$F_{Ratio} < F_{Critical} = Not\ Significant$$

$$F_A = 4.50 < 5.12 \quad Not\ Significant$$

$$F_B = 14.60 > 5.12 \quad Significant$$

Source of Variation	Sum of the Squares	Degrees of Freedom	Mean Squares	F Ratio	F′ Critical	% Contribution
A	18.75	1	18.75	4.50	5.12	% A
B	60.75	1	60.75	14.60	5.12	% B
Within	37.420	9	4.16			% Within
Total	116.920	11				

Table 4.20 ANOVA Table with Critical Value of *F* Applied

This information can then be applied to ANOVA Table 4.20.

Step 6. Calculate the % Contribution

The % Contribution element of the ANOVA table is the quotient of the sum of the squares for the treatment and within, and the sum of the squares for Total as indicated in Eqs (4.15) and (4.16):

$$\%Contribution\ Treatments = \frac{SS_{Treatments}}{SS_{Total}} \tag{4.15}$$

$$\%Contribution\ Within = \frac{SS_{Within}}{SS_{Total}} \tag{4.16}$$

Using this equation, we can calculate the % Contribution using the information from our calculations of the sum of the squares.

$$\%Contribution\ A = \frac{SS_A}{SS_{Total}} = \frac{18.75}{116.92} = .16$$

$$\%Contribution\ B = \frac{SS_B}{SS_{Total}} = \frac{60.75}{116.92} = .52$$

$$\%Contribution\ Within = \frac{SS_{Within}}{SS_{Total}} = \frac{37.42}{116.92} = .32$$

The final step can now be completed by applying the data to the ANOVA Calculation and Decision Table as indicated in Table 4.21:

Evaluating the Results of ANOVA

We can now use the Two-Way ANOVA table we just completed as a fact-based decision-making tool. These data now indicate the following:

Source of Variation	Sum of the Squares	Degrees of Freedom	Mean Squares	F Ratio	F' Critical	% Contribution
A	18.75	1	18.75	4.50	5.12	16.00
B	60.75	1	60.75	14.60	5.12	52.00
Within	37.420	9	4.16			32.00
Total	116.920	11				

Table 4.21 Completed ANOVA Table

- The Product variance caused by the treatment *A* (Supplier) is not significant.
- The Product variance caused by the treatment *B* (Test Set) is significant.
- The test set treatment is contributing 52% to the overall product variability.
- Thirty-seven percent of the product variability is not accounted for in this analysis.

LINEAR CONTRASTS

Accepting or rejecting the hypothesis in the ANOVA analysis implies that there is a difference in means. Using ANOVA, the exact nature of this difference is not specified. A further understanding may be useful. Linear contrasts can provide this additional understanding and also provide for a graphical representation of the data. A contrast is the sum of the high-level mean minus the sum of the low-level mean.

One-Way Linear Contrasts

Determining that the data from an ANOVA are significant implies that there is a difference between treatment means (as measured by their variance). But the exact nature of these differences is not specified. The effect of the difference of the means is not measured; rather, it is only an assessment that they are "significantly different." Using linear contrasts will provide a tool to measure that difference and also provide a visualization of that difference. This approach to measuring the effects empirically and graphically is demonstrated in Figure 4.1. The contrast between the two input variables ($A-$ and $A+$) is measured as the difference between the means of each treatment $\bar{y}_{A-} - \bar{y}_{A+}$. In this way we measure the effects as indications of the linear contrasts; it also provides an assessment of the effects of the different treatments and provides a graph of these effects.

The analysis of linear contrasts can be accomplished in four steps.

1. Calculate the effects.
2. Graph the effects.
3. Determine the significance of the linear contrast.
4. Evaluate the results.

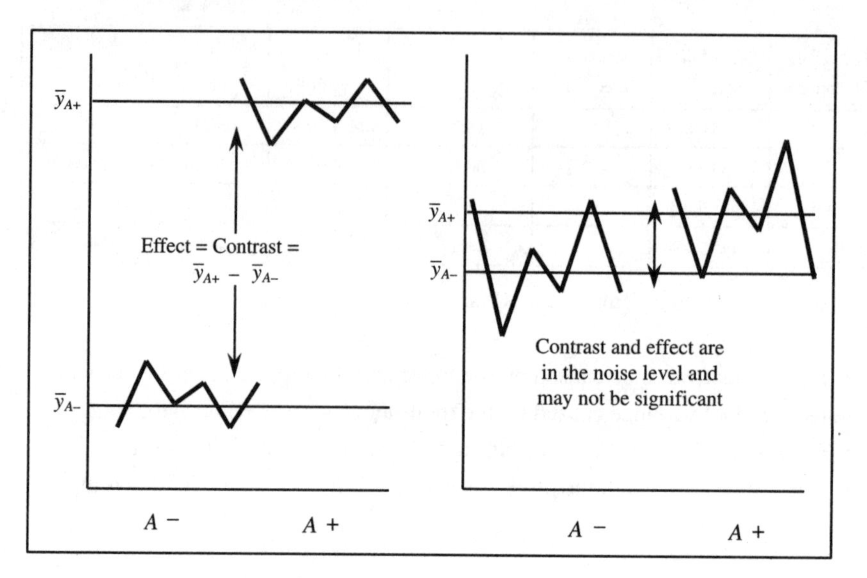

Figure 4.1 Linear Contrasts

To demonstrate these steps, we will use the data in Table 4.22. You will recognize these data as the test set data from the evaluation of two-way ANOVA. Using this same data will assist you in comprehending the differences between ANOVA and linear contrasts. Additionally, we have introduced the different types of notation used in DOE to define levels of a treatment. These levels can be defined as Level 1 and Level 2 ($B1$, $B2$) or as Minus and Plus ($B-$, $B+$).

Step 1. Calculate the Effects of the different treatments using the following equation. The effect is the sum of the treatment data at the plus level ($\Sigma K+$) divided by the total number of samples at that level ($n+$), minus the sum of the treatment data at the plus level ($\Sigma K-$) divided by the total number of samples at that level ($n-$).

Test Set	
*B*1 (–)	*B*2 (+)
98	98
94	97
97	100
89	99
99	92
94	96

Table 4.22 Linear Contrast Data

Test Set				
*B*1 (–)		*B*2 (+)		
	98		98	$Effect_k = \sum \frac{K+}{n_{K+}} - \sum \frac{K-}{n_{K-}}$
	94		97	
	97		100	$Effect_B = \frac{582}{6} - \frac{571}{6}$
	89		99	
	99		92	$Effect_B = 1.83$
	94		96	
$\sum K_{B-}$	571	$\sum K_{B+}$	582	
$\sum K_{B-}/n_{B-}$	95.17	$\sum K_{B+}/n_{B+}$	97.00	
		$Effect_B$	1.83	

Table 4.23 Effects Calculation Spreadsheet

$$Effect_k = \sum \frac{K+}{n} - \sum \frac{K-}{n}$$

Using this equation, we can now calculate the effects. This equation is also easily translated to a spreadsheet, as indicated in Table 4.23. The spreadsheet is used to calculate the effect.

$$Effect_k = \sum \frac{K+}{n} - \sum \frac{K-}{n}$$

$$Effect_B = \frac{582}{6} - \frac{571}{6}$$

$$Effect_B = 1.83$$

Step 2. Graph the Effects We can now graph the effect of the treatment levels to evaluate what effect the levels are having on the quality characteristic. Figure 4.2 displays the graphs associated with the effect we have calculated.

The slope of the line indicates the significance of the effect. The steeper the slope, the more significant the effect is. The graph also indicates which treatment level is producing the desired effect of optimizing the process. This graph of linear effects indicates that *B*+ would optimize yield. The calculation of the effects indicates that taking this action would increase yield by 1.83 units.

Step 3. Determine the Significance of the Linear Contrast Determine the significance of the linear contrasts by using Fisher's *F* Statistic (ANOVA). This can be accomplished using the same first steps as One-Way ANOVA. Since we have covered

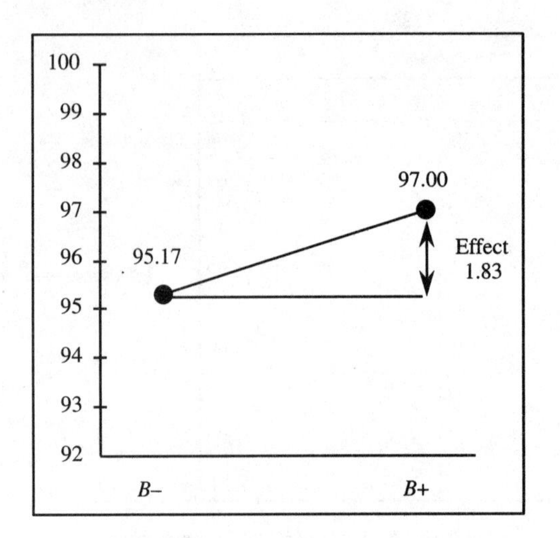

Figure 4.2 Linear Contrast Graph

these functions in detail previously, we will use the statistical function from the Excel program to perform the evaluation, as indicated in Table 4.24.

Step 4. Evaluating the Results Now that we have completed our one-way linear contrast evaluation, we can make several *fact-based* determinations:

- $B-$ provides a yield of 95.17.
- $B+$ provides a yield of 97.00.

ANOVA: Single-Factor

Summary

Groups	Count	Sum	Average	Variance
*B*1 –	6.00	571.00	95.17	13.37
*B*2 +	6.00	582.00	97.00	8.00

ANOVA

Source of Variation	*SS*	*df*	*MS*	*F*	*p*-value	*F* crit
Between Groups	10.08	1.00	10.08	0.94	0.35	4.96
Within Groups	106.83	10.00	10.68			
Total	116.92	11				

Table 4.24 Test for Significance Using ANOVA

- The effect is 1.83.
- The value of the F_{Ratio} (F) does not exceed the critical value of F_{Crit}.
- Therefore, the contrast may be caused by the data or some other factor not taken into account in this analysis.
- The p-value in this printout indicated that the confidence level is 0.35.

Two-Way Linear Contrasts

Just as with the relationship of one-way ANOVA to one-way linear contrasts, so does the two-way linear contrast provide a method to measure the effect of the differences of two treatments simultaneously. This is very important because different treatments can affect the results of your analysis by their individual effects and their interactions. In two-way ANOVA, we will be measuring the effects of interactions for the first time. This is a good introduction to the methods of performing the analysis of Classical and Taguchi DOE. Figure 4.3 illustrates the linear relationship of two variables ($A+, A-$ and $B+, B-$).

The analysis of two-way linear contrasts can be accomplished in the same four steps as one-way linear contrasts.

1. Calculate the effects.
2. Graph the effects.
3 Determine the significance of the linear contrast.
4. Evaluate the results.

To demonstrate these steps, we will use the data in Table 4. 25. Your will recognize this data as the test set data from the evaluation of two-way ANOVA. Using this same data will assist you in comprehending the differences between ANOVA and linear

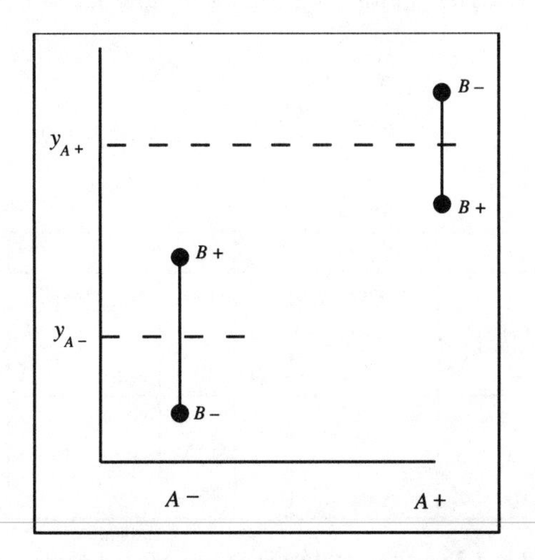

Figure 4.3 Two-Way Linear Contrasts

	A1 (–)	A2 (+)
B1 (–)	97	89
	94	92
	97	94
B2 (+)	98	99
	98	99
	100	96

Table 4.25 Linear Contrast Data

contrasts. Additionally, we have introduced the different types of notation used in DOE to define levels of treatment.

Step 1. Calculate the Effects of the different treatments using the same equation as used for one-way linear contrasts. The effect is the sum of the treatment data at the plus level ($\Sigma K+$) divided by the total number of samples at that level (n_{K+}), minus the sum of the treatment data at the plus level ($\Sigma K-$) divided by the total number of samples at that level (n_{K-}), for each treatment (A, B). This gives us an opportunity to demonstrate the relationship between ANOVA, Linear Contrasts, and the DOE Orthogonal Arrays. The data from Table 4.25 translate directly into a test matrix, as indicated in Table 4.26.

The test matrix describes how treatments within each test cell are allocated, as indicated in cells 1, 2, 3, and 4 in Table 4.27. That matrix can then be translated directly into an orthogonal array.

Recall the following equation, which is used to determine effects. It requires that we summarize the treatment effects at the plus level and at the minus level (Σ+, Σ–). This is called the sum plus–sum minus approach to evaluating treatment data.

$$Effect_k = \sum \frac{K+}{n_{K+}} - \sum \frac{K-}{n_{K-}}$$

	A1 (–)	A2 (+)
B1 (–)	97	89
	94	92
	97	94
B2 (+)	98	99
	98	99
	100	96

→

	A1 (–)	A2 (+)
B1 (–)	A(–) B (–) 1	A (+) B (–) 3
B1 (+)	A (–) B (+) 2	A (+) B (–) 4

Table 4.26 Translating a Data Table into a Test Matrix

	A1 (–)	A2 (+)
B1 (–)	A(–) B (–) 1	A (+) B (–) 3
B1 (+)	A (–) B (+) 2	A (+) B (–) 4

	A	B	AB
1	–	–	+
2	–	+	–
3	+	–	–
4	+	+	+

Table 4.27 Translating Test Cells into an Orthogonal Array

This is a very simplified approach and does not require software or advanced mathematics. In Table 4.28 we have applied the data from the test matrix onto the orthogonal array. The data from test cell 1 are applied to Run 1, and so on.

Once this is accomplished, we need to summarize the test data in the $\bar{y}$ column of the orthogonal array. Then we can summarize the plus and minus treatment conditions for each treatment and interaction. This is accomplished at the end of Table 4.28. Now we have the data we need to evaluate and plot the effects.

$$Effect_A = \sum \frac{A+}{n_{A+}} - \sum \frac{A-}{n_{A-}} = \frac{189.67}{2} - \frac{194.67}{2}$$

$$Effect_A = 94.83 - 97.33 = -2.5$$

$$Effect_B = \sum \frac{B+}{n_{B+}} - \sum \frac{B-}{n_{B-}} = \frac{196.67}{2} - \frac{187.67}{2}$$

$$Effect_B = 98.33 - 93.83 = -4.5$$

$$Effect_{AB} = \sum \frac{AB+}{n_{AB+}} - \sum \frac{AB-}{n_{AB-}} = \frac{194.00}{2} - \frac{190.33}{2}$$

$$Effect_A = 97.00 - 95.16 = -1.84$$

	A	B	AB	y_1	y_2	y_3	$\bar{y}$
1	–	–	+	97	94	97	96.00
2	–	+	–	98	98	100	98.67
3	+	–	–	89	92	94	91.67
4	+	+	+	99	99	96	98.00
Σ+	189.67	196.67	194.00				
Σ–	194.67	187.67	190.33				

Table 4.28 Orthogonal Array with Data Applied for Analysis

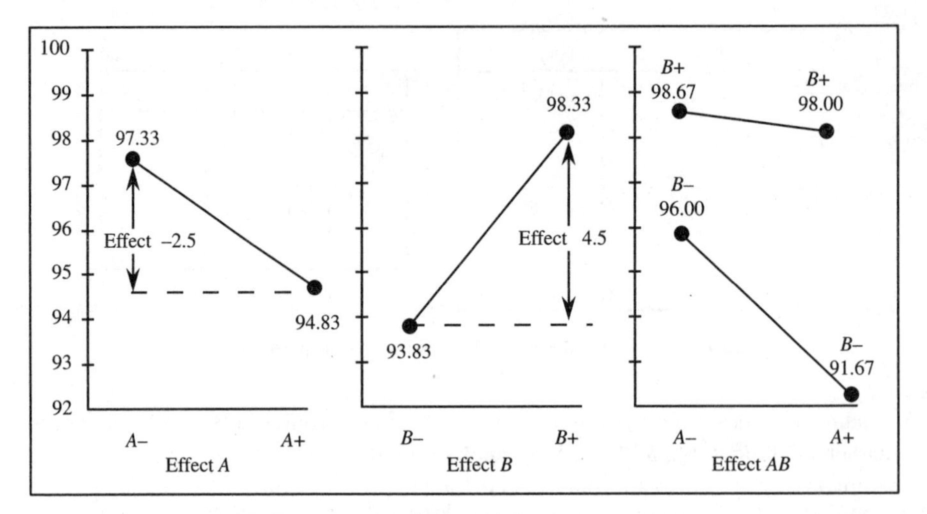

Table 4.29 Two-Way Linear Contrast Graph

Step 2. Graph the Effects We can now graph the effect of the treatment levels to evaluate what effect they are having on the quality characteristic. Table 4.29 displays the graphs associated with the effect we have calculated.

Here also the slope of the line indicates the significance of the effect. The steeper the slope, the more significant the effect. The graph also indicates which treatment and treatment level are producing the desired effect of optimizing the process. This graph of linear effects indicates that $A-$ and $B+$ would optimize yield. In the graph of interactions, where the lines run parallel, there is no or only a small interaction; when the lines cross, there is a clear interaction. The graph of the interactions indicates that there are no or a very small interaction effect.

Step 3. Determine the Significance of the Linear Contrast Determine the significance of the linear contrasts by using Fisher's F Statistic (ANOVA). This can be accomplished using the same first steps from one-way ANOVA. Since we have covered these functions in detail previously, we will use the statistical function from the Excell program to perform the evaluation. This information is presented in Table 4.30.

Source of Variation	*SS*	*df*	*MS*	*F*	F_{crit}
Rows (B)	69.42	1.00	69.42	21.73	6.61
Columns (A)	18.75	1.00	18.75	5.87	6.61
Error	28.75	9.00	3.19		
Total	116.92	11.00			

Table 4.30 Test for Significance Using a Spreadsheet Program

Step 4. Evaluating the Results Now that we have completed our one-way linear contrast evaluation, we can make several *fact-based* determinations:

- Treatment A is optimized at the negative treatment level $A-$.
- Treatment B is optimized at the positive treatment level $B+$.
- The effect of treatment A is 12.5.
- The effect of treatment B is 4.5.
- The value of the F_{Ratio} (F) for Treatment A does not exceed the critical value of F_{Crit}.
- The value of the F_{Ratio} (F) for Treatment B does exceed the critical value of F_{Crit}.

MULTIVARIANT ANOVA

Table 4.31 also establishes the sequence of steps to perform the multivariant ANOVA as we implement it for the evaluation of process data and for use in designed experiments. There are six steps in our approach to the ANOVA:

1. Calculate the sum of the squares.
2. Determine the degrees of freedom.
3. Calculate the mean squares.
4. Calculate the F ratio.
5. Look up the critical value of F.
6. Calculate the % contribution.

These steps follow directly the construction of the ANOVA computation and decision table, as indicated in Table 4.31.

To demonstrate these steps, we will use the data from Table 4.32. To provide a reasonable level of confidence in this data, we have selected a minimum sample size of three as discussed in Chapter 2.

Step 1. Sum of the Squares

The sum of the squares is the first calculation on our ANOVA table. We must calculate the sum of the squares for total variation (SS_{Total}) for the variation attributed to treatment main effects ($SS_{Main\ Effects}$), and interactions ($SS_{Interactions}$), and the variation

Step	1	2	3	4	5	6
Source of Variation	Sum of the Squares	Degrees of Freedom	Mean Squares	F Ratio	F' Critical	% Contribution

Table 4.31 ANOVA Table Process Steps

		A							
		A1			A2				
B	B1	10	12	11	10	11	10	C1	C
	B1	8	11	9	12	9	10	C2	
	B2	11	11	11	13	16	15	C1	
	B2	13	14	13	16	15	16	C2	

Table 4.32 ANOVA Data

attributed to error (SS_{Error}). The classic view of the calculation of the sums of the squares is given in Eqs. (4.17) through (4.19).

$$SS_{Total} = \sum_{i=1}^{a} \sum_{j=1}^{b} \sum_{k=1}^{c} \sum_{1=1}^{n} y_{ijkl}^2 - \frac{y_{...}^2}{abcn} \tag{4.17}$$

$$SS_{Treatment} = \sum_{i=1}^{a} \frac{y_{i...}^2}{bcn} - \frac{y_{...}^2}{abcn} \tag{4.18}$$

$$SS_{Interactions} = \sum_{i=1}^{a} \sum_{j=1}^{b} \frac{y_{ij...}^2}{cn} - \frac{y_{...}^2}{abcn} \tag{4.19}$$

We have significantly simplified this approach and designed statistical notation that is comprehensible and easily translated to a spreadsheet. For this reason, we will not discuss the intricacies of this dot and summary notation. The following calculations for the sums of the squares follow this approach.

Sum of the Squares Total (SS_{Total}) The sum of the squares for the total variation SS_{Total} is calculated by subtracting the squared value of the summation of y $((\Sigma y)^2)$ divided by the total number of data points (N) from the summation of the squared individual values of y (Σy^2), as indicated in Eq. (4.20).

$$SS_{Total} = \sum y^2 - \frac{\left(\sum y\right)^2}{N} \tag{4.20}$$

This equation is easily translated into a simple spreadsheet, as indicated in Table 4.33. The spreadsheet is used to calculate the SS_{total}.

$$SS_{Total} = \sum y^2 - \frac{\left(\sum y\right)^2}{N}$$

$$SS_{Total} = 3561 - \frac{82369}{24}$$

$$SS_{Total} = 3561.000 - 3432.042$$

$$SS_{Total} = 128.958$$

n	y	y^2	n	y	y^2
1	10	100	13	10	100
2	8	64	14	12	144
3	11	121	15	13	169
4	13	169	16	16	256
5	12	144	17	11	121
6	11	121	18	9	81
7	11	121	19	16	256
8	14	196	20	15	225
9	11	121	21	10	100
10	9	81	22	10	100
11	11	121	23	15	225
12	13	169	24	16	256
$SS_{Total} = \sum y^2 - \frac{(\sum y)^2}{N}$			$\sum y$	287	$\sum y^2$ 3561
			$(\sum y)^2$	82369	
$SS_{Total} = 128.958$			$(\sum y)^2/N$	3432.042	

Table 4.33 SS_{Total} Calculation Spreadsheet

Sum of the Squares Treatment ($SS_{Treatments}$) To calculate the sum of the squares for treatment variation ($SS_{Treatments}$) for Treatment Main Effects (SS_A, SS_B), first summarize the squared values of each level of our data for each treatment ($A1$, $A2$, etc.) divided by the number of data points in each level (n) and subtract the squared value of the summation of y ($(\Sigma y)^2$) divided by the total number of data points (N), as indicated in Eq. (4.21).

$$SS_{Main\ Effects} = \left(\frac{\sum y_i^2}{n} + \frac{\sum y_j^2}{n} \right) - \frac{(\sum y)^2}{N} \tag{4.21}$$

Applying this equation to our data from the data table we can use a spreadsheet to calculate the values for the Treatment Main Effects Sum of the Squares for treatment A SS_A, as indicated in Table 4.34.

n	y_{A1}	n	y_{A2}
1	10	1	10
2	8	2	12
3	11	3	13
4	13	4	16
5	12	5	11
6	11	6	9
7	11	7	16
8	14	8	15
9	11	9	10
10	9	10	10
11	11	11	15
12	13	12	16
$\sum y_{A1}$	134	$\sum y_{A2}$	153
$\left(\sum y_{A1}\right)^2$	17956.000	$\left(\sum y_{A2}\right)^2$	23409.000
$\left(\sum y_{A1}\right)^2 / n$	1496.333	$\left(\sum y_{A2}\right)^2 / n$	1950.750
$SS_A = \left(\frac{\sum y_{A1}^2}{n_{A1}} + \frac{\sum y_{A2}^2}{n_{A2}}\right) - \frac{\left(\sum y\right)^2}{N} = 15.042$			

Table 4.34 Spreadsheet Calculations for SS_A

$$SS_A = \left(\frac{\sum y_{A1}^2}{n_{A1}} + \frac{\sum y_{A2}^2}{n_{A2}}\right) - \frac{\left(\sum y\right)^2}{N}$$

$$SS_A = \left(\frac{17956.000}{12} + \frac{23409.000}{12}\right) - \frac{82369.000}{24}$$

$$SS_A = (1496.333 + 1950.750) - 3432.042$$

$$SS_A = 15.042$$

In the same way SS_B and SS_C can be calculated using the same equation and a spreadsheet, as demonstrated in Table 4.35.

n	y_{B1}	n	y_{B2}	n	y_{C1}	n	y_{C2}
1	10	1	11	1	10	1	8
2	12	2	11	2	12	2	11
3	11	3	11	3	11	3	9
4	10	4	13	4	10	4	12
5	11	5	16	5	11	5	9
6	10	6	15	6	10	6	10
7	8	7	13	7	11	7	13
8	11	8	14	8	11	8	14
9	9	9	13	9	11	9	13
10	12	10	16	10	13	10	16
11	9	11	15	11	16	11	15
12	10	12	16	12	15	12	16
$\sum y_{B1}$	123	$\sum y_{B2}$	164	$\sum y_{C1}$	141	$\sum y_{C2}$	146
$\left(\sum y_{B1}\right)^2$	155129.000	$\left(\sum y_{B2}\right)^2$	26896.000	$\left(\sum y_{C1}\right)^2$	19881.000	$\left(\sum y_{C2}\right)^2$	21316.000
$\left(\sum y_{B1}\right)^2 / n$	1260.750	$\left(\sum y_{B2}\right)^2 / n$	2241.333	$\left(\sum y_{C1}\right)^2 / n$	1656.750	$\left(\sum y_{C2}\right)^2 / n$	1776.333
$SS_B = \left(\frac{\sum y_{B1}^2}{n_{B1}} + \frac{\sum y_{B2}^2}{n_{B2}}\right) - \frac{\left(\sum y\right)^2}{N} = 70.042$				$SS_C = \left(\frac{\sum y_{C1}^2}{n_{C1}} + \frac{\sum y_{C2}^2}{n_{C2}}\right) - \frac{\left(\sum y\right)^2}{N} = 1.042$			

Table 4.35 Spreadsheet Calculations for SS_B and SS_C

$$SS_B = \left(\frac{\sum y_{B1}^2}{n_{B1}} + \frac{\sum y_{B2}^2}{n_{B2}}\right) - \frac{\left(\sum y\right)^2}{N}$$

$$SS_B = \left(\frac{15129.000}{12} + \frac{26896.000}{12}\right) - \frac{82369.000}{24}$$

$$SS_B = (1260.750 + 2241.333) - 3432.042$$

$$SS_B = 70.042$$

$$SS_C = \left(\frac{\sum y_{C1}^2}{n_{C1}} + \frac{\sum y_{C2}^2}{n_{C2}}\right) - \frac{\left(\sum y\right)^2}{N}$$

$$SS_C = \left(\frac{19881.00}{12} + \frac{21316.000}{12}\right) - \frac{82369.000}{24}$$

$$SS_C = (1656.750 + 1776.333) - 3432.042$$

$$SS_C = 1.042$$

Sum of the Squares for Interactions ($SS_{Interactions}$) For Treatment Interactions (SS_{AxB}, SS_{AxC}, etc.), we summarize the squared values of each level of our data for each treatment ($A1$, $A2$, etc.) and subtract the squared value of the summation of y ($(\Sigma y)^2$) divided by the total number of data points (N), and subtract the sum of the squares for the associated main effects (SS_A, SS_B, etc.), as indicated in Eq. (4.22)

$$SS_{Interactions} = \left(\frac{\sum y_{i1}^2}{n} + \frac{\sum y_{i2}^2}{n} + \frac{\sum y_{j1}^2}{n} + \frac{\sum y_{j2}^2}{n} \right) - \frac{\left(\sum y\right)^2}{N} - SS_i - SS_j \qquad (4.22)$$

Applying this equation to our data from Table 4.32 we can use a spreadsheet to calculate the values for the Treatment Interactions of treatments AxB, as indicated in Table 4.36.

$$SS_{AB} = \left(\frac{\sum y_{A1B1}^2}{n_{A1B1}} + \frac{\sum y_{A1B2}^2}{n_{A1B2}} + \frac{\sum y_{A2B1}^2}{n_{A2B1}} + \frac{\sum y_{A2B2}^2}{n_{A2B2}} \right) - \frac{\left(\sum y\right)^2}{N} - SS_A - SS_B$$

$$SS_{AB} = \left(\frac{3721}{6} + \frac{5329}{6} + \frac{3844}{6} + \frac{8281}{6} \right) - \frac{82369}{24} - 15.042 - 70.042$$

$$SS_{AB} = (620.167 + 888.167 + 640.667 + 1380.167) - 3432.042 - 15.042 - 70.042$$

$$SS_{AB} = 12.041$$

Using this same method, we can calculate the interactions for the interactions of AxC, BxC, and $AxBxC$ as indicated in Tables 4.37 through 4.39.

n	y_{A1B1}	n	y_{A1B2}	n	y_{A2B1}	n	y_{A2B2}
1	10	1	11	1	10	1	13
2	12	2	11	2	11	2	16
3	11	3	11	3	10	3	15
4	8	4	13	4	12	4	16
5	11	5	14	5	9	5	15
6	9	6	13	6	10	6	16
$\sum y_{A1B1}$	61	$\sum y_{A1B2}$	73	$\sum y_{A2B1}$	62	$\sum y_{A2B2}$	91
$\left(\sum y_{A1B1}\right)^2$	3721	$\left(\sum y_{A1B2}\right)^2$	5329	$\left(\sum y_{A2B1}\right)^2$	3844	$\left(\sum y_{A2B2}\right)^2$	8281
$\left(\sum y_{A1B1}\right)^2/n$	620.167	$\left(\sum y_{A1B2}\right)^2/n$	888.167	$\left(\sum y_{A2B1}\right)^2/n$	640.667	$\left(\sum y_{A2B2}\right)^2/n$	1380.167

$$SS_{AB} = \left(\frac{\left(\sum y_{A1B1}\right)^2}{n} + \frac{\left(\sum y_{A1B2}\right)^2}{n} + \frac{\left(\sum y_{A2B1}\right)^2}{n} + \frac{\left(\sum y_{A2B2}\right)^2}{n} \right) - \frac{\left(\sum y\right)^2}{N} - SS_A - SS_B = 12.041$$

Table 4.36 Spreadsheet for Calculating Interaction of Treatment AxB (SS_{AxB})

n	y_{A1C1}	n	y_{A1C2}	n	y_{A2C1}	n	y_{A2C2}
1	10	1	8	1	10	1	12
2	12	2	11	2	11	2	9
3	11	3	9	3	10	3	10
4	11	4	13	4	13	4	16
5	11	5	14	5	16	5	15
6	11	6	13	6	15	6	16
$\sum y_{A1C1}$	66	$\sum y_{A1C2}$	68	$\sum y_{A2C1}$	75	$\sum y_{A2C2}$	78
$\left(\sum y_{A1C1}\right)^2$	4356	$\left(\sum y_{A1C2}\right)^2$	4624	$\left(\sum y_{A2C1}\right)^2$	5625	$\left(\sum y_{A2C2}\right)^2$	6084
$\left(\sum y_{A1C1}\right)^2/n$	726.000	$\left(\sum y_{A1C2}\right)^2/n$	770.667	$\left(\sum y_{A2C1}\right)^2/n$	937.500	$\left(\sum y_{A2C2}\right)^2/n$	1014.000

$$SS_{AC} = \left(\frac{\left(\sum y_{A1C1}\right)^2}{n} + \frac{\left(\sum y_{A1C2}\right)^2}{n} + \frac{\left(\sum y_{A2C1}\right)^2}{n} + \frac{\left(\sum y_{A2C2}\right)^2}{n}\right) - \frac{\left(\sum y\right)^2}{N} - SS_A - SS_C = 0.041$$

Table 4.37 Spreadsheet for Calculating Interaction of Treatments AxC (SS_{AxC})

$$SS_{AC} = \left(\frac{\sum y_{A1C1}^2}{n_{A1C1}} + \frac{\sum y_{A1C2}^2}{n_{A1C2}} + \frac{\sum y_{A2C1}^2}{n_{A2C1}} + \frac{\sum y_{A2C2}^2}{n_{A2C2}}\right) - \frac{\left(\sum y\right)^2}{N} - SS_A - SS_C$$

$$SS_{AC} = \left(\frac{4356}{6} + \frac{4624}{6} + \frac{5625}{6} + \frac{6084}{6}\right) - \frac{82369}{24} - 15.042 - 1.042$$

$$SS_{AC} = (726.000 + 770.667 + 5625.000 + 1 - 14.000) - 3432.042 - 15.042 - 1.042$$

$$SS_{AC} = 0.041$$

n	y_{B1C1}	n	y_{B1C2}	n	y_{B2C1}	n	y_{B2C2}
1	10	1	8	1	11	1	13
2	12	2	11	2	11	2	14
3	11	3	9	3	11	3	13
4	10	4	12	4	13	4	16
5	11	5	9	5	16	5	15
6	10	6	10	6	15	6	16
$\sum y_{B1C1}$	64	$\sum y_{B1C2}$	59	$\sum y_{B2C1}$	77	$\sum y_{B2C2}$	87
$\left(\sum y_{B1C1}\right)^2$	4096	$\left(\sum y_{B1C2}\right)^2$	3481	$\left(\sum y_{B2C1}\right)^2$	5929	$\left(\sum y_{B2C2}\right)^2$	7569
$\left(\sum y_{B1C1}\right)^2/n$	682.667	$\left(\sum y_{B1C2}\right)^2/n$	580.167	$\left(\sum y_{B2C1}\right)^2/n$	988.167	$\left(\sum y_{B2C2}\right)^2/n$	1261.500

$$SS_{BC} = \left(\frac{\left(\sum y_{B1C1}\right)^2}{n} + \frac{\left(\sum y_{B1C2}\right)^2}{n} + \frac{\left(\sum y_{B2C1}\right)^2}{n} + \frac{\left(\sum y_{B2C2}\right)^2}{n}\right) - \frac{\left(\sum y\right)^2}{N} - SS_B - SS_C = 9.374$$

Table 4.38 Spreadsheet for Calculating Interaction of Treatment BxC (SS_{BxC})

$$SS_{BC} = \left(\frac{\sum y_{B1C1}^2}{n_{B1C1}} + \frac{\sum y_{B1C2}^2}{n_{B1C2}} + \frac{\sum y_{B2C1}^2}{n_{B2C1}} + \frac{\sum y_{B2C2}^2}{n_{B2C2}} \right) - \frac{\left(\sum y\right)^2}{N} - SS_B - SS_C$$

$$SS_{BC} = \left(\frac{4096}{6} + \frac{3481}{6} + \frac{5929}{6} + \frac{7569}{6} \right) - \frac{82369}{24} - 70.042 - 1.042$$

$$SS_{BC} = (682.667 + 580.167 + 988.167 + 1261.500) - 3432.042 - 70.042 - 1.042$$

$$SS_{BC} = 9.374$$

Sum of the Squares Error (SS_{Error}) The sums of the squares for main effects and interactions are cumulative; therefore, we can calculate the sum of the squares for error as the total sum of the squares minus the sum of the squares for treatments and interactions as indicated in Eq.(4.23).

$$SS_{Error} = SS_{Total} - (SS_{Main\ Effects} + SS_{Interactions}) \tag{4.23}$$

Using this equation with the data we have used in developing the calculation of the sum of the squares element of the ANOVA table, we can determine the SS_{Error} as follows:

$$SS_{Error} = SS_{Total} - (SS_{Main\ Effects} + SS_{Interactions})$$

$$SS_{Error} = 128.958 - (86.126 + 16.456)$$

$$SS_{Error} = 26.332$$

Please note that although this is called the sum of the squares for error, it should not imply that an error has been made in your data acquisition, calculations, or

n	A1B1C1	A1B1C2	A1B2C1	A1B2C2	A2B1C1	A2B1C2	A2B2C1	A2B2C2
1	10	8	11	13	10	12	13	16
2	12	11	11	14	11	9	16	15
3	11	9	11	13	10	10	15	16
$\sum y_{ABC}$	33	28	33	40	31	31	44	47
$\left(\sum y_{ABC}\right)^2$	1089	784	1089	1600	961	961	1936	2209
$\left(\sum y_{ABC}\right)^2 / n$	363	261.333	363.000	533.333	320.333	320.333	645.333	736.333

$$SS_{ABC} = \left(\frac{\left(\sum y_{A1B1C1}\right)^2}{n} + \frac{\left(\sum y_{A1B1C2}\right)^2}{n} + \frac{\left(\sum y_{A1B2C1}\right)^2}{n} + \frac{\left(\sum y_{A1B2C2}\right)^2}{n} + \frac{\left(\sum y_{A2B1C1}\right)^2}{n} + \frac{\left(\sum y_{A2B1C2}\right)^2}{n} + \frac{\left(\sum y_{A2B2C1}\right)^2}{n} + \frac{\left(\sum y_{A2B2C2}\right)^2}{n} \right)$$

$$- \frac{\left(\sum y\right)^2}{N} - SS_A - SS_B - SS_C - SS_{AB} - SS_{AC} - SS_{BC} = 3.376$$

Table 4.39 Spreadsheet for Calculating Interaction of Treatment $AxBxC$ (SS_{AxBxC})

Step	1	2	3	4	5	6
Source of Variation	Sum of the Squares	Degrees of Freedom	Mean Squares	F Ratio	F' Critical	% Contribution
A	15.042					
B	70.042					
C	1.042					
AB	12.041					
AC	0.041					
BC	9.374					
ABC	3.376					
Error	26.332					
Total	128.958					

Table 4.40 ANOVA Table with Sum of the Squared Data Applied

analysis. It is also not related to sampling error or any other statistic. Rather, this is the variance that has not been accounted for by your main effects or interactions. That variance can be attributed to the uncontrolled environment, treatments that you have not selected for evaluation or other variance other than between treatment data variance.

Step 1 can now be completed by applying the sum of the squares data to the ANOVA calculation and decision table as indicated in Table 4.40.

Step 2. Degrees of Freedom

Degrees of freedom (*df*) are the number of independent comparisons available to estimate a specific treatment or level of a treatment. Therefore, if you have N treatments, you would have $N-1$ degrees of freedom. This is also an element of the critical value of F. Degrees of freedom must be found for both the numerator and denominator of the F statistic. The number of degrees of freedom for treatment main effects is T_{A-1} or the number of applicable treatment levels minus one. The degrees of freedom for treatment interactions is $(T_a-1)\ (T_b-1)$ or the degrees of freedom for each of the effects in the interaction. The total degrees of freedom are the total number of independent comparisons for all treatments, treatment levels, and replicates $N-1$ or the total number of data elements in the sample minus one. The degrees of freedom applicable to variance within treatments (error) is the total degrees of freedom minus the degrees of freedom for treatments and treatment interactions. This can be demonstrated using the data in Table 4.32.

Treatment Main Effects The degrees of freedom derived from this data, for the treatment main effects, are the number of treatment levels for each treatment minus one. In this example, there are two treatment levels for each treatment; $df_{Main\ Effects}$ are:

$$df_{Treatment\ A} = T_A - 1 = 2 - 1 = 1$$

$$df_{Treatment\ B} = T_B - 1 = 2 - 1 = 1$$

$$df_{Treatment\ C} = T_C - 1 = 2 - 1 = 1$$

Treatment Interactions The degrees of freedom derived from this data, for treatment interactions, are the multiplied degrees of freedom for the associated treatments. In this example, there are four possible treatment interactions, A x B, A x C, B x C, and A x B x C; therefore, $df_{Interactions}$ are:

$$df_{Interaction\ A \times B} = (T_A - 1)(T_B - 1) = 1 \times 1 = 1$$

$$df_{Interaction\ A \times C} = (T_A - 1)(T_C - 1) = 1 \times 1 = 1$$

$$df_{Interaction\ B \times C} = (T_B - 1)(T_C - 1) = 1 \times 1 = 1$$

$$df_{Interaction\ A \times B \times C} = (T_A - 1)(T_B - 1)(T_C - 1) = 1 \times 1 \times 1 = 1$$

The degrees of freedom for within treatments (or error) are derived by subtraction of the $df_{\text{Main Effects}}$ and $df_{Interactions}$ from the df_{Total}:

$$df_{Error} = df_{Total} - (df_{Treatments} - df_{Interactions}) = 23 - 3 - 4 = 16$$

The total degrees of freedom for this figure are the total number of samples (24) minus one.

$$df_{Total} = N - 1 = 24 - 1 = 23$$

Step 2 can now be completed by applying the degrees of freedom data to the ANOVA calculation and decision table as indicated in Table 4.41

Step 3. Mean Squares

The Mean Squares (MS) element of the ANOVA table is the quotient of the sum of the squares for main effects and interactions and the degrees of freedom (*df*) as indicated in Eqs. (4.24) through (4.26).

$$MS_{Main\ Effects} = \frac{SS_{Main\ Effects}}{df_{Main\ Effects}} \tag{4.24}$$

$$MS_{Interactions} = \frac{SS_{Interactions}}{df_{Interactions}} \tag{4.25}$$

$$MS_{Error} = \frac{SS_{Error}}{df_{Error}} \tag{4.26}$$

Step	1	2	3	4	5	6
Source of Variation	Sum of the Squares	Degrees of Freedom	Mean Squares	F Ratio	F' Critical	% Contribution
A	15.042	1				
B	70.042	1				
C	1.042	1				
AB	12.041	1				
AC	0.041	1				
BC	9.374	1				
ABC	3.376	1				
Error	26.332	16				
Total	128.958	23				

Table 4.41 ANOVA Table with Degrees of Freedom Data Applied

Applying this equation to our data from Table 4.41, we can calculate the values for the mean squares as follows:

$$MS_A = \frac{SS_A}{df_A} = 15.042 \qquad MS_{AB} = \frac{SS_{AB}}{df_{AB}} = 12.041$$

$$MS_B = \frac{SS_B}{df_B} = 70.042 \qquad MS_{AC} = \frac{SS_{AC}}{df_{AC}} = 0.041$$

$$MS_C = \frac{SS_C}{df_C} = 1.042 \qquad MS_{BC} = \frac{SS_{BC}}{df_{BC}} = 1.042$$

$$MS_{ABC} = \frac{SS_{ABC}}{df_{ABC}} = 3.376$$

$$MS_{Error} = \frac{SS_{Error}}{df_{Error}} = \frac{26.332}{16} = 1.646$$

Step 3 can now be completed by applying the mean squares data to the ANOVA calculation and decision table as indicated in Table 4.42:

Step 4. *F* Ratio

The F_{Ratio} is the quotient of the $MS_{Main\ Effects}$, $MS_{Interactions}$, and MS_{Within} and can be calculated using Eq. (4.27).

$$F_{Ratio\ Main\ Effects} = \frac{MS_{Main\ Effects}}{MS_{Within}} \qquad F_{Ratio\ Interactions} = \frac{MS_{Interactions}}{MS_{Within}} \tag{4.27}$$

Step	1	2	3	4	5	6
Source of Variation	Sum of the Squares	Degrees of Freedom	Mean Squares	F Ratio	F' Critical	% Contribution
A	15.042	1	15.042			
B	70.042	1	70.042			
C	1.042	1	1.042			
AB	12.041	1	12.041			
AC	0.041	1	0.041			
BC	9.374	1	9.374			
ABC	3.376	1	3.376			
Error	26.332	16	1.646			
Total	128.958	23				

Table 4.42 ANOVA Table with Mean Squares Data Applied

Applying this equation to our data from Table 4.42, we can calculate the values for the F ratio as follows:

$$F_{Ratio\ A} = \frac{MS_A}{MS_{Within}} = \frac{15.042}{1.646} = 9.139$$

$$F_{Ratio\ B} = \frac{70.042}{1.646} = 42.553 \qquad F_{Ratio\ C} = \frac{1.042}{1.646} = 0.663$$

$$F_{Ratio\ AB} = \frac{12.041}{1.646} = 7.315 \qquad F_{Ratio\ AC} = \frac{0.041}{1.646} = 0.025$$

$$F_{Ratio\ BC} = \frac{1.042}{1.646} = 0.663 \qquad F_{Ratio\ ABC} = \frac{3.376}{1.646} = 2.051$$

Step 4 can now be completed by applying the F ratio data to the ANOVA calculation and decision table as indicated in Table 4.43:

Step 5. Look Up the Critical Value of F'

We compare the F_{Ratio} to the critical value of F' ($F_{Critical}$) to determine if the variance demonstrated is significant. The critical value of F' is determined by referring to the F table (Table A.2 in Appendix A) for the applicable degrees of freedom and significance level selected for your evaluation.

$$F': df_{Numerator}, df_{Denominator}, \alpha$$

The level of significance applied to the analysis can be a very subjective choice where no specific standards exist. It is important that some standard for selection of

Step	1	2	3	4	5	6
Source of Variation	Sum of the Squares	Degrees of Freedom	Mean Squares	F Ratio	F' Critical	% Contribution
A	15.042	1	15.042	9.139		
B	70.042	1	70.042	42.553		
C	1.042	1	1.042	0.633		
AB	12.041	1	12.041	7.315		
AC	0.041	1	0.041	0.025		
BC	9.374	1	9.374	5.695		
ABC	3.376	1	3.376	2.051		
Error	26.332	16	1.646			
Total	128.958	23				

Table 4.43 ANOVA Table with the *F* Ratio Data Applied

the significance level be implemented and applied to analysis uniformly throughout. The selection of the level of significance often reflects the consequences of the decision that will result from the analysis. A typical decision table for level of significance is provided in Table 4.9. We will select a significance level (α) of .10. Therefore, our $F'_{Critical}$ will be:

$$F': df^{1}, df^{16}, .10$$

The degrees of freedom for the numerator are the degrees of freedom for treatment main effects and interactions. The degrees of freedom for the denominator are the degrees of freedom for error. From our continuing example we can determine that the degrees of freedom for the numerator is one (1) for all main effects and interactions. The degrees of freedom for error are 16.

From the *F* distribution table (Table A.2) we can determine that the critical value of *F* is 3.05 for each interaction and main effect. The variance is significant when the F_{Ratio} exceeds the $F_{Critical}$ value:

$$F_{Ratio} > F_{Critical} = Significant$$

$$F_{Ratio} < F_{Critical} = Not\ Significant$$

Step 5 can now be completed by applying the critical value of F' to the ANOVA calculation and decision table as indicated in Table 4.44:

Step 6. Calculate the % Contribution

The % contribution element of the ANOVA table is the quotient of the sum of the squares for main effects, interactions, and error and the sum of the squares for total as indicated in Eq. (4.28).

Step	1	2	3	4	5	6
Source of Variation	Sum of the Squares	Degrees of Freedom	Mean Squares	*F* Ratio	*F'* Critical	% Contribution
A	15.042	1	15.042	9.139	3.050	
B	70.042	1	70.042	42.553	3.050	
C	1.042	1	1.042	0.633	3.050	
AB	12.041	1	12.041	7.315	3.050	
AC	0.041	1	0.041	0.025	3.050	
BC	9.374	1	9.374	5.695	3.050	
ABC	3.376	1	3.376	2.051	3.050	
Error	26.332	16	1.646			
Total	128.958	23				

Table 4.44 ANOVA Table with the Critical Value of *F* Applied

$$\%Contribution = \frac{SS_{Treatments}}{SS_{Total}},$$

$$\%Contribution = \frac{SS_{Interactions}}{SS_{Total}}, \tag{4.28}$$

$$\%Contribution = \frac{SS_{Error}}{SS_{Total}}$$

Using this equation, we can calculate the % Contribution using the information from Table 4.15.

$$\%_{Contibution\ A} = \frac{SS_A}{SS_{Total}} = \frac{15.042}{128.958} = .12 \qquad \%_{Contibution\ AB} = \frac{SS_{AB}}{SS_{Total}} = \frac{12.042}{128.958} = .09$$

$$\%_{Contribution\ B} = \frac{SS_B}{SS_{Total}} = \frac{70.042}{128.958} = .54 \qquad \%_{Contibution\ AC} = \frac{SS_{AC}}{SS_{Total}} = \frac{0.041}{128.958} = .00$$

$$\%_{Contribution\ C} = \frac{SS_C}{SS_{Total}} = \frac{1.042}{128.958} = .01 \qquad \%_{Contibution\ BC} = \frac{SS_{BC}}{SS_{Total}} = \frac{1.042}{128.958} = .01$$

$$\%_{Contibution\ ABC} = \frac{SS_{ABC}}{SS_{Total}} = \frac{3.376}{128.958} = .03$$

The final step can now be completed by applying the *F* ratio data to the ANOVA calculation and decision table as indicated in Table 4.45:

Step	1	2	3	4	5	6
Source of Variation	Sum of the Squares	Degrees of Freedom	Mean Squares	*F* Ratio	*F*′ Critical	% Contribution
A	15.042	1	15.042	9.139	3.050	12%
B	70.042	1	70.042	42.553	3.050	54%
C	1.042	1	1.042	0.633	3.050	1%
AB	12.041	1	12.041	7.315	3.050	9%
AC	0.041	1	0.041	0.025	3.050	0%
BC	9.374	1	9.374	5.695	3.050	7%
ABC	3.376	1	3.376	2.051	3.050	3%
Error	26.332	16	1.646			14%
Total	128.958	23				

Table 4.45 Completed ANOVA Computation and Decision Table

Evaluating the Results of ANOVA

We can now use the completed Table 4.45 as a fact-based decision-making tool. Evaluating the data in this form becomes almost intuitive. A few of the important facts that we can extract from the table are listed here. These *facts* form the basis for informed business and engineering decisions concerning such things as design, materials selection and procurement, training, variability reduction programs, process selection, and the management of further designed experiments.

- The variability is significant for treatments *A* and *B* and for the interaction of treatments *A* x *B*.
- The main effect of treatment *B* is contributing 54% to the total process, product, or system variability
- The main effect of treatment *A* is contributing 12% to the total variability.
- The interaction of treatments *A* x *B* is contributing 6% to the total variability.
- 20% of the total variability in this process or product is comprised of factors we did not evaluate.

WHAT HAVE YOU LEARNED?

- The Analysis of Variance (ANOVA) can be used to evaluate existing data and data from designed experiments.
- The ANOVA can be accomplished in six steps:
  - Calculate the sum of the squares.
  - Determine the degrees of freedom.

- Calculate the mean squares.
- Calculate the F ratio.
- Look up the critical value of F.
- Calculate the % contribution.

- Calculating the sum of the squares can be accomplished using the following equations:

$$SS_{Total} = \sum y^2 - \frac{\left(\sum y\right)^2}{N} \tag{4.20}$$

$$SS_{Main\ Effects} = \left(\frac{\sum y_i^2}{n} + \frac{\sum y_j^2}{n} \right) - \frac{\left(\sum y\right)^2}{N} \tag{4.21}$$

$$SS_{Interactions} = \left(\frac{\sum y_{i1}^2}{n} + \frac{\sum y_{i2}^2}{n} + \frac{\sum y_{j1}^2}{n} + \frac{\sum y_{i2}^2}{n} \right) - \frac{\left(\sum y\right)^2}{N} - SS_i - SS_j \tag{4.22}$$

$$SS_{Error} = SS_{Total} - \left(SS_{Main\ Effects} + SS_{Interactions}\right) \tag{4.23}$$

- The degrees of freedom (*df*) are the number of independent comparisons available to estimate a specific treatment or level of treatment. Determining the degrees of freedom can be accomplished as follows:

$$df_{Total} = T - 1$$

$$df_{Treatments} = T_A - 1$$

$$df_{Interactions} = (T_A - 1)(T_B - 1)$$

$$df_{Error} = df_{Total} - (df_{Treatments} + df_{Interactions})$$

- Calculating the Mean Squares (MS) can be accomplished using the following equations:

$$MS_{Main\ Effects} = \frac{SS_{Main\ Effects}}{SS_{Main\ Effects}} \tag{4.24}$$

$$MS_{Interactions} = \frac{SS_{Interactions}}{SS_{Interactions}} \tag{4.25}$$

$$MS_{Error} = \frac{SS_{Error}}{SS_{Error}} \tag{4.26}$$

- The F ratio (F) can be calculated using the following equation:

$$F_{Ratio\ Main\ Effects} = \frac{MS_{Main\ Effects}}{MS_{Within}} \qquad F_{Ratio\ Interaction} = \frac{MS_{Interactions}}{MS_{Within}} \tag{4.27}$$

- The critical value of F (F') is determined from the F table in the statistical appendix, depending on the degrees of freedom in the numerator ($df_{Numerator}$), the degrees of freedom in the denominator ($\text{df}_{Denominator}$), and the selected level of significance (α).

$$F': (df_{Numerator}), (\text{df}_{Denominator}), \alpha$$

- The percent contribution evaluates the proportion of variance attributable to each of the main effects, interactions, and error based on the ratio of each of these factors to the total sum of the squares.

$$\%Contribution = \frac{SS_{Treatments}}{SS_{Total}},$$
$$\%Contribution = \frac{SS_{Interactions}}{SS_{Total}}, \tag{4.28}$$
$$\%Contribution = \frac{SS_{Error}}{SS_{Total}}$$

- The evaluation of ANOVA is determined by the criticality of the data determined by the F statistic and by the evaluation of the percent contribution to variation.

TERMINOLOGY

The following are some of the terms you may have first encountered in this chapter:

Analysis of Variance: A statistical technique for determining the significance of the variation among several treatments using Fisher's F Statistic. ANOVA measures the criticality of the variation based on the comparison of the calculated value of F (F ratio) to the critical value of F (F') based on a specific level of significance (α) selected for your evaluation.

Degrees of Freedom: The number of independent values associated with a given factor, treatment, or treatment level.

Main Effect: The measured variation of a specific treatment.

Interaction: The measured variation as a result of the combined effect of two or more treatments, *AB*, *AC*, *BC*, and *ABC*.

Treatment: The controllable factors used as inputs to the products and processes under evaluation. These are the input variables (also called the independent variables) that can be varied to change the effect on the output or dependent variable (main effects and interactions).

Treatment Level: Levels are the value of the treatments being studied. In most instances of designed experiments, we can use two levels of the treatments: a high level, symbolized by a (+) sign or the number 1, and a low level, symbolized by a (–) sign or the number 2.

Percent Contribution (% *A*): The proportion of variation each factor contributes.

EXERCISES

1. Using the data table provided, complete the ANOVA calculation and decision table using the six steps described in this chapter. To accomplish this you may use a spreadsheet program, such as Excel or Lotus, or you can perform the calculations manually. Do not use a canned statistical program. Demonstrate your work at each step.

Data Matrix

		A							
			*A*1			*A*2			
B	*B*1	107	117	108	115	113	113	*C*1	*C*
	*B*1	115	107	117	118	123	123	*C*2	
	*B*2	108	107	108	117	118	116	*C*1	
	*B*2	111	112	109	120	118	117	*C*2	

ANOVA Table and Six Steps

Step	1	2	3	4	5	6
Source of Variation	Sum of the Squares	Degrees of Freedom	Mean Squares	*F* Ratio	F' Critical	% Contribution

2. The Martin Baker Marketing Research Company of Chicago conducts market surveys for various industries. These surveys are conducted by telephone and by personal interview. Because of the time differences across the country, the company employs two shifts—a morning shift starting at 7:00 A.M. and an afternoon shift starting at 1:00 P.M. The survey takers are divided between new employees who have recently graduated with a marketing degree and part-time employees of various qualifications. The company is attempting to determine which survey approach yields the most responses. They have collected data indicating the percentages of complete responses received, over a five-day work week:

Treatment Combinations

*A*1 First Shift	*B*1 Marketing Graduates	*C*1 Interviewing
*A*2 Second Shift	*B*2 Part-Time Employees	*C*2 Telephone

Data Table

		A											
		*A*1					*A*2						
B	*B*1	0.42	0.44	0.44	0.44	0.46	0.32	0.29	0.30	0.27	0.25	*C*1	*C*
	*B*1	0.44	0.45	0.45	0.49	0.44	0.27	0.28	0.29	0.26	0.27	*C*2	
	*B*2	0.47	0.47	0.44	0.42	0.46	0.33	0.31	0.33	0.30	0.33	*C*1	
	*B*2	0.45	0.49	0.47	0.46	0.48	0.31	0.33	0.31	0.33	0.31	*C*2	

Using this data, complete the ANOVA calculation and decision table using the six steps described in this chapter. What is the most significant contributor to complete marketing responses?

ANOVA Table and Six Steps

Step	1	2	3	4	5	6
Source of Variation	Sum of the Squares	Degrees of Freedom	Mean Squares	*F* Ratio	*F*′ Critical	% Contribution

Note: Check your calculation of the total degrees of freedom carefully for this exercise.

3. Evaluate the following ANOVA analysis and decision table.

Step	1	2	3	4	5	6
Source of Variation	Sum of the Squares	Degrees of Freedom	Mean Squares	*F* Ratio	*F*′ Critical	% Contribution
A	15.777	1	15.777	9.585	3.050	17%
B	13.568	1	13.568	8.243	3.050	15%
C	1.042	1	1.042	0.633	3.050	1%
AB	9.762	1	9.762	5.931	3.050	11%
AC	0.041	1	0.041	0.025	3.050	0%
BC	1.042	1	1.042	0.633	3.050	1%
ABC	3.376	1	3.376	2.051	3.050	4%
Error	45.756	16	2.860			51%
Total	90.364	23				

a. Which sources of variation are critical?

b. Based on the % contribution, which items should be selected for change?

c. What is the % contribution for error?

d. What does the % contribution for error indicate?

4. Evaluate the following ANOVA analysis and decision table.

Step	1	2	3	4	5	6
Source of Variation	Sum of the Squares	Degrees of Freedom	Mean Squares	F Ratio	F' Critical	% Contribution
A	16.721	1	16.721	10.159	3.050	19%
B	3.667	1	3.667	2.228	3.050	4%
C	39.456	1	39.456	23.971	3.050	46%
AB	5.987	1	5.987	3.637	3.050	7%
AC	1.076	1	1.076	0.654	3.050	1%
BC	1.042	1	1.042	0.633	3.050	1%
ABC	3.376	1	3.376	2.051	3.050	4%
Error	14.538	16	0.909			17%
Total	85.863	23				

a. Which sources of variation are critical?

b. Based on the % contribution, which items should be selected for change?

c. What is the % contribution for error?

d. What does the % contribution for error indicate?

5

FULL FACTORIAL EXPERIMENTS

Now that we have studied the planning and management necessary to execute a designed experiment, the basic statistical decision-making tools used to make intelligent decisions, and the basic analytical tool of Analysis of Variance (ANOVA), we are ready for the main event. Factorial experiments are widely used in business, industry, and services to define, describe, optimize, and improve products and processes. They are used in simultaneously evaluating a number of factors effecting the response of a product or process. This class of designs is of very great practical importance as the most frequently used technical decision-making tool. In this chapter, we will discuss the basic concepts of factorial experiments, and the uses, generation, and evaluation of 2^k experimental designs.

BASIC CONCEPTS OF FACTORIAL EXPERIMENTATION

Design of Experiments (DOE) is one of the most powerful tools available for the design, characterization, and improvement of products and services. DOE is a group of techniques used to organize and evaluate testing so that it provides the most valuable data and makes efficient use of assets. This chapter explains how to carry out the calculations needed to identify the factors that have the most influence on your process; hence the name *factorial analysis*. The intent here is to present a simple technique that can withstand day-to-day use in a variety of industries while providing a basic understanding that will ease further studies. A few concepts are therefore discussed in depth, rather than displaying a myriad of options. With this new technique, you will be able to decide whether a change in your process is worth the added cost, how to improve the yield or other quality characteristic, and how to reduce the variability of your product or service. Variability breeds dissatisfaction, both in the unlucky customer who got the poorer item and in the lucky one who then wonders whether his luck will hold the next time he buys. Customers *will* talk to each other! By using your design parameters (dimensions, materials, etc.) as factors in the analysis, you can make your quality characteristic insensitive ("robust") to an outside influence such as ambient temperature.

A designed experiment is simply a test or trial project that has been well structured to gain the most information from the response or yield caused by certain inputs and treatments. Full factorial experimental designs provide a comprehensive analysis of all treatments, their levels, and their interactions. Evaluating full factorial experiments includes:

- The analysis of the effects of the treatments and treatment levels.
- Graphing these effects.
- Performing an ANOVA to determine the significance of the treatments.
- Determining the percent contribution of the variance at each level.

Optimizing the quality characteristic depends on the metric selected and the nature of the process. If we are measuring a process yield, maximum is best; for failure rates, minimum is best; for the deviation from a variable standard. nominal is best.

Experimental Matrix Figure 5.1 displays a typical DOE experimental matrix or test matrix. The matrix indicates the number of treatments, levels of the treatments, and the treatment combinations for each experimental run (trial or test).

The DOE experimental matrix translates directly to an array suitable for computation. This matrix lays out the DOE in runs, indicating the levels for each run applied to the main effects and interactions of the treatments. In Figure 5.2 the treatment combination $A2B2C2$ is represented as experimental run 1, with main effects for A set as low ($-$), B set as low ($-$), and C set as low ($-$). The interactions for this experimen-

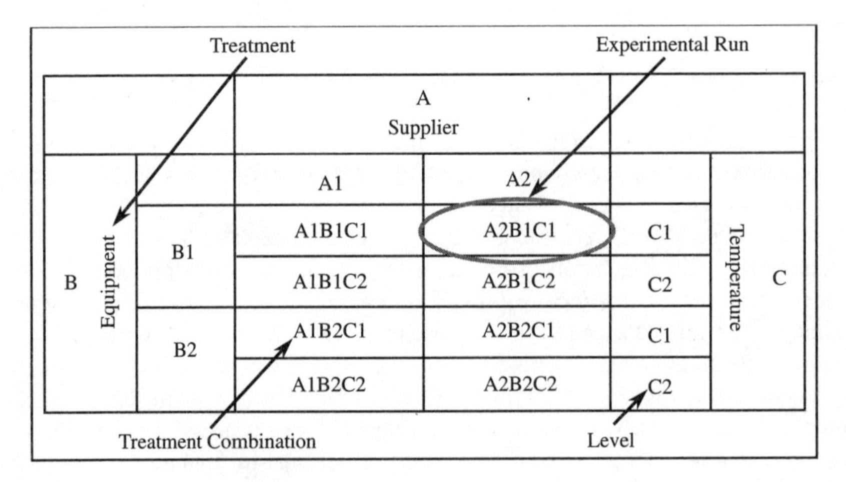

Figure 5.1 DOE Experimental Matrix

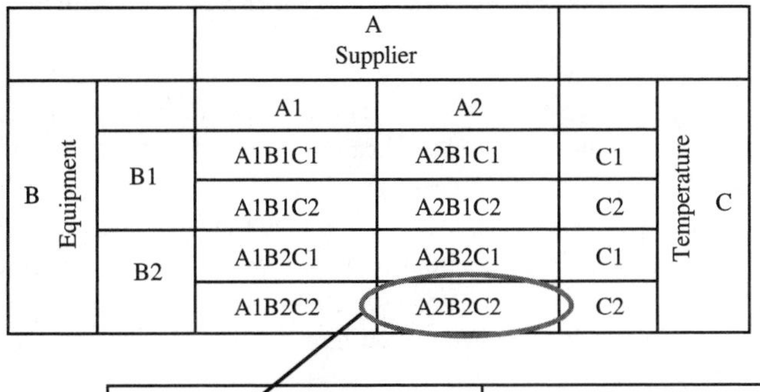

		A Supplier			
		A1	A2		
B Equipment	B1	A1B1C1	A2B1C1	C1	Temperature C
		A1B1C2	A2B1C2	C2	
	B2	A1B2C1	A2B2C1	C1	
		A1B2C2	A2B2C2	C2	

	Main Effects			Interactions				Sample	
Run	A	B	C	AB	AC	BC	ABC	y_1	y_2
1	−	−	−	+	+	+	−		
2	+	−	−	−	−	+	+		
3	−	+	−	−	+	−	+		
4	+	+	−	+	−	−	−		
5	−	−	+	+	−	−	+		
6	+	−	+	−	+	−	−		
7	−	+	+	−	−	+	−		
8	+	+	+	+	+	+	+		

Figure 5.2 DOE Analytical Matrix

tal run are *AB* set as high (+), *AC* set as high (+), *BC* set as high (+), and the three-way interaction *ABC* set as low (−).

Main Effect The main effect of a treatment is the measured change in the response as a result of changing that specific treatment. In Figures 5.1 and 5.2, the main effects are for treatments *A*, *B*, and *C*.

Interaction Interaction is the measured change in the response as a result of the combined effect of two or more treatments. In Figures 5.1 and 5.2, the interactions are for treatment combinations *AB*, *AC*, *BC*, and *ABC*.

Treatment Treatments are the controllable factors used as inputs to the products and processes under evaluation. These are the input variables (also called the independent variables) that can be varied to change the effect on the output or dependent variable. In Figures 5.1 and 5.2, the treatments are (*A*) the supplier, (*B*) the equipment used in the process, and (*C*) the temperature.

Level Levels are the values of the treatments being studied. In most instances of designed experiments, we can use two levels of the treatments: a high level, symbolized by a (+) sign or the number 1, and a low level, symbolized by a (−) sign or the number 2. In Figure 5.2, there are two levels for each treatment, Supplier *A*1 (+) and *A*2 (−), Equipment *B*1 (+) and *B*2 (−), and Temperature *C*1 (+) and *C*2 (−). There can be more levels than two; however, we will discuss these in the chapter on advanced topics.

Treatment Combination A treatment combination is the set of treatments and the associated levels used for an individual experimental run. In Figure 5.2, the treatment combinations are displayed for each experimental run. For instance, $A1B2C1$ indicates that the experimental run will be accomplished using material from Treatment A (Supplier) at Level 1 (+), Treatment B (Equipment) at Level 2 (−), and Treatment C (Temperature) at Level 1 (+). These treatment combinations also describe the treatments and levels of an experiment and determine the number of experimental runs required.

Experimental Run An experimental run is the accomplishment of a treatment combination. It may consist of one or more trials. In the example given in *Treatment Combination*, there are three treatments at two levels (2^3), or eight experimental runs in this designed experiment. The number of experimental runs needed in a designed experiment can be determined by the number of treatments (k) and the levels (L) of each treatment. These are stated as an exponential expression (L^k), with the levels being the base number and the treatments being the exponent. We therefore can calculate the number of experimental runs required for any combination of treatments and levels as follows:

A designed experiment can have as few as four experimental runs (2^2) or as many as 2187 (three levels with seven treatments 3^7). This clearly demonstrates that DOE can cover a wide range of experimental combinations and levels of data. As we progress through this chapter, we will describe DOE methods for dealing with large experiments.

$$
\begin{aligned}
2^2 &= 4 & 3^2 &= 9 \\
2^3 &= 8 & 3^3 &= 27 \\
2^4 &= 16 & 3^4 &= 81 \\
2^5 &= 32 & 3^5 &= 243 \\
2^6 &= 64 & 3^6 &= 729 \\
2^7 &= 128 & 3^7 &= 2187
\end{aligned}
$$

Orthogonal Array Analyzing the results of a designed experiment includes calculating the mean effect of the levels of each treatment. To accomplish this result, it is necessary for the experiment to be balanced. A balanced set of experiments contains an equal number of experiments for each level of each treatment. The example in Table 5.1 demonstrates this principle of a balanced experiment.

There are three criteria that can be applied to determine whether a test array is orthogonal:

1. Sum all the levels in each column. If there is an equal number of levels in each column, then we have passed the first test for an orthogonal array. In each col-

Run	A	B	C	AB	AC	BC	ABC
1	−	−	−	+	+	+	−
2	+	−	−	−	−	+	+
3	−	+	−	−	+	−	+
4	+	+	−	+	−	−	−
5	−	−	+	+	−	−	+
6	+	−	+	−	+	−	−
7	−	+	+	−	−	+	−
8	+	+	+	+	+	+	+
+ −	4 4	4 4	4 4	4 4	4 4	4 4	4 4

Table 5.1 Orthogonal Array

umn of Table 5.1, the treatment has 4 plus-level (+) and 4 minus-level (−) values. The orthogonal array forms the basis for many designed experiments, so it is important to recognize when a proposed test array is or is not orthogonal.

2. All rows with a certain symbol (−) in a given column must have an equal number of occurrences of all symbols in each other column. For example, in Table 5.1, when B is minus (−), there are two minuses and two pluses in C.
3. The selected matrix must have the least number of rows that satisfy the first two criteria for the selected number of treatments.

Sample Size The sample size is the number of times all experimental runs are accomplished as a repeat or replicate. A repeat sample is used when the experiment is simply duplicated for each experimental run. A replicate sample is used when the experimental run is a measurement that is sensitive to setup, environment, or some other factor outside the sample treatments. The minimum sample size for a designed experiment is 2. This minimum is required to establish a variance about the mean of the response variable. The sample size then becomes dependent on the levels of confidence needed in the experiment, as indicated during our discussion on ANOVA. The sample size for each experimental run is indicated by a lowercase n, and the total sample size for the complete experiment is indicated by an uppercase N. The samples for a designed experiment then can be described by $N = nL^k$.

$$(2)(2^2) = 8 \qquad (2)(3^2) = 18$$
$$(2)(2^3) = 16 \qquad (2)(3^3) = 54$$
$$(2)(2^4) = 32 \qquad (2)(3^4) = 162$$
$$(2)(2^5) = 64 \qquad (2)(3^5) = 486$$
$$(2)(2^6) = 128 \qquad (2)(3^6) = 1{,}458$$
$$(2)(2^7) = 256 \qquad (2)(3^7) = 4{,}374$$

It is apparent that we can use the sample size and number of experimental runs to plan our data management needs. If we run the simplest experiment with the minimum number of samples, there will be only eight resulting data points. In a (3^7) full factorial experiment with the minimum number of samples, there would be 4,374 data points.

Sample size is always a critical decision in any experimental design. This decision is based on the need for experimental data—the smaller the differences you are trying to detect, the larger your sample size needs to be—the economics of the situation, and the resources being used to perform the experiment. Although it is true that the minimum sample size required to evaluate a designed experiment is two (2), you would not want to evaluate any kind of data based on a sample of two. The sampling tables in the statistical appendix will assist you with selecting the proper sample size.

Response A response is a result of an experimental trial. It is the dependent variable (also called the response variable). It is the measured effect, on the product or process, of using the specific combination of treatments and levels. In a 2^7 full factorial experiment with two repeats or replicates, there are 256 responses.

Randomization We assign a treatment combination to an experimental run by random chance, using a randomization program or randomization table. Randomization prevents the influence of data due to any uncontrolled environmental variables in any test run. Randomization must always be used when the experimenter does not have total control over the environment, or when there are input variables outside the experiment that may affect the process. Table 5.2 has randomized the experimental runs first demonstrated in Table 5.1. A possible random order of the experimental runs is presented in the second column of the figure.

GENERATING FULL FACTORIAL DESIGNS AT TWO LEVELS

To perform a factorial experiment, you must first select the treatments and treatment levels. Since the experimenter always has the option of selecting these treatments and variables, we always use the fixed effects models for analysis. Then, based on the number of treatments and levels, you must select an existing design format that fits your needs or generate a design format. The number of test runs is determined by the number of treatments and treatment levels as previously described. Therefore, a 2^3 full factorial experiment will have $2 \times 2 \times 2$ or 8 runs.

These runs are applied to the test format (orthogonal array) as indicated in Figure 5.3. The levels of treatments are designated with a plus sign (+) for the higher levels and a minus sign (−) for the lower levels. From time to time in other texts you will also see these designated as +1 and −1, 1 and 2, or 0 and 1. It makes no difference which signs are associated with the higher and lower levels, as long as the labeling is consistent. From the test matrix we can apply the signs to each test run. We then have placed the symbols in *standard order*, the order in which they appear most frequently. We have applied that order here for consistency with general texts and general usage.

Run	Rand	A	B	C	AB	AC	BC	ABC
1	3	–	–	–	+	+	+	–
2	5	+	–	–	–	–	+	+
3	4	–	+	–	–	+	–	+
4	7	+	+	–	+	–	–	–
5	2	–	–	+	+	–	–	+
6	8	+	–	+	–	+	–	–
7	6	–	+	+	–	–	+	–
8	1	+	+	+	+	+	+	+

Table 5.2 Randomized Experimental Runs

Next we must add the signs for evaluating interactions. The coefficients for evaluating interactions are the product of the corresponding coefficients for the treatment main effects. Algebraically this is simply calculated as the product of the treatments for each interaction. The following equations are the products of the interaction of

Test Matrix

*A*1				*A*2			
*B*1		*B*2		*B*1		*B*2	
*C*1	*C*2	*C*1	*C*2	*C*1	*C*2	*C*1	*C*2
1	2	3	4	5	6	7	8
*A*1*B*1*C*1	*A*1*B*1*C*2	*A*1*B*2*C*1	*A*1*B*2*C*2	*A*2*B*1*C*1	*A*2*B*1*C*2	*A*2*B*2*C*1	*A*2*B*2*C*2

Test Format from Test Matrix → Test Format at Standardized Form

	Treatments		
Runs	*A*	*B*	*C*
1	+	+	+
2	+	+	–
3	+	–	+
4	+	–	–
5	–	+	+
6	–	+	–
7	–	–	+
8	–	–	–

	Treatments		
Runs	*A*	*B*	*C*
1	–	–	–
2	+	–	–
3	–	+	–
4	+	+	–
5	–	–	+
6	+	–	+
7	–	+	+
8	+	+	+

Figure 5.3 Establishing the Test Design Format

Treatment A and Treatment B (AB) and the three-level interaction of treatments A, B, and C (ABC). Multiplying these symbols is simply performed directly on the design format, as indicated in Table 5.3.

$$A \cap B = A \times B = AB$$

$$-1 \times -1 = 1$$

$$- \times - = +$$

Negative Times Negative Equals Positive

$$A \cap B \cap C = A \times B \times C = ABC$$

$$-1 \times -1 \times -1 = 1$$

$$- \times - \times - = -$$

Negative Times Negative Times Negative Equals Negative

2^K FULL FACTORIAL DOE

Full factorial experimental designs provide a comprehensive analysis of all treatments, levels, and interactions related to a selected quality characteristic. The full factorial designs we will discuss are commonly called the 2^k designs, where 2 is the number of levels and k represents the number of treatments. The orthogonal arrays for other designs such as 3^k are provided in Appendix A. The analytical principles for all orthogonal arrays are essentially the same and can be applied to all factorial experiments. Full factorial experiments can be accomplished in eight steps.

1. Identify a problem or opportunity for improvement.
2. Perform a cause-and-effect analysis to select treatments.
3. Select treatments and treatment levels.
4. Select a full factorial experimental format.
5. Conduct the experiment and acquire data.
6. Determine the effects.
7. Graph the results.
8. Perform ANOVA.

Step 1. Identify a Problem or an Opportunity for Improvement

The first step in any designed experiment is to identify the opportunity for improvement. The most effective way to accomplish this is through a thorough understanding

	Treatments						
Runs	A	B	C	AB	AC	BC	ABC
1	–1	–1	–1	1	1	1	–1
2	1	–1	–1	–1	–1	1	1
3	–1	1	–1	–1	1	–1	1
4	1	1	–1	1	–1	–1	–1
5	–1	–1	1	1	–1	–1	1
6	1	–1	1	–1	1	–1	–1
7	–1	1	1	–1	–1	1	–1
8	1	1	1	1	1	1	1

	Treatments						
Runs	A	B	C	AB	AC	BC	ABC
1	–	–	–	+	+	+	–
2	+	–	–	–	–	+	+
3	–	+	–	–	+	–	+
4	+	+	–	+	–	–	–
5	–	–	+	+	–	–	+
6	+	–	+	–	+	–	–
7	–	+	+	–	–	+	–
8	+	+	+	+	+	+	+

Table 5.3 Completed Full Factorial Design Format with Interactions

of your processes and products. These opportunities can be derived from many sources in existing and new processes, products, and services. Examples include:

- Opportunities for variability reduction to improve the efficiency and effectiveness of existing processes.
- Opportunities to improve products and services based on customer requirements.
- Opportunities to reduce reject and scrap rates.
- Opportunities to improve the development of new products and services.

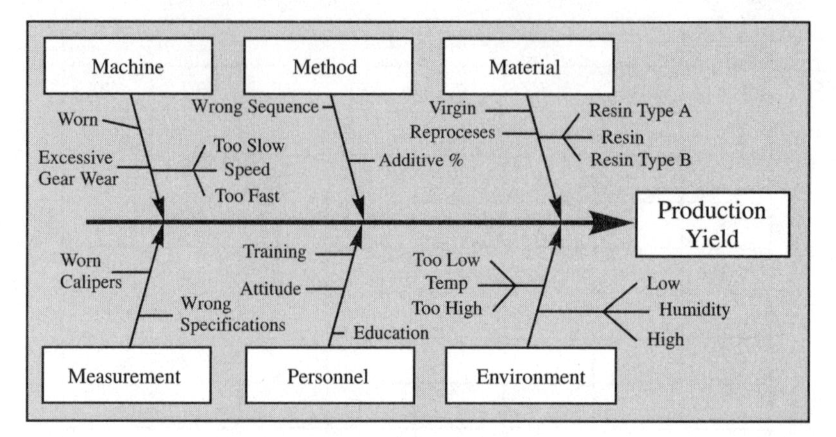

Figure 5.4 Potential Parameter Design Factors

Step 2. Perform a Cause-and-Effect Analysis

Figure 5.4 is an example of a fishbone diagram for cause-and-effect analysis. We will follow this example through our development of parameter design. To identify the opportunities for improvement, four subject experts spent approximately one hour of their time to brainstorm the problem and create the cause-and-effect analysis.

Step 3. Select Treatments, Levels, and Values

Based on the cause and effect analysis, select which factors will be tested. Determine the levels of the factors and assign a test value for each level. From our previous example, the factors and levels are as indicated in Table 5.4. Two of the factors are qualitative equipment and suppliers, and the third factor is quantitative temperature. The ability to mix these distinctively different types of data into a single experiment is an important feature of designed experiments.

Step 4. Selecting a 2^k Full Factorial Design

To select the appropriate full factorial design for your experiment, use the table of 2^k factorial designs given in Table 5.5. The complete table is provided in Appendix A.

Treatments	Levels	
	+	−
A Equipment	*A*1	*A*2
B Supplier	*B*1	*B*2
C Temperature	*C*1 High	*C*2 Low

Table 5.4 Selected Parameter Design Factors, Levels, and Values

Run	Design 2^t	Treatment Combinations														
		A	B	AB	C	AC	BC	ABC	D	AD	BD	ABD	CD	ACD	BCD	ABCD
1	2^2	−	−	+	−	+	+	−	−	+	+	−	+	−	−	+
2		+	−	−	−	−	+	+	−	−	+	+	+	+	−	−
3		−	+	−	−	+	−	+	−	+	−	+	+	−	+	−
4		+	+	+	−	−	−	−	−	−	−	−	+	+	+	+
5	2^3	−	−	+	+	−	−	+	−	+	+	−	−	+	+	−
6		+	−	−	+	+	−	−	−	−	+	+	−	−	+	+
7		−	+	−	+	−	+	−	−	+	−	+	−	+	−	+
8		+	+	+	+	+	+	+	−	−	+	−	−	−	−	−
9	2^4	−	−	+	−	+	+	−	+	−	+	+	−	+	+	−
10		+	−	−	−	−	+	+	+	+	+	−	−	−	+	+
11		−	+	−	−	+	−	+	+	−	−	−	−	+	−	+
12		+	+	+	−	−	−	−	+	+	−	+	−	−	−	−
13		−	−	+	+	−	−	+	+	−	+	+	+	−	−	+
14		+	−	−	+	+	−	−	+	+	+	−	+	+	−	−
15		−	+	−	+	−	+	−	+	−	−	−	+	−	+	−
16		+	+	+	+	+	+	+	+	+	−	+	+	+	+	+

Table 5.5 2^k Factorial Designs

The selected design is based on the number of treatments and levels selected for the experiment. The table demonstrates the full factorial designs for two-treatment (2^2), three-treatment (2^3), and four-treatment (2^4) experimental designs at two levels. If we were to select a design for three treatments (*ABC*) at two levels from this table, the resulting designed-experiment format would take the form of Table 5.6, a full factorial for three treatments at two levels (2^3).

Step 5. Conduct the Experiment and Acquire Data

Using the design selected in Step 4, conduct the experiment. Measure the quality characteristic using the measure determined in Step 2. Record the results on your worksheet along with any notes on related circumstances that might provide infor-

	Treatments						
Runs	A	B	C	AB	AC	BC	ABC
1	−	−	−	+	+	+	−
2	+	−	−	−	−	+	+
3	−	+	−	−	+	−	+
4	+	+	−	+	−	−	−
5	−	−	+	+	−	−	+
6	+	−	+	−	+	−	−
7	−	+	+	−	−	+	−
8	+	+	+	+	+	+	+

Table 5.6 2^3 Experimental Format

	Treatments							Test Results				
Runs	A	B	C	AB	AC	BC	ABC	y_1	y_2	y_3	$\bar{y}$	Range
1	–	–	–	+	+	+	–	81	79	74	78.00	7
2	+	–	–	–	–	+	+	78	72	76	75.33	6
3	–	+	–	–	+	–	+	76	71	74	73.67	5
4	+	+	–	+	–	–	–	75	71	75	73.67	4
5	–	–	+	+	–	–	+	79	80	76	78.33	4
6	+	–	+	–	+	–	–	76	71	74	73.67	5
7	–	+	+	–	–	+	–	75	76	77	76.00	2
8	+	+	+	+	+	+	+	74	77	78	76.33	4

Table 5.7 Experimental Format with Data Applied

mation concerning the test results. The resulting data can be recorded as indicated in Table 5.7.

Step 6. Determine the Effects

To calculate the effects of the treatments and treatment levels on the quality characteristic, we will use Eq. (5.1). This can also be accomplished directly on the spreadsheet as indicated in Table 5.8. First sum the experimental data for each treatment level ($\Sigma k+$ *and* $\Sigma k-$) at the bottom of the spreadsheet. Then divide each of the treatment summaries by the total number of runs at that level ($n+$ *and* $n-$). The effect is then calculated by subtracting the quotient at the plus level from the quotient at the minus level.

$$Effect_k = \frac{\sum y_+}{\sum n_+} - \frac{\sum y_-}{\sum n_-} \tag{5.1}$$

The effects were calculated in Table 5.8 using Eq. (5.1) as follows:

$$Effect_A = \frac{\sum y_{A+}}{\sum n_{A+}} - \frac{\sum y_{A-}}{\sum n_{A-}} = \frac{299.00}{4} - \frac{306.00}{4} = -1.75$$

$$Effect_B = \frac{299.67}{4} - \frac{305.33}{4} = -1.42 \qquad Effect_C = \frac{304.33}{4} - \frac{300.67}{4} = 0.92$$

$$Effect_{AB} = \frac{306.33}{4} - \frac{298.67}{4} = 1.92 \qquad Effect_{AC} = \frac{301.67}{4} - \frac{303.33}{4} = -0.42$$

$$Effect_{BC} = \frac{305.67}{4} - \frac{299.33}{4} = 1.58 \qquad Effect_{ABC} = \frac{303.67}{4} - \frac{301.33}{4} = 0.58$$

	Treatments							Test Results				
Runs	A	B	C	AB	AC	BC	ABC	y_1	y_2	y_3	$\bar{y}$	Range
1	–	–	–	+	+	+	–	81	79	74	78.00	7
2	+	–	–	–	–	+	+	78	72	76	75.33	6
3	–	+	–	–	+	–	+	76	71	74	73.67	5
4	+	+	–	+	–	–	–	75	71	75	73.67	4
5	–	–	+	+	–	–	+	79	80	76	78.33	4
6	+	–	+	–	+	–	–	76	71	74	73.67	5
7	–	+	+	–	–	+	–	75	76	77	76.00	2
8	+	+	+	+	+	+	+	74	77	78	76.33	4
$\sum +$	299.00	299.67	304.33	306.33	301.67	305.67	303.67					
$\sum -$	306.00	305.33	300.67	298.67	303.33	299.33	301.33					
$\sum + / n_+$	74.75	74.92	76.08	76.58	75.42	76.42	75.92					
$\sum - / n_-$	76.50	76.33	75.17	74.67	75.83	74.83	75.33					
Effect	–1.75	–1.42	0.92	1.92	–0.42	1.58	0.58					

Table 5.8 Summary of Experimental Data and Effects

Step 7. Graph the Results

We can now graph the main effects and interactions to evaluate what effect the treatments and treatment levels are having on the quality characteristic. Figure 5.5 displays the graphs associated with the treatment main effects. This is simply a graph of the values of $\Sigma+/n+$ *and* $\Sigma-/n-$ from Table 5.8. The slope of the line for main effects indicates the significance of the effect. The steeper the slope, the more significant the effect is. The graph also indicates which treatment level is producing the desired effect of optimizing the process.

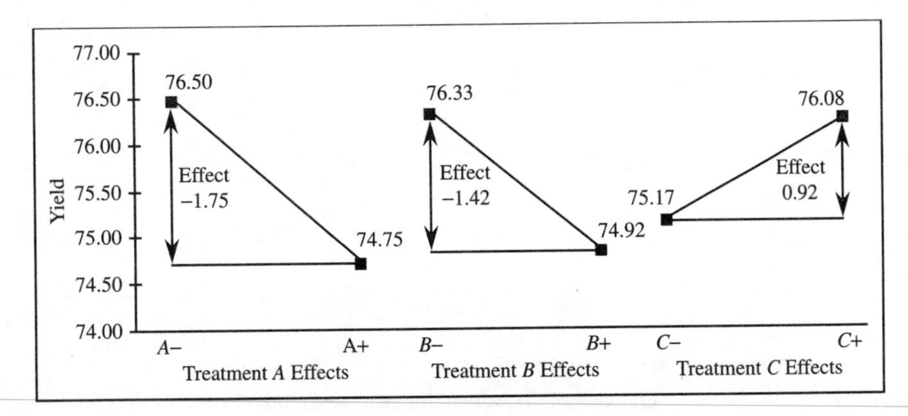

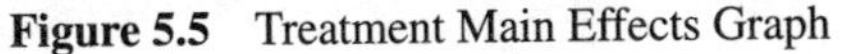
Figure 5.5 Treatment Main Effects Graph

Determining the treatment interactions requires one additional calculation. It is necessary to break down the effects of the treatments into the effects of the treatment interactions. To accomplish this, we calculate the effects of *A*- when *B* is at the plus and minus levels and the effects of *A*+ when *B* is at the plus and minus levels. This is accomplished using Eq. (5.2).

$$Interaction_{AB} = \frac{\sum y_{AB}}{n_{AB}} \tag{5.2}$$

$$Interaction_{AB} = \frac{\sum y_{AB}}{n_{AB}} = \frac{78.00 + 78.33}{2} = 78.17$$

These calculations can be accomplished directly on the same spreadsheet used to calculate treatment main effects in Table 5.8 as indicated in the following. We are now ready to plot the treatment interactions as indicated in Figure 5.6. When the lines of the graph are intersecting or converging, there is an indication of an interaction. If the lines run parallel, there is no interaction.

This graph of the main effects and interactions indicates that the yield for this process can be maximized by selecting Treatment *A* at the low setting, Treatment *B* at

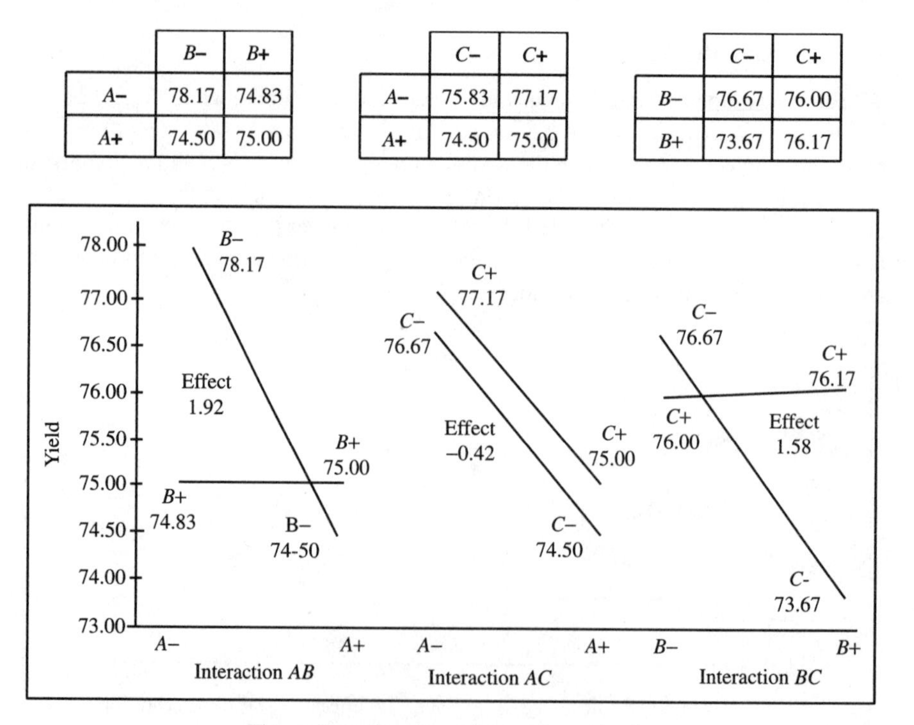

	B–	*B*+
A–	78.17	74.83
A+	74.50	75.00

	C–	*C*+
A–	75.83	77.17
A+	74.50	75.00

	C–	*C*+
B–	76.67	76.00
B+	73.67	76.17

Figure 5.6 Treatment Interactions Graph

Source of Variation	Sum of the Squares	Degrees of Freedom	Mean Squares	*F* Ratio	*F'* Critical	% Contribution
A	45.375	1	45.375	27.567	3.050	20
B	0.375	1	0.375	0.228	3.050	0
C	2.042	1	2.042	1.240	3.050	1
AB	7.042	1	7.042	4.278	3.050	3
AC	0.375	1	0.375	0.228	3.050	0
BC	7.042	1	7.042	4.278	3.050	3
ABC	12.041	1	12.041	7.316	3.050	5
Error	150.667	16	9.417			67
Total	224.958	23				

Table 5.9 ANOVA Analysis and Decision Table

the low setting, and Treatment *C* at the high setting. There are apparent interactions between *A* x *B* and *B* x *C* but no apparent interaction between *A* x *C*.

Step 8. Analysis of Variance

ANOVA, applied as part of a designed experiment, is used to measure the significance of the main effects and interactions, and the level of contribution. ANOVA decision table, Table 5.9, is based on our design of experiments and indicates that Treatment A is significant and contributed 20% to the overall process variation. This clearly indicates which treatment is to be the target of improvement processes such as CMI and Variability Reduction Programs.

It also indicates which of the treatments in our experimental design has the most effect on our outcome. Additionally, it is equally important to notice that the percent contribution from error is 67% of the total! This indicates that 67% of the variability is attributable to factors we have not evaluated in our designed experiment and/or to the uncontrolled environment.

NONLINEARITY

One of the assumptions made in using a two-level factorial design is linearity, which is that the model of your function is linear or essentially linear. Perfect linearity is not necessary. As discussed previously, if the treatment levels are selected sufficiently apart, any nonlinearity between the points will be accounted for in the magnitude of the effects. This relationship is demonstrated in Figure 5.7. In this example, the nonlinear is extreme, a cubic function. In this and in most cases where the levels of the treatments are properly selected, the nonlinearity of the data distribution is accounted for and the proper decision to maximize or minimize the treatment would have been made.

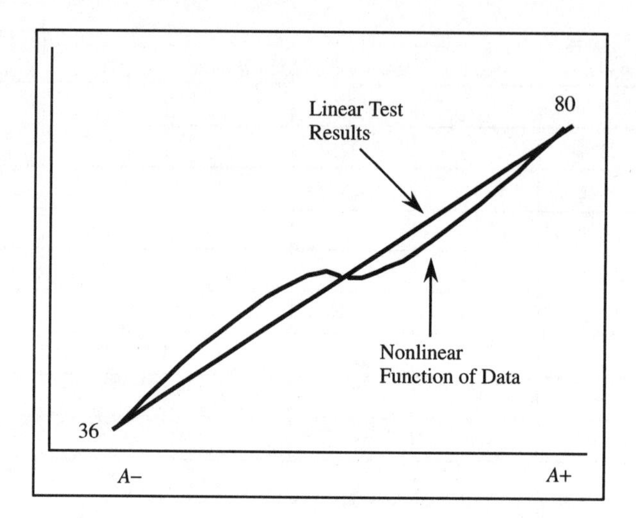

Figure 5.7 The Effects of a Nonlinear Function

There are two methods of obtaining information concerning the linearity of your experimental data and that will allow you to understand the effects of nonlinearity if and where they exist. This can be done by adding midpoints to some of your test runs as replicates or adding midpoints to the confirmation run repeats or replicates.

ESTIMATING THE RESULTS

The most effective methods of predicting the results of optimization based on your designed experiment is multiple linear regression. We will explain and use multiple linear regression as the standard method to predict the outcome of your designed experiment, given that the assumption of linearity exists as indicated previously. This method of regression analysis is used almost exclusively to study the relationship of one or more independent variables ($X_1, X_2, X_3 ...... X_1$) and a dependent response variable (y). Therefore, we can think of y as a function of x.

$$y = f\,(X)$$

Knowing that your dependent variable y is a function of X, you can use this relationship to understand the relationship between the multiple effects of multiple independent variables X_1, X_2, $X_1 ....X_i$ on the estimated value of your dependent variable $\hat{y}$. Therefore, you can view the function as follows.

$$\hat{y} = f_{X_1 + X_2 + X_3 ...... X_i}$$

With this relationship and with understanding how the independent variables (treatments) of a designed experiment affect the output variable (Quality Characteristic), you can evaluate the estimated results of your designed experiment using the following equation:

$$\hat{y} = \bar{\bar{X}} + \left(\bar{X}_A - \bar{\bar{X}}\right) + \left(\bar{X}_B - \bar{\bar{X}}\right) + \ldots + \left(\bar{X}_i - \bar{\bar{X}}\right) \tag{5.3}$$

where:

$\hat{y}$ = Estimated value of the response variable

$\bar{\bar{X}}$ = Grand average

$\bar{X}$ = Average of the treatment at the optimal level

Since the value of $\bar{X}_A$ is simply the effect of Treatment *A* the optimal level (+ *or* −), you can substitute the effect of *A* (E_A) you have already calculated for the $\bar{X}_A$. This step saves recalculating the value for $\bar{X}_A$ and simply substitutes the already calculated value of E_A, as indicated:

$$\hat{y} = \bar{\bar{X}} + \left(E_A - \bar{\bar{X}}\right) + \left(E_B - \bar{\bar{X}}\right) + \ldots + \left(E_i - \bar{\bar{X}}\right)$$

Using the data from Table 5.7 in our continuing example, we can calculate the optimum results. First we need to calculate the value of $\bar{\bar{X}}$, the grand average. This can be accomplished using a spreadsheet by totaling the values of each test result and dividing by the total number of samples or by totaling the average of each run and dividing by the total number of runs, as indicated in Eqs. (5.4) and (5.5).

$$\bar{\bar{X}} = \sum \bar{y} \Big/ k = \frac{\bar{y}_{Run\,1} + \bar{y}_{Run\,2} \cdots \bar{y}_{Run\,i}}{k} \tag{5.4}$$

where

$\bar{y}$ = Average value of each run

k = Number of runs

$$\bar{\bar{X}} = \sum X \Big/ N = \frac{X_1 + X_2 + X_3 + \ldots + X_i}{N} \tag{5.5}$$

where

X = Value of each data point

N = Total number of data points

Applying these equations to the data using a spreadsheet as indicated in Table 5.10, calculate the value of $\bar{\bar{X}}$.

	Test Results			
Runs	y_1	y_2	y_3	$\bar{y}$
1	81	79	74	78.00
2	78	72	76	75.33
3	76	71	74	73.67
4	75	71	75	73.67
5	79	80	76	78.33
6	76	71	74	73.67
7	75	76	77	76.00
8	74	77	78	76.33
	$\sum X$ 1815			
	$\sum X/n$ 75.63			

	Test Results			
Runs	y_1	y_2	y_3	$\bar{y}$
1	81	79	74	78.00
2	78	72	76	75.33
3	76	71	74	73.67
4	75	71	75	73.67
5	79	80	76	78.33
6	76	71	74	73.67
7	75	76	77	76.00
8	74	77	78	76.33
	$\sum \bar{y}$ 605.00			
	$\sum \bar{y}/k$ 75.63			

Table 5.10 Calculating the Grand Mean

$$\bar{\bar{X}} = \sum \bar{y} \Big/ k = \frac{78.00 + 75.33 + 73.67 + 73.67 + 78.33 + 73.67 + 76.00 + 76.33}{8} = 75.63$$

To predict the optimum results based on this evaluation insert the optimum treatment level effect and the grand mean into Eq. (5.5) as shown. The expected yield from the optimum treatment combinations is 77.65.

$$\hat{y} = \bar{\bar{X}} + \left(E_{A-} - \bar{\bar{X}}\right) + \left(E_{B-} - \bar{\bar{X}}\right) + \left(E_{C+} - \bar{\bar{X}}\right)$$

$$\hat{y} = 75.63 + (76.50 - 75.63) + (76.33 - 75.63) + (76.08 - 75.63)$$

$$\hat{y} = 75.63 + 0.87 + 0.70 + 0.45$$

$$\hat{y} = 77.65$$

BLOCKING

For many reasons it may be unreasonable to perform a complete full factorial experiment in a single block, that is, to run all test runs in the test format or randomized sequence. This can be true because of many factors—the length of each run may require an entire day, the raw materials for the evaluation may come in specific sized

	Treatments						
Runs	A	B	C	AB	AC	BC	ABC
1	–	–	–	+	+	+	–
2	+	–	–	–	–	+	+
3	–	+	–	–	+	–	+
4	+	+	–	+	–	–	–
5	–	–	+	+	–	–	+
6	+	–	+	–	+	–	–
7	–	+	+	–	–	+	–
8	+	+	+	+	+	+	+

Table 5.11 2^3 Full Factorial Experimental Format Blocked by the Higher Order Interaction

batches, the time required to heat or cool conditioning chambers may be long and expansive. Confounding is an experimental design technique that will allow you to arrange a complete full factorial experiment into blocks. This treatment may cause information about higher level treatment interactions to be confounded. That is indistinguishable from other effects within the block.

The individual blocks of a block design are incomplete full factorial experiments and cannot be evaluated separately. The orthogonality of factorial design will allow for the resulting data to be evaluated in exactly the same way as any experimental design. Full factorial experimental formats may be blocked in two distinct ways. First, for the purpose of running your experiment in incomplete blocks due to size or time considerations, you can block the design by using the higher order interaction to generate two or more blocks. In the case of a 2^3 design format, the higher order interaction is *ABC* and the experimental format is divided into two blocks by separating the levels of the *ABC* interaction as indicated in Table 5.11.

Blocking this 2^3 experimental design format by the levels of the higher order interaction will produce two blocked designs that can be executed in four runs each. The runs can then be randomized within the blocks as indicated in Table 5.12.

The second approach to blocking your experiment is used when there is a specific treatment or treatment levels that are problematic. This often occurs when a specific treatment level requires some very specialized equipment, is separated by a significant distance, or is costly or time-consuming to change. In our 2^3 full factorial experimental format, let's assume that factor *A* is temperature and that conditioning to high and low temperature is expensive and time-consuming. Therefore, we would desire to block by the levels in Treatment *A* as indicated in Table 5.13.

This action would result in two separate test blocks, as shown in Table 5.14, each running a single level of Treatment *A* (Temperature). Please note that just as in block-

Block 1

		Treatments						
Runs	Random	A	B	C	AB	AC	BC	ABC
1	2	+	–	–	–	–	+	+
2	4	–	+	–	–	+	–	+
3	1	–	–	+	+	–	–	+
4	3	+	+	+	+	+	+	+

Block 2

		Treatments						
Runs	Random	A	B	C	AB	AC	BC	ABC
1	4	–	–	–	+	+	+	–
2	2	+	+	–	+	–	–	–
3	3	+	–	+	–	+	–	–
4	1	–	+	+	–	–	+	–

Table 5.12 2^3 Full Factorial Experimental Format in Two Blocks

ing by higher order interactions, the effects of *A* are confounded within the individual blocks. After running the blocks separately, the data may be consolidated and the blocked experiment evaluated just as any other experimental design.

	Treatments						
Runs	A	B	C	AB	AC	BC	ABC
1	–	–	–	+	+	+	–
2	+	–	–	–	–	+	+
3	–	+	–	–	+	–	+
4	+	+	–	+	–	–	–
5	–	–	+	+	–	–	+
6	+	–	+	–	+	–	–
7	–	+	+	–	–	+	–
8	+	+	+	+	+	+	+

Table 5.13 2^3 Full Factorial Experimental Format Blocked by Treatment Level

Block 1

		Treatments						
Runs	Random	A	B	C	AB	AC	BC	ABC
1	1	+	–	–	–	–	+	+
2	4	+	+	–	+	–	–	–
3	3	+	–	+	–	+	–	–
4	2	+	+	+	+	+	+	+

Block 2

		Treatments						
Runs	Random	A	B	C	AB	AC	BC	ABC
1	3	–	–	–	+	+	+	–
2	4	–	+	–	–	+	–	+
3	2	–	–	+	+	–	–	+
4	1	–	+	+	–	–	+	–

Table 5.14 2^3 Full Factorial Experimental Format in Two Blocks

WHAT HAVE YOU LEARNED?

- A designed experiment can be planned, managed, and executed using the experimental matrix, trial formats, and sample size requirements.
- Full factorial experiments can be accomplished in eight steps.
  - Identify a problem or opportunity for improvement.
  - Perform a cause-and-effect analysis to select treatments.
  - Select treatments and treatment levels.
  - Select a full factorial experimental format.
  - Conduct the experiment and acquire data.
  - Determine the effects.
  - Graph the results.
  - Perform ANOVA.
- Identify a problem or opportunity for improvement: The most effective way to accomplish this is through a thorough understanding of your processes and products. These opportunities can be derived from many sources; they are:
  - Opportunities for variability reduction.
  - Opportunities to improve products and services.
  - Opportunities to reduce reject and scrap rates.
  - Opportunities to improve the development of new products and services.

- Perform a cause-and-effect analysis to select treatments: Form a cross-functional team to perform cause-and-effect analysis. Use the results of the cause-and-effect analysis to select treatments (input variables).
- Select treatments and treatment levels: Select the treatment levels for the designed experiment. These should be broad enough to clearly distinguish between the levels.
- Select a full factorial experimental format: Based on the number of treatments and treatment levels, select a designed experimental format.
- Conduct the experiment and acquire data: Using the design you have selected, measure the quality characteristic. Record the results on your worksheet along with any notes on related circumstances that might provide information concerning the test results.
- Determine the effects: To calculate the effects of the treatment and treatment levels on your quality characteristic, using the following equation:

$$Effect_k = \frac{\sum y_+}{\sum n_+} - \frac{\sum y_-}{\sum n_-}$$

- Graph the results: Graph the main effects and interactions to evaluate what effect the treatments and treatment levels are having on the quality characteristic. The slope of the line for main effects indicates the significance of the effect. The steeper the slope, the more significant the effect is. The graph also indicates which treatment level is producing the desired effect of optimizing the process. If the lines of the graph are intersecting or converging, there is an interaction. If the lines run parallel, there is no interaction.
- Perform ANOVA: ANOVA is used to measure the significance of the main effects and interactions, and the level of contribution.

TERMINOLOGY

The following are some of the terms you may have first encountered in this chapter:

Array (or Matrix): The DOE experimental (or test) array indicates the number of treatments and the levels of treatments and treatment combinations in a format convenient for computation and "bookkeeping."

Effects: To calculate the effects of the treatment and treatment levels on our quality characteristic, we must determine the difference that each of these is making in the response.

Experimental Run: The accomplishment of each treatment combination.

Factor: Any quantity that can influence the quality characteristic of your process.

Interaction: The measured change in the response as a result of the combined effect of two or more treatments.

Level: The values of the treatments being studied. In most instances of designed experiments, two levels are enough to identify an improvement: a high level, symbolized by a (+) sign or the number 2, and a low level, symbolized by a (−) sign or the number 1.

Main Effect: The measured change in the response as a result of changing a specific treatment.

Noise: A factor contributing to variance that is not controlled during the experiment.

Repeat: To obtain a repeat sample, the experiment is simply duplicated for each run.

Replicate: To obtain a replicate sample, the experimental run is repeated after all important treatment settings have been changed and then adjusted to the proper values for the run. Do not assume that simply doing another treatment combination between repeats is good enough. Make sure that every treatment setting has been changed and then changed back. If it appears to be part of the problem, disassembly and reassemble of the system may be warranted.

Response: A response is a result of an experimental run. It is the dependent variable (also called the response variable). It is the measured effect on the product or process of using the specific combination of treatments and levels. In a 2^7 full factorial experiment, with the minimum number of samples, there are 256 responses.

Sample Size: The number of experimental trials.

Treatment: The controllable factors used as inputs to the products and processes under evaluation. These are the input variables (also called the independent variables) that can be varied to change the effect on the output or dependent variable.

Treatment Combination: The set of treatments and the associated levels used for an individual experimental run.

Trial: A single measurement of a response.

EXERCISES

1. A small accounting firm needed to optimize its timelines for producing customer quarterly financial statements. A team consisting of management, lead accountants, and accounting clerks was formed. They determined that the following factors contributed most to the timelines of quarterly financial statements.

- Level of systems automation
- Training of accounting clerks
- Experience of accountants
- Timelines of receipt of customer records (debits and credits)
- Staffing level

a. If each of these factors could be evaluated at two levels, which full factorial experimental design format would be used?

b. If ten samples or replicates were accomplished, how many total (N) samples would there be?

c. One of these input variables cannot be controlled by the experimenters. Which is it?

d. Given there is one treatment that is part of the uncontrollable environment, how many treatments are there? Which experimental format should be used?

e. How many total data elements would there be with 10 samples with the reduced number of runs?

2. In the following 2^3 experimental format, calculate the proper treatment levels for the interactions.

	Treatments						
Runs	A	B	C	AB	AC	BC	ABC
1	−	−	−				
2	+	−	−				
3	−	+	−				
4	+	+	−				
5	−	−	+				
6	+	−	+				
7	−	+	+				
8	+	+	+				

3. An engineer is evaluating the effects of Cooling Medium (A), Tool Angle (B), and Tool Age (C) on the quality of a machine part. The quality is measured by the acceptance inspection of a sample of 25 randomly selected parts from each lot of 250 produced. Four complete lots were produced using the selected treatments and levels. The experiment was conducted at two levels with three factors 2^3. The results of the acceptance inspection follow:

	Treatments							Test Results			
Runs	A	B	C	AB	AC	BC	ABC	y_1	y_2	y_3	y_4
1	−	−	−	+	+	+	−	19	20	20	21
2	+	−	−	−	−	+	+	24	25	24	23
3	−	+	−	−	+	−	+	23	22	21	21
4	+	+	−	+	−	−	−	23	24	23	23
5	−	−	+	+	−	−	+	23	21	23	21
6	+	−	+	−	+	−	−	25	24	25	24
7	−	+	+	−	−	+	−	23	22	24	23
8	+	+	+	+	+	+	+	23	22	23	22

a. Calculate and graph the effects. Which treatment levels provide the optimum effects?

b. Perform an ANOVA on the results. Which treatments are significant? What is the percentage of contribution for each main effect?

c. What combination of treatments and levels would you recommend to optimize the yield?

d. Estimate the expected yield from your recommended changes.

4. There are hundreds of small vendors in the mall area of the U.S. capitol in Washington, D.C. They set up their carts by the roadside or next to the mall itself and sell a wide variety of items, from hot dogs and pizza to T-shirts and souvenirs. A small-business owner employed six mobile hot dog vending stands in this area. The owner needed to optimize the profits from these stands and we decided to perform a designed experiment. The experiment consisted of each of the six vending carts selling a combination of items of different types; in addition, the vendor was interested in determining if the age of the vendor had any effect on profitability. The test results (Quality Characteristic) represent the net profit for seven uniform periods of time.

	Treatments				Test Results						
Runs	A	B	C	D	y_1	y_2	y_3	y_4	y_5	y_6	y_7
1	–	–	–	–	765	721	732	779	764	791	805
2	+	–	–	–	732	717	725	732	760	781	800
3	–	+	–	–	647	680	696	649	683	694	702
4	+	+	–	–	790	805	886	775	783	820	813
5	–	–	+	–	604	615	609	611	620	635	649
6	+	–	+	–	645	632	621	670	666	645	671
7	–	+	+	–	625	612	601	650	646	625	651
8	+	+	+	–	715	652	704	715	690	704	687
9	–	–	–	+	710	647	699	710	685	699	682
10	+	–	–	+	690	711	725	709	732	760	754
11	–	+	–	+	645	678	694	647	681	692	700
12	+	+	–	+	810	790	805	811	797	825	811
13	–	–	+	+	620	622	617	594	632	647	650
14	+	–	+	+	635	647	678	628	661	672	690
15	–	+	+	+	645	678	694	647	681	692	700
16	+	+	+	+	645	655	672	655	648	681	700

Treatment	Low (−)	High (+)
Hot Dog	Standard Hot Dog	Premium Hot Dog
Drinks	Canned Soda Only	Canned Soda and Other Drinks
Location	Capital End of Mall	Lincoln Memorial End of Mall
Vendor	Students	Older

a. Calculate and graph the effects and treatment combinations, and provide the optimum profit.

b. Perform an ANOVA on the results. Which treatments are significant? What is the percentage of contribution for each main effect?

c. Estimate the expected yield from your recommended changes.

5. The experiment conducted in Exercise 4 is large and may be difficult to run. If the experimenter desires to block this experiment by location, what would the new test format look like?

Block 1

	Treatments			
Runs	A	B	C	D
1				
2				
3				
4				
5				
6				
7				
8				

Block 2

	Treatments			
Runs	A	B	C	D
1				
2				
3				
4				
5				
6				
7				
8				

6

FRACTIONAL FACTORIAL EXPERIMENTS

As the number of treatments to be evaluated in a full factorial experiment grows large, the costs in time, resources, and budget grow. The number of runs required to complete a full factorial quickly outgrows the resources of most experimenters. As an example, completing a 2^3 full factorial experiment requires eight runs; completing a 2^8 experiment requires 256 runs. The runs required to execute these full factorial experiments contain higher order interactions as well as main effects. The number of runs required to evaluate these interactions is a significant portion of the full factorial experiment. Where these higher order interactions are not clearly of concern, as is true in most experimental cases, information on main effects only or main effects and lower order interactions can be obtained from fractional factorial experiments with great savings in resources. Several key points contribute to our ability to fractionalize and use fractional factorial experiments:

- When there are several variables to be evaluated, the system being studied is most likely driven by some subset of main effects and possibly a few lower order interactions.
- The subset of significant factors identified in a fractional factorial experiment can be projected into a full factorial experiment and fully evaluated.
- Fractional factorial experiments can be used in a sequence of experiments used to refine and identify significant factors.
- Fractional factorial experiments can themselves be used to optimize systems where there is little or no direct concern about interactions.

Using fractional factorial designs leads directly to improved efficiency and effectiveness of experimental design. The most effective use of fractional factorials is when runs are made sequentially. That is when fractional factorial experiments are used as screening and characterizing purposes leading to a full factorial or a fractional factorial that can be used to optimize a system. For example, if you needed to understand the relationship of 6 treatments (2^6) on a quality characteristic, that would require 64 test runs. Running a half fractional factorial (2^{6-1}) experiment would require only 32 runs, and a one quarter fractional factorial (2^{6-2}) would require only 16 runs. From this fractional factorial data you could determine which subset of the original 6 factors was significant. You could then conduct a full factorial on that subset, thereby saving significant resources. Alternatively, you could learn enough from the first fractional factorial to change levels, add or remove treatments, or change the response characteristic.

All experiments should be sequential, beginning with fractional factorials to screen larger numbers of treatments and refine other experimental parameters. This will lead to effective use of full factorials to finalize your analysis and optimize the system or process under study. As a basic rule no more than 25% of your experimental budget should be used to accomplish your first experiment.

FRACTIONALIZING 2^{k-1} FACTORIAL EXPERIMENTS

One half fractional factorial experiments are called 2^{k-1} designs. This is because you are going to perform the experiment in $k-1$ runs. As an example, in a 2^3 experimental design there are 8 runs. In a 2^{3-1} fractional factorial experimental design there are 2^2 or 4 runs. The basic premise of separating experimental runs for fractional factorial experiments is the same as for blocking a full factorial, that is, to derive from the full factorial a smaller subset that is orthogonal. In a fractional factorial the blocks are not likely to be rejoined into a full factorial, so we must separate them by their higher order interaction only.

This method is easier to visualize if we use the already familiar 2^3 full factorial. The transition for this full factorial design to a fractional factorial is demonstrated in Table 6.1. The higher order interaction (*ABC*) here is the defining factor for confounding. The eight-run experiment can be separated into two four-run designs that separate the treatment combinations according to the sign (+ or $-$) conversion of interaction *ABC* as shown in Table 6.2. The primary fraction is the identity column where *ABC* is equal to plus (+), $I=ABC$. This then would yield a fractional factorial experiment in four runs as indicated in Table 6.3.

This method will separate this full factorial into two fractional factorials, each with four runs. Four runs designated by *ABC* are equal to minus, and four runs designated by *ABC* are equal to plus. The resulting two fractional factorial arrays are demonstrated in Table 6.3. This method also creates aliases, that is, more than one factor with the same treatment combination. The aliases result when more than one factor has the same treatment levels in the same run. In Table 6.2 the aliases are:

$$A=BC \qquad B=AC \qquad C=AB$$

	Treatments						
Runs	A	B	C	AB	AC	BC	ABC
1	−	−	−	+	+	+	−
2	+	−	−	−	−	+	+
3	−	+	−	−	+	−	+
4	+	+	−	+	−	−	−
5	−	−	+	+	−	−	+
6	+	−	+	−	+	−	−
7	−	+	+	−	−	+	−
8	+	+	+	+	+	+	+

Table 6.1 2^3 Full Factorial Separated by the Higher Level Interaction

	Treatments						
Runs	A	B	C	AB	AC	BC	ABC
2	+	−	−	−	−	+	+
3	−	+	−	−	+	−	+
5	−	−	+	+	−	−	+
8	+	+	+	+	+	+	+
Alias	BC	AC	AB				Identity

Table 6.2 2^{3-1} Fractional Factorial Format with Identity and Confounding

		Treatments		
Runs	Random	A	B	C
1	3	+	−	−
2	2	−	+	−
3	1	−	−	+
4	4	+	+	+

Table 6.3 Fractional Factorial Design

	Treatments		
Runs	A	B	C
1	–	–	–
2	+	–	–
3	–	+	–
4	+	+	–
5	–	–	+
6	+	–	+
7	–	+	+
8	+	+	+

Table 6.4 Standard Format 2^3 *A*, *B*, and *C* Main Effects Columns

An alternative method for fractionalizing a full factorial design which can be quicker and more expedient, especially if you do not have a full factorial experimental format, is to follow the pattern of the basic design. We will use a 2^{4-1} fractional factorial design format as the example.

A 2^{4-1} fractional factorial experiment is a four-factor (*A*, *B*, *C*, and *D*) experiment that is performed in the same number of runs as a 2^3 factorial experiment. Therefore, the number of runs required will be eight. Write down the standard formal experimental treatments and levels for the eight runs of a 2^3 experiment as written for treatments *A*, *B*, and *C*, indicated in Table 6.4.

The column for factor *D* will be generated by $D=ABC$, that is, the *D* treatment column will be generated as the product of the *A*, *B*, and *C* columns. This can be verified by the fact that when the identity column is calculated for this design, it is the primary fraction or $I - ABCD$ all at the plus level, as indicated in Table 6.5.

2^{k-p} FACTORIAL EXPERIMENTS

When the number of treatments in a factorial experiment becomes very large and/or the experimental runs are very expensive, it may be necessary to reduce the number of runs further than the one half fraction of the 2^{k-1} design. The 2^{k-p} design is constructed in much the same way as a 2^{k-1} design. At each level there are specific generators that define the relationship between the basic design and the fractional factorial design format. The 2^{k-p} design may be constructed by first writing down a basic design consisting of the runs for a full factorial in $k-p$ runs. Then the additional columns needed can be generated using the appropriate generators.

Using the format generators listed in Table 6.6, we will construct a quarter fractional factorial design format for a five treatment experiment (2^{5-2}). A complete

	Treatments				
Runs	A	B	C	D	ABCD
1	–	–	–	–	+
2	+	–	–	+	+
3	–	+	–	+	+
4	+	+	–	–	+
5	–	–	+	+	+
6	+	–	+	–	+
7	–	+	+	–	+
8	+	+	+	+	+
				D=ABC	Identity

Table 6.5 2^{4-1} Format with Identity Column

listing of these generators is contained in Appendix A. They can be used to generate any fractional factorial format required to perform a designed experiment.

As an example, a 2^{5-2} fractional factorial design uses a 2^3 factorial format as the basic design. First draw a standard format experimental format for the *A*, *C*, and *D* factors of a three-treatment experiment. Then choose the generators for treatments *D* and *E* from the table of fractional factorial generators in Table 6.6. In this case they are *D*=*AB* and *E*=*AC*. These generators are used to define the levels for treatments *D* and *E* as indicated in Table 6.7.

Number of Treatments	Fraction	Number of Runs	Design Generators
3	2^{3-1}	4	C = +/– AB
4	2^{4-1}	8	D = +/– ABC
5	2^{5-1}	16	E = +/– ABCD
	2^{5-2}	8	D = +/– AB E = +/– AC
6	2^{6-1}	32	F = +/– ABCDE
	2^{6-2}	16	E = +/– ABC F = +/– BCD
	2^{6-3}	8	D = +/– AB E = +/– AC F = +/– BC

Table 6.6 Fractional Factorial Format Generators

	Treatments				
Runs	A	B	C	D	E
1	–	–	–	+	+
2	+	–	–	–	–
3	–	+	–	–	+
4	+	+	–	+	–
5	–	–	+	+	–
6	+	–	+	–	+
7	–	+	+	–	–
8	+	+	+	+	+
				D=AB	E=AC

Table 6.7 2^{5-2} Fractional Factorial Format with Generators

CONFOUNDING AND ALIASING

In the act of fractionalizing a designed experiment, a phenomenon occurs that is both a concern and a potential help to the experimenter. This is known as *confounding* and *aliasing*. Confounding occurs when you reduce the number of runs required in your experiment by fractionalizing a full factorial design format. This causes certain interactions to be confounded with main effects and other interactions. That is that some of the information acquired for main effects may contain the effects of interactions. The interactions that are confounded with main effects or other interactions are called aliases, that is, the main effect is an alias for the interaction ($A=BC$).

In performing fractional factorial experiments and determining design formats for fractional factorial orthogonal arrays, it is important to be able to obtain the confounding and aliasing for any selected format. The following is a simple algebraic method for calculating the confounding patterns and determining the aliases. We will use a 2^{4-1} fractional factorial to demonstrate this method. Construct a 2^{4-1} fractional factorial using the *ABC* generator from Table 6.6. First write down a basic design for a 2^{4-1} or a 2^3 design format. This basic design has the correct number of runs but only three treatments (A,B,C). To find the fourth factor (D) set $D=ABC$ as indicated in Table 6.8.

Whenever fractionalizing a full factorial design format, the highest level interaction is called the identity (I). This is due to the fact that the primary fraction as we previously discussed is always the highest level interaction separated by its plus level runs. Therefore, in our example from Table 6.8, the defining relationship is $I=ABCD$. Multiplying any main effect by the defining relationship yields the alias for that effect. Using the basic defining relationship we can calculate the confounding effects on the main effects and therefore identify their aliases. The confounding and alias

	Treatments			
Runs	A	B	C	D
1	–	–	–	–
2	+	–	–	+
3	–	+	–	+
4	+	+	–	–
5	–	–	+	+
6	+	–	+	–
7	–	+	+	–
8	+	+	+	+
Column D Generator ⟶				D = ABC

Table 6.8 Standards 2^{4-1} Fractional Factorial Format

structure is easily determined by multiplying each column defining a treatment by the identity. Remember that in our plus and minus two-level experiment, any column multiplied by itself will be all pluses, therefore equal to the defining relationship, and can be canceled out. These calculations for the treatment main effects of our 2^{4-1} fractional factorial example are:

A	B	C	D
$I=ABCD$	$I=ABCD$	$I=ABCD$	$I=ABCD$
$A \bullet I=A \bullet ABCD$	$B \bullet I=B \bullet ABCD$	$C \bullet I=C \bullet ABCD$	$D \bullet I=D \bullet ABCD$
$A \bullet I=A^2BCD$	$B \bullet I=AB^2CD$	$C \bullet I=ABC^2D$	$D \bullet I=ABCD^2$
$A=BCD$	$B=ACD$	$C=ABD$	$D=ABC$

Each treatment main effect is confounded with a three-level interaction, and the alias for that interaction is $A=BCD$, $B=ACD$, $C=ABD$, and $D=ABC$. Notice that the alias for treatment main effect D is the same as the design format generator. This will hold true throughout fractional factorial experiments. Additionally, two-level interactions are confounded with other two-level interactions as indicated here; these are calculated in the same way.

$$AB=CD \qquad AC=BD \qquad \text{and} \qquad BC=AD$$

Understanding how confounding occurs and what the aliases for each fractional factorial are provides you with a tool you can use in designing your experiments. To make this effort much easier, a complete table of confounding and alias relationships is provided in Appendix A.

EXECUTING A FRACTIONAL FACTORIAL EXPERIMENT

Fractional factorial experiments are conducted and evaluated exactly like full factorial experiments. A complete set of fractional factorial design formats is provided in Appendix A. Fractional factorial experiments can be accomplished in eight steps.

1. Identify a problem or opportunity for improvement.
2. Perform a cause-and-effect analysis to select treatments.
3. Select treatments and treatment levels.
4. Select a fractional factorial experimental format.
5. Conduct the experiment and acquire data.
6. Determine the effects.
7. Graph the results.
8. Perform ANOVA.

Step 1. Identify a Problem or Opportunity for Improvement

Selecting the opportunity for improvement in a fractional factorial is accomplished in exactly the same way as for full factorials. A fractional factorial can be either a screening experiment used to determine which of many factors are critical, a system characterization experiment used to determine how a system will react to several treatments and levels, or as an optimizing experiment in cases where interactions may be of little concern. Therefore, the opportunity for improvement may relate to any of these factors. The basic idea remains constant; however, your output variable that relates to the opportunity for improvement will relate to quality, cost, or schedule in some way. Some examples include:

- Identifying the treatments and levels affecting the reliability of a system (Quality).
- Determining the treatments and levels that reduce cycle time (Schedule).
- Opportunities to reduce rework (Cost).
- Opportunities to improve process effectiveness and efficiency (Quality, Cost, and Schedule).

Step 2. Perform a Cause-and-Effect Analysis to Select Treatments

Figure 6.1 is another example of an Ishakowa diagram used to select treatments and levels. This is accomplished using a cross-functional team during a brainstorming session. Other methods for selecting treatments and levels are quality function deployment, failure modes and effects analysis for systems and processes, and process analysis. In using any of these methods, remember the importance of using a cross-functional team. The team approach will bring a broad spectrum of knowledge, dis-

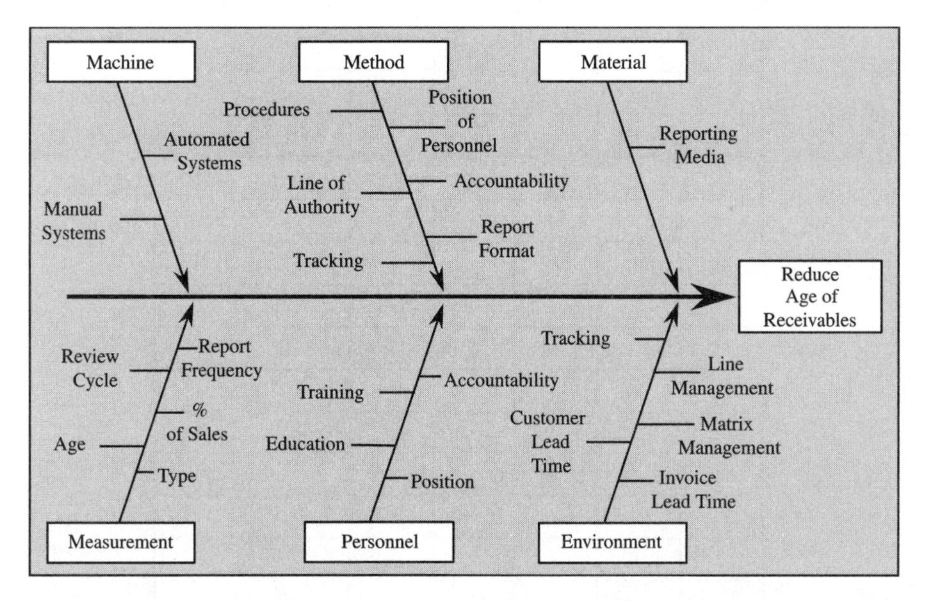

Figure 6.1 Cause-and-Effect Analysis

ciplines, and experience to the evaluation and make your designed experiment more effective and efficient.

Step 3. Select Treatments, Levels, and Values

Based on the cause-and-effect analysis, select which treatments and treatment levels will be selected for your designed experiment. The treatments and treatment levels selected are indicated in Table 6.9.

Treatment	Level	
	+	–
A Training	Clerk	Technician
B Follow-up	30-day	15-Day
C Contractor	Contract	Company
D Procedure	Yes	No
E Automated	Tracking	Billing Tracking Follow-up
F Level of Authority	Executive	Management

Table 6.9 Treatments and Treatment Levels

	A	B	C	D	E = ABC	F = BCD
1	–	–	–	–	–	–
2	+	–	–	–	+	–
3	–	+	–	–	+	+
4	+	+	–	–	–	+
5	–	–	+	–	+	+
6	+	–	+	–	–	+
7	–	+	+	–	–	–
8	+	+	+	-	+	–
9	–	–	–	+	–	+
10	+	–	–	+	+	+
11	–	+	–	+	+	–
12	+	+	–	+	–	–
13	–	–	+	+	+	–
14	+	–	+	+	–	–
15	–	+	+	+	–	+
16	+	+	+	+	+	+
	BCE	ACE	ABE	BCF	ABC	BCD
	DEF	CDF	BDF	AEF	ADF	ADE

Table 6.10 2^{6-2} Fractional Factorial Format

Step 4. Select or Create a 2^{k-p} Fractional Factorial Experiment

Since the number of runs required of a fractional factorial experiment for a 2^6 experiment would be very large, it was determined that a fractional factorial experiment would be used. The resources available and the fact that interactions may not be of concern led the experimenters to use a 2^{6-2} fractional factorial. The resulting orthogonal array is demonstrated in Table 6.10.

The applicable confounding and aliasing are listed across the bottom of the table. In this fractional factorial format, the treatment main effects are confounded with three-level interactions. These confounding and aliasing patterns can be calculated using the previous algebraic example or more simply by referring to the table of confounding and alias relationships provided in Appendix A.

Step 5. Conduct the Experiment and Acquire Data

Using the design selected in Step 4, conduct the experiment. Measure the quality characteristic using the measure determined in Step 2. Record the results on your

	A	B	C	D	E	F	y^1	y^2	y^2	$\bar{y}$
1	–	–	–	–	–	–	76	70	69	71.67
2	+	–	–	–	+	–	100	96	81	92.33
3	–	+	–	–	+	+	121	105	117	114.33
4	+	+	–	–	–	+	96	84	95	91.67
5	–	–	+	–	+	+	101	96	115	104.00
6	+	–	+	–	–	+	67	72	76	71.67
7	–	+	+	–	–	–	96	106	100	100.67
8	+	+	+	–	+	–	115	122	119	118.67
9	–	–	–	+	–	+	96	89	98	94.33
10	+	-	-	+	+	+	121	122	135	126.00
11	–	+	–	+	+	–	165	173	181	173.00
12	+	+	–	+	–	–	119	127	132	126.00
13	–	–	+	+	+	–	112	124	99	111.67
14	+	–	+	+	–	–	86	82	84	84.00
15	–	+	+	+	–	+	100	109	111	106.67
16	+	+	+	+	+	+	176	184	195	185.00

Table 6.11 Fractional Factorial Format with Data Applied

worksheet along with any notes on related circumstances that might provide information concerning the test results. The resulting data can be recorded as indicated in Table 6.11.

Step 6. Determine the Effects

To determine the effects of a fractional factorial designed experiment, we will use the same equation as we did for determining the effects of a full factorial experiment. Eq. (6.1) is the same as Eq.(5.1); it is presented again here for the purpose of continuity with this example.

$$Effect_k = \frac{\sum y_+}{n_+} - \frac{\sum y_-}{n_-} \tag{6.1}$$

The Effects are calculated using the information from Table 6.11 and Eq. (6.1) as follows:

$$Effect_A = \frac{\sum y_{A+}}{n_{A+}} - \frac{\sum y_{A-}}{n_{A-}} = \frac{895.33}{8} - \frac{876.33}{8} = 2.38$$

$$Effect_B = \frac{1016.00}{8} - \frac{755.67}{8} = 32.54 \quad Effect_C = \frac{882.33}{8} - \frac{889.33}{8} = -0.87$$

$$Effect_D = \frac{100.67}{8} - \frac{765.00}{8} = 30.21 \qquad Effect_E = \frac{1025.00}{8} - \frac{746.67}{8} = 34.79$$

$$Effect_F = \frac{893.67}{8} - \frac{878.00}{8} = 1.96$$

These calculations can be most easily accomplished and understood using the standard spreadsheet format and example in Table 6.12. First sum the experimental data for each treatment level ($\Sigma k+$ and $\Sigma k-$) at the bottom of the spreadsheet. Then divide each of these treatment summaries by the total number of runs at that level ($n+$ and $n-$). The effect is then calculated by subtracting the quotient at the plus level from the quotient at the minus level.

Step 7. Graph the Results

Graph the main effects to evaluate what effect the treatments and treatment levels are having on the quality characteristic. Figure 6.2 displays the graphs associated with the

	A	B	C	D	E	F	y^1	y^2	y^2	$\bar{y}$
1	−	−	−	−	−	−	76	70	69	71.67
2	+	−	−	−	+	−	100	96	81	92.33
3	−	+	−	−	+	+	121	105	117	114.33
4	+	+	−	−	−	+	96	84	95	91.67
5	−	−	+	−	+	+	101	96	115	104.00
6	+	−	+	−	−	+	67	72	76	71.67
7	−	+	+	−	−	−	96	106	100	100.67
8	+	+	+	−	+	−	115	122	119	118.67
9	−	−	−	+	−	+	96	89	98	94.33
10	+	−	−	+	+	+	121	122	135	126.00
11	−	+	−	+	+	−	165	173	181	173.00
12	+	+	−	+	−	−	119	127	132	126.00
13	−	−	+	+	+	−	112	124	99	111.67
14	+	−	+	+	−	−	86	82	84	84.00
15	−	+	+	+	−	+	100	109	111	106.67
16	+	+	+	+	+	+	176	184	195	185.00
$\sum +$	895.33	1016.00	882.33	1006.67	1025.00	893.67				
$\sum -$	876.33	755.67	889.33	765.00	746.67	878.00				
$\sum + / n_+$	111.92	127.00	110.29	125.83	128.13	111.71				
$\sum - / n_-$	109.54	94.46	111.17	95.63	93.33	109.75				
Effect	2.38	32.54	-0.87	30.21	34.79	1.96				

Table 6.12 Summary of Experimental Data and Effects

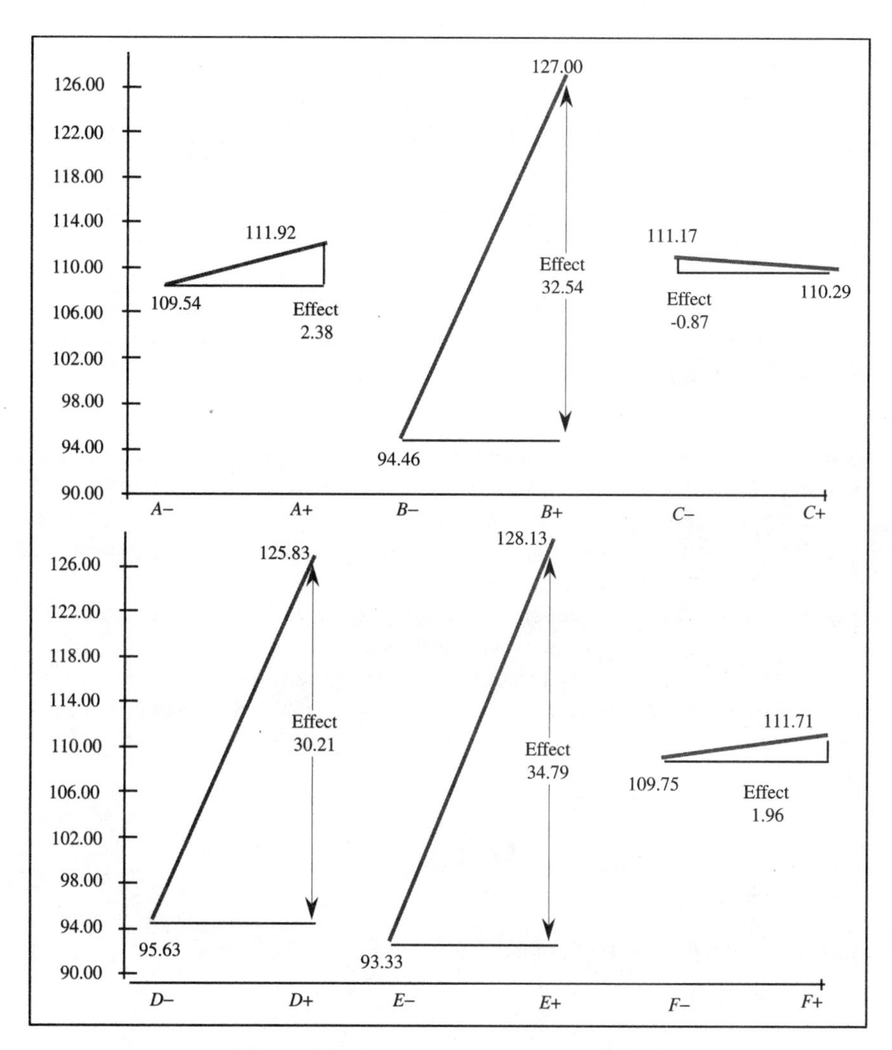

Figure 6.2 Graphing the Treatment Effects

treatment main effects. This is simply a graph of the values of $\Sigma+/n+$ and $\Sigma-/n-$ from Figure 6.2. The slope of the line for main effects indicates the significance of the effect. The steeper the slope, the more significant the effect is. The graph also indicates which treatment level is producing the desired effect of optimizing the process.

Step 8. Analysis of Variance

Analysis of Variance, applied as part of a fractional factorial designed experiment, is used to measure the significance of the treatments and the level of contribution. The ANOVA decision table in Table 6.13 is based on the techniques developed in Chapter 4. It is calculated using the experimental results from Table 6.12.

Variation	Squares	Freedom	MS	Ratio	Critical	% Contribution
A	67.21	1	67.21	0.35	7.310	0
B	12707.00	1	12707.00	65.84	7.310	27
C	8.71	1	8.71	0.05	7.310	0
D	10950.00	1	10950.00	56.74	7.310	24
E	14525.04	1	14525.04	75.26	7.310	31
F	45.54	1	45.54	0.24	7.310	0
Error	7925.50	41	193.30			17
Total	46229.00	47				

Table 6.13 ANOVA Calculation and Decision Table

This fractional factorial experiment has indicated that Treatments B, D, and E are clearly significant. Treatments A, C, and F are not significant. The determination of significance using Fisher's F Statistic in an ANOVA table indicates that, from a statistical point of view, the variance may have been caused by the sampling or some other factor not attributed to the treatment under study. The percent contribution indicates what you may gain by taking action and optimizing the treatment under study. In this study the percent contribution for Treatment B is 27%, for treatment D it is 24%, and for treatment E it is 31%. This indicates that 83% of the variance may be eliminated by optimizing treatments B, D, and E. What would this mean for this process. That can be demonstrated by predicting the optimum results using the following equation:

$$\hat{y} = \frac{T}{n} \pm \left(A_X - \frac{T}{n}\right) + \left(B_X - \frac{T}{n}\right) + \ldots + \left(F_X - \frac{T}{n}\right) \tag{6.2}$$

where

T = Total of all experimental results

n = Number of samples

$\frac{T}{n}$ = Grand average

$A_X,\ B_X \ldots F_X$ = Performance values at optimum level for each factor

When utilizing the effects factors to estimate the optimum results for a designed experiment, you should not include the treatments that were found to be insignificant. Therefore, the optimized process can be estimated as follows:

Optimized at B_{-}, D_{-}, E_{-}

$$\hat{y} = \frac{T}{n} - \left(B_{-} - \frac{T}{n}\right) + \left(D_{-} - \frac{T}{n}\right) + \left(E_{-} - \frac{T}{n}\right)$$

$$\hat{y} = 110.73 - (94.46 - 110.73) + (95.63 - 110.73) + (93.33 - 110.73) = 61.96$$

TAGUCHI QUALITY ENGINEERING

Quality Engineering is a design, development, and improvement strategy to optimize the design of a product, service or process. Basically, it deals with activities that reduce the variability of products and processes through a series of optimization analyses, using design of experiments as a basic tool. There are two phases of quality engineering: on-line and off-line quality engineering.

Noise and Robustness

Variability in products and processes is the enemy of quality and the cause of unnecessary cost and scheduling problems. The primary cause of variability is the uncontrolled or uncontrollable environment that a system is subjected to in production and use. Dr. Taguchi calls variation coming from these sources *noise*. Just as the factors that the design team uses to optimize a system are called control factors, these uncontrolled factors are called noise factors.

Noise factors can be classified by their source. Outer noises are factors that affect product and process performance from an external source. Outer noise factors are such things as ambient temperature, training, operating conditions, or anything that affects the system that is external to the product or process. Inner noises are the system conditions that cause variation from within the product or process. Inner noise factors are such things as tool wear, process procedures, component capability, different materials, and any other factor that is part of the system, product, or process that causes variability. Between-product noises are the factors that cause variability between products when produced by exactly the same process. This variability is related very closely to the common-cause variation of process control. It is caused by such factors as machine capability, process capability, operator competency, and others. The effect of these noise factors is demonstrated in Figure 6.3.

It is not critical to identify and categorize each source of noise. It is imperative, however, that the design team create products and processes to be insensitive to all

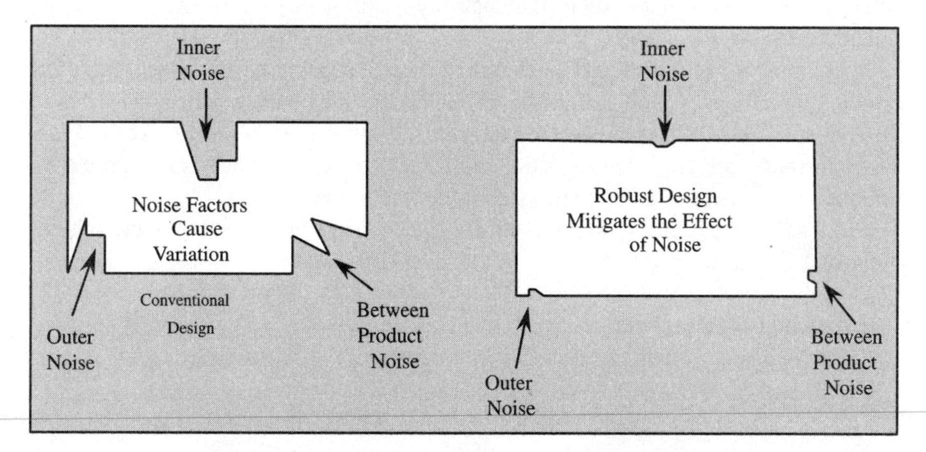

Figure 6.3 Noise Effects on Products and Processes

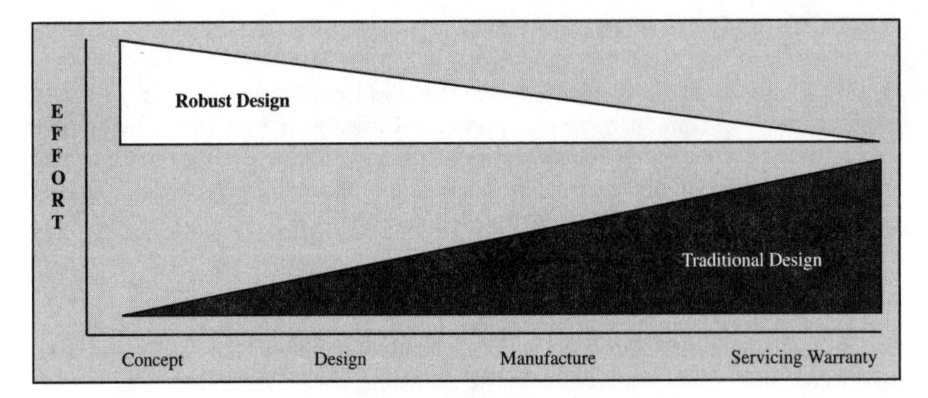

Figure 6.4 Comparative Effort (Cost and Time)

types of noise. A system that is designed to perform at a high level of quality despite being subjected to uncontrollable environmental conditions in manufacture and use is said to be *robust*.

Robustness is synonymous with high-quality, low-cost, and on-schedule processes. High quality, low cost, and on schedule is the optimum state for any system. Therefore, it is the goal of the design team to search for and find robust designs for products and processes. The best method for achieving this goal is through Taguchi Off-Line Quality Engineering during the design and development process. If robustness is achieved for systems, products, and processes early in the life cycle of the product, fewer problems will occur during the manufacture process and when the product is in the hands of the customer for use. The relationship of these efforts is demonstrated in Figure 6.4.

Quality Characteristics

One of the first steps in performing quality engineering is to identify the quality characteristics that we are going to measure. These are the output or dependent variables of our designed experiment. The basic types of quality characteristics are presented here.

A nominal-is-best characteristic is one that has a specific target value with a tolerance limit around the nominal value. A smaller-is-better characteristic always has an ideal value of zero. A larger-is-better characteristic has infinity as its ideal value. Classified attributes are characteristics that are not amenable to measurement on a continuous scale. These characteristics are graded on a qualitative scale. The dynamic characteristic measures the direct relationship between a system's input and output; more fuel equals more speed.

- Nominal-Is-Best (NIB)
  - Dimensions
  - Clearances
  - RPM

- Smaller-Is-Better(SIB)
  - Scrap
  - Warranty Returns
  - MTBF
- Larger-Is-Better (LIB)
  - Expected Life
  - Process Yield
  - Miles Per Gallon
- Classified Attribute
  - On/Off
  - Grade I/II/III
  - Pass/Fail

Understanding what type of characteristic to measure is an important first step in measuring your response variable. The characteristic you select must be relevant to the product or process you are measuring. There must be a functional link between your selected characteristic and the quality, cost, or schedule of the product or process under analysis. Do not select a measure simply because it is available and easy to obtain and understand. Selecting the proper quality characteristic to measure is not always an easy task. The characteristic you select should relate the functional inputs directly to the output. Remember what your goal is and that you are designing and developing a robust system. Some measures of yield, for instance, do not relate directly to customer satisfaction; therefore, they are poor measures. To quote Taguchi from the 1990 Taguchi Symposium, "To get quality, don't measure quality." This means that to optimize the goal, do not measure the goal itself, measure the characteristics that make up the goal.

On-Line Quality Engineering

On-line quality engineering activities are utilized as process and product improvement strategies. The purpose of on-line quality engineering is to control the process and products to ensure that they meet the robust design standards established during the design and development process and to improve those products and processes. Figure 6.5 demonstrates the elements of this process. It is the aim of this process to balance design, control, and improvement with quality, cost, and schedule.

Off-Line Quality Engineering

Off-line quality engineering is performed early in the design and development process of a product. The intent of off-line quality engineering is to optimize the functionality of a system through a design strategy that will render a robust system. The product, or process, is designed to be robust against the environments of use, production, materials, and other pertinent factors selected by the design team.

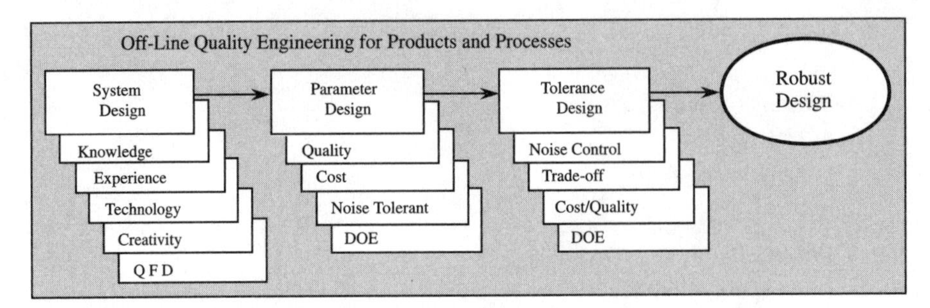

Figure 6.5 Off-Line Quality Engineering

This method of design focuses on how to design a product that can withstand the environment rather than one that is designed to meet every specific related problem. This guides the engineer in developing efficient processes and effective products that enhance capabilities rather than just meet needs. The process is efficient because the materials, methods, and processes are optimized to enhance the controllable factors of quality, cost, and schedule during the design and production process. The product is effective because the design engineer has made the design to be robust to operate in the user's environment. As we first indicated in Figure 1.1, these activities take place during the design-and-development phase of a product life cycle. Off-line quality engineering is performed using the three phases demonstrated in Figure 6.5.

Taguchi describes the engineering design and improvement cycle in three phases: System Design, Parameter Design, and Tolerance Design. The system design phase occurs when the fundamental design and concept for a product or service are established. This phase is central to the engineering and subject expert technologies.

System Design The system design phase requires the design engineer and the design team to use their knowledge and experience to build a system that is technically able to meet the customer's needs, requirements, and wants. The best tools to use in developing the system, its requirements and the associated assemblies are the product development strategies presented in Chapter 9.

In this phase the basic concept of the system is formed and the system technologies are selected. Based on the voice of the customer (VOC), the desired output parameters of the system are defined and the assemblies, components, and parts are selected that will meet these requirements. The intent of this effort is to develop the best technical approach to producing the desired function. The What/How method of Quality Function Deployment (QFD) is the best approach to structuring this effort successfully.

Parameter Design In the parameter design phase we design the specifications of the system. That is, we improve quality without controlling or eliminating the sources of variance that occur in the customer or production environment. These sources influence target dimensions, material properties, voltages, and other parameters. We set these parameters to make the product less sensitive to the causes of variation. Parameter design makes the product more robust to the causes of variation.

Parameter design is the most important step in off-line quality engineering. In this phase, the system's capability to best perform its function is developed and the system is made to be robust against the production and user's environments. During this phase the design team will select the specific levels for the system factors that will lead to a high-quality, low-cost, robust product that satisfies the VOC. The goal of parameter design is to combine levels for the product or process parameters that achieve robustness at the least cost. Taguchi DOE is the best tool for performing parameter design and selecting the optimized system.

Parameter design studies typically involve two basic types of factors: *control factors* and *noise factors*. Control factors are the treatments that can be controlled by the engineer and are typically referred to as treatments. These can be set, specified, and maintained during the test runs and throughout the life cycle of the system, production, and use. Noise factors are the factors that cannot be controlled from a technical, practical, or economical standpoint; these are the factors that appear as the error factors in ANOVA. As you can see from the ANOVA tables previously used in this chapter, noise (error) can constitute a significant proportion of the variability of a system. Parameter design allows you to design your factors/treatments so that they can overcome the effects of the environment.

Control and noise factors must be included in any DOE for parameter design. The inclusion of these two types of factors in required to assess their relationship. This relationship between the levels of the control factors and the associated signal-to-noise (S/N) ratio is the primary selection factor for designing robust systems. We will utilize a parameter design to demonstrate the Taguchi method of performing a design of experiments later in this section.

Tolerance Design Tolerance design is used when a further optimization of a system is required. If you have accomplished all you can by performing parameter design and the system still does not meet the requirements of the customer, you must now attempt to control the noise (environment). This phase focuses on the trade-off between the cost of controlling the environment and the quality obtained, which is measured using the loss function we previously introduced. Tolerance design is the judicious manipulation of the levels of tolerances, materials and components. The design team judiciously considers upgrading materials and components, using higher grade processes, and tightening tolerances only for those factors having a significant effect on quality. The strategy for tolerance design is to reduce noise by increasing cost. Tolerance design should be considered only after a complete parameter design has been completed. The Taguchi method for performing DOE is the best method for performing tolerance design.

Taguchi Design of Experiments

The Taguchi method of DOE uses orthogonal arrays that are forms of the fractional factorial arrays previously presented in this chapter. Taguchi indicates that there are no higher order interactions (three interactions or more) that are critical. Key two-way interactions can be considered, based on engineering and technical assessment, but they are assumed to be less significant than main effects. The focus of the Taguchi DOE methods is clearly main effects, with interactions as a secondary consideration for the experimenter. We will review the Taguchi methods of DOE that are the most widely applicable to the design and development of new products and services. We will use a parameter design DOE to demonstrate the Taguchi method of DOE. Parameter values for each product or process element of the system should be designed to provide consistent performance with little or no variability due to process or use noise. We will develop the Taguchi method of DOE in eight steps.

1. Identify an element of the system design for analysis.
2. Perform a cause-and-effect analysis.
3. Select treatments, levels, and values.
4. Determine how experimental results will be expressed.
5. Select a designed experiment.
6. Conduct the experiment.
7. Perform data analysis.
8. Graph the results.

Step 1. Identify an Element of the System Design for Analysis The first step in parameter design is to identify the process or product element of the system design that we need to evaluate. The most effective way to accomplish this is by using Process Analysis, 7-MP tools, and 7-QC tools. These opportunities can be derived from many sources in existing and new processes, products, and services. Examples include:

- Opportunities for variability reduction to improve the efficiency and effectiveness of existing processes.
- Opportunities to improve products and services based on customer requirements.
- Opportunities to reduce reject and scrap rates.
- Opportunities to improve the development of new products and services.

Step 2. Perform a Cause-and-Effect Analysis Figure 6.6 is an example of a fishbone diagram for cause-and-effect analysis. This is an example based on the production of a composite material to be used in signs. We will follow this example through our development of parameter design. To identify the opportunities for improvement, four subject experts spent approximately one hour of their time to brainstorm the problem and create the cause-and-effect analysis.

Step 3 Select Treatments, Levels, and Values Based on the cause-and-effect analysis, the team selected which factors will be tested, determined the levels of the factors, and assigned a test value for each level. From our previous example, the factors and levels are indicated in Table 6.14.

The team determined that the composite must be designed to be robust against three noise factors Those factors and their levels are indicated in Table 6.15.

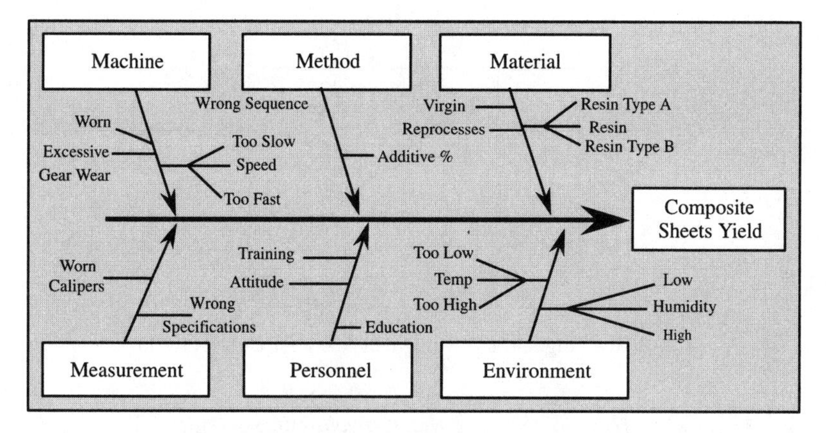

Figure 6.6 Potential Parameter Design Factors

Control Factors	Levels	
	1	2
A. Plastic Material	Virgin	Reprocesses
B. Fiberglass Additive	40%	20%
C. Binder Resin	A	B

Table 6.14 Selected Parameter Design Factors, Levels, and Values

Noise Factors	Levels	
	1	2
D. Humidity	High Amb.	Low Amb.
E. Temperature	High	Low
F. Operator Training	Company	OJT

Table 6.15 Selected Noise Factors, Levels, and Values

Inner Array			
	Control Factors		
Run	A	B	C
1	1	1	1
2	1	2	2
3	2	1	2
4	2	2	1

Table 6.16 Taguchi L_4 Orthogonal Array for Control Factors

Step 4. Determine How to Express the Experimental Results Determine how the experimental results will be expressed. What quality characteristic will we try to measure? This is the output variable for the designed experiment. From the cause-and-effect analysis we have completed the quality characteristic, which will be quantified as follows:

Yield as Acceptable 4×8 Sheets Per Hour

Step 5. Select a Designed Experiment Based on the number of factors and factor levels required to evaluate the output variable, select the appropriate Taguchi designed experiment from the statistical appendix. Remember the basic rules for selecting an experimental design: The design must be orthogonal and must contain the minimum number of runs to satisfy the requirements.

In our continuing example, we have three factors at two levels each 2^3. In the list of Taguchi orthogonal arrays in the appendix, our experiment, with three levels and two factors, is most closely represented by the L_4 orthogonal array. The array is presented in Tables 6.16 and 6.17. The L_4 orthogonal array indicates that we will require four test runs to evaluate the three factors at two levels. The letters across the top of the table represent column headings and correspond to the factors we selected in Figure 6.15. This array is called the inner array because it represents the control factors of our designed experiment.

Outer Array			
	Noise Factors		
Run	D	E	F
1	1	1	1
2	1	2	2
3	2	1	2
4	2	2	1

Table 6.17 Taguchi L_4 Orthogonal Array for Noise Factors

					Outer Array			
				Noise Factor	Run			
					1	2	3	4
				D	1	1	2	2
Inner Array				E	1	2	1	2
	Control Factor			F	1	2	2	1
Run	A	B	C					
1	1	1	1		y_1	y_2	y_3	y_4
2	1	2	2		y_5	y_6	y_7	y_8
3	2	1	2		y_9	y_{10}	y_{11}	y_{12}
4	2	2	1		y_{13}	y_{14}	y_{15}	y_{16}

Table 6.18 Test Layout for Taguchi L_4 Array with Noise Factors

We must now select an array for the evaluation of the noise factors (outer array). The same rules apply to the selection of an orthogonal array to evaluate noise factors. Therefore, a Taguchi L_4 array will be selected to evaluate the effects of noise on the process of producing composite sheets. The L_4 orthogonal array in Table 6.16 represents the noise factors from Table 6.17.

The inner and outer arrays must now be arranged to provide for the distribution of the noise factors over the range of response data from the control factors. This is accomplished by rearranging the standard L_4 array as indicated in Table 6.18. Note that to accomplish this it is necessary to test the number of replicates or repeats equal to the number of runs in the outer noise array. In this example it will be necessary to accomplish sixteen test runs (y_1 through y_{16}) to satisfy all the test requirements of the control and noise factors.

In this example the number of factors for control and noise were equal to the factors and levels for the L_4 array. It is not necessary to model your experiment to fit any specific Taguchi orthogonal arrays. The array can be customized to fit any experimental requirement.

Step 6. Conduct the Experiment Using the experimental design formulated in Step 5, conduct the series of tests. Measure the quality characteristic using the measure determined in Step 4. Record the results on your work sheet along with any notes on related circumstances that might provide information concerning the test results. The resulting data can be recorded as indicated in Table 6.19, or on a separate spreadsheet if the experiment is large and there are many data points. In any case, use automated data acquisition whenever possible and always use software or a spreadsheet program to perform your analysis.

Step 7. Perform Data Analysis To perform data analysis of your designed experiment, it will be necessary to calculate several factors. Using the data from the experimental results, calculate the following:

	Inner Array			Outer Array				
				Noise Factor	Run			
					1	2	3	4
				D	1	1	2	2
				E	1	2	1	2
	Control Factor			F	1	2	2	1
Run	A	B	C					
1	1	1	1		28	27	28	29
2	1	2	2		22	24	23	21
3	2	1	2		23	24	21	22
4	2	2	1		26	25	25	24

Table 6.19 Record the Results of the Designed Experiment

- S/N Ratio
- Level Averages
- Grand Average
- Level Effects

These calculations can then be applied directly to your spreadsheet, as indicated in Table 6.20.

	Inner Array			Outer Array						
				Noise Factor	Run					
					1	2	3	4		
				D	1	1	2	2		
				E	1	2	1	2		
	Control Factor			F	1	2	2	1		
Run	A	B	C						$\bar{y}$	*S/N*
1	1	1	1		28	27	28	29	28.00	28.93
2	1	2	2		22	24	23	21	22.50	27.01
3	2	1	2		23	24	21	22	22.50	27.01
4	2	2	1		26	25	25	24	25.00	27.95
LA1	27.97	27.97	28.44	Notes:						
LA2	27.48	27.48	27.01	LA = Level Average						
GA	27.73	27.73	27.73	GA = Grand Average						
LE1	0.24	0.24	0.71	LE = Level Effects						
LE2	−0.24	−0.24	−0.71	TE = Total Effects						
TE	0.49	0.49	1.43							

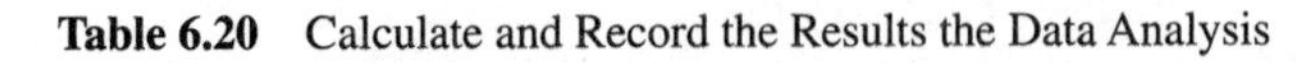
Table 6.20 Calculate and Record the Results the Data Analysis

S/N Ratio Use the experimental results to calculate the Signal-to-Noise Ratio (S/N). The Signal-to-Noise Ratio calculation is dependent on the characteristic of the response, (SIB), (NIB), and (LIB), using the appropriate equations as indicated:

$$SIB\text{: S / N ratio } \eta = -10\log\frac{1}{n}\left(y_1^2 + y_2^2 + \ldots y_n^2\right)$$

$$NIB\text{: S / N ratio } \eta = -10\log\frac{1}{n}\left[\left(y_1 - y\right)^2 + \left(y_2 - y\right)^2 + \ldots\left(y_n - y\right)^2\right]$$

$$LIB\text{: S / N ratio } \eta = -10\log\frac{1}{n}\left(\frac{1}{y_1^2} + \frac{1}{y_2^2} + \ldots\frac{1}{y_n^2}\right)$$

Since the quality characteristic we are measuring (yield) is a LIB characteristic, the LIB equation is used to calculate the S/N Ratio for each of the four test runs in our example:

$$\text{Run 1: } \eta = -10\log\frac{1}{4}\left(\frac{1}{28^2} + \frac{1}{27^2} + \frac{1}{28^2} + \frac{1}{29^2}\right)$$

$$= -10\log\frac{1}{4}(.0013 + .0014 + .0013 + .0012) = 28.93$$

$$\text{Run 2: } \eta = 27.01$$

$$\text{Run 3: } \eta = 27.01$$

$$\text{Run 4: } \eta = 27.01$$

These data can now be applied to the DOE data table and spreadsheet as was indicated in Table 6.20. All the other calculations required to complete the analysis of your designed experiment can be accomplished directly on the spreadsheet.

Level Average Obtain the level means by adding all the observations at that level for each main effect and interaction and dividing by the total number of observations for that level. In our example, the level averages are calculated as follows:

$$Level\ Average = LA = \frac{Observation\ Level_X}{n_X}$$

$$LA_{A1} = \frac{28.93 + 27.01}{2} = 27.97 \qquad LA_{A2} = \frac{27.01 + 27.95}{2} = 27.48$$

$$LA_{B1} = 27.97 \qquad LA_{B2} = 27.48 \qquad LA_{C1} = 28.44 \qquad LA_{C2} = 27.01$$

Grand Average Obtain the grand average by adding the S/N from each control factor run and dividing by the total of control factor runs. In our example, the grand average is calculated as follows:

$$Grand\ Average = \frac{S/N\ Ratio\ Run\ 1 + S/N\ Ratio\ Run\ 2 + \ldots S/N\ Ration\ Run\ i}{Control\ Factor\ Runs}$$

$$Grand\ Average = \frac{28.93 + 27.01 + 27.01 + 27.95}{4} = 27.73$$

Level Effects The level effect is calculated by subtracting the grand average (7.23) from each level mean.

$$Level\ Effects\ (LE) = Level\ Average - Grand\ Average$$

$$LE_{A1} = 27.97 - 27.73 = 0.247 \qquad LE_{A2} = 27.65 - 27.73 = -0.247$$

$$LE_{B1} = 0.247 \qquad LE_{B2} = -0.247 \qquad LE_{C1} = 0.715 \qquad LE_{C2} = -.715$$

Total Effect The total effect is calculated by subtracting the least level effect from the greatest level effect (the range).

$$Total\ Effect\ (TE) = Greatest\ Level\ Effect - Least\ Level\ Effect$$

$$TE_A = 27.97 - 27.48 = 0.49 \qquad TE_B = 0.49 \qquad TE_C = 1.43$$

Step 8. Graph the Results To confirm the information provided by the calculations, and to provide a graphical understanding of the effects on the product and process, graph the results as shown in Figure 6.7. Based on the Lib criterion, the optimized

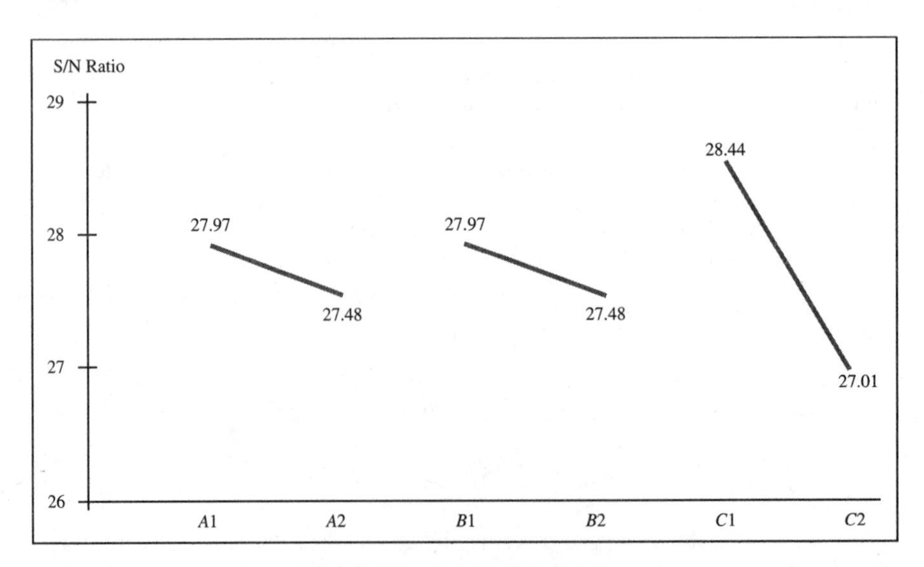

Figure 6.7 Graphing the Results

process inputs for the manufacture of plastic sheets is $A1$, $B1$, and $C1$, with the $C1$ factor clearly being the most significant.

Predicting Optimum Results

To predict the optimum results based on our evaluation in Step 8, insert the optimum treatment and level input variables into the equation to estimate the value of the optimized quality characteristic:

$$\hat{y} = \frac{T}{n} + \left(A_X - \frac{T}{n}\right) + \left(B_X - \frac{T}{n}\right) + \ldots + \left(F_X - \frac{T}{n}\right)$$

where

T = Total of all experimental results

n = Number of samples

$\frac{T}{n}$ = Grand average

A_X, $B_X \ldots F_{X=}$Performance values at optimum level for each factor

Applying the data from our example to the optimization equation, we obtain:

$$\hat{y} = \frac{T}{n} + \left(A_1 - \frac{T}{n}\right) + \left(B_1 - \frac{T}{n}\right) + \ldots + \left(C_1 - \frac{T}{n}\right)$$

$$\hat{y} = \frac{392}{16} + \left(25.25 - \frac{392}{16}\right) + \left(25.25 - \frac{392}{16}\right) + \left(26.25 - \frac{392}{16}\right)$$

$$\hat{y} = 24.5 + .75 + .75 + 2 = 28$$

The predicted yield of the optimized production process is 28 acceptable sheets per hour. Please note that we used the actual data to perform this prediction. We must now verify it. The confirmation experiment is run using the optimum factor levels, based on the LIB criteria.

Taguchi-Style ANOVA

Taguchi-style ANOVA uses several methods developed to facilitate the calculation and understanding of ANOVA. The basic rules of ANOVA apply here also, with regard to hypothesis testing, partitioning of variance, and measures of significance. One of the important elements of Taguchi-style ANOVA is the integration of the percent contribution directly into the ANOVA calculations. The clear advantages of this method are the ease in performing the calculations and the integration of the percent contribution. We will use the data from our DOE example to construct the

ANOVA. This may be accomplished using the transformed S/N data or the raw data. For the purposes of clarity, we will calculate the ANOVA using the raw data from Table 6.20.

1. Calculate the grand average
2. Calculate the level mean, level effects, and total effects.
3. Compute the sum of the squares.
4. Compute the mean squares.
5 Calculate the F_{Ratio}.
6. Determine the critical value of F (F_{Crit})
7. Calculate the pure sum of the squares.
8. Determine the percent contribution.

Step 1. Calculate the Grand Average The grand average is calculated by adding the S/N from each control factor run and dividing by the total number of runs.

$$Grand\ Average = \frac{Sum\ of\ Individual\ Observations}{Total\ Number\ of\ Observations}$$

Using the three levels of data from the parameter design DOE as an example, we obtain the following:

$$Grand\ Average = \frac{28+27+28+\ldots+24+21+22+\ldots+24}{16} = 24.50$$

Step 2. Calculate the Level Mean, Level Effects, and Total Effects The level mean is calculated by summarizing the observations at each level of treatment and dividing by the number of observations in that level.

$$Level\ Mean = \frac{\sum(All\ Observations\ at\ Level\ i)}{Number\ of\ Observations\ at\ Level\ i}$$

Using the two levels of data from the parameter design DOE with the Plastic Material at level 1, we obtain:

$$Level\ A1 = \frac{28+27+28+29+22+24+23+21}{8} = 25.25$$

The level effect is calculated by subtracting the grand average from the level mean. The total effect is calculated by subtracting the least level mean from the greatest level mean:

$$Level\ Effects = Level\ Mean - Grand\ Average$$

$$Total\ Effect = Greatest\ Level\ Mean - Least\ Level\ Mean$$

These data are consolidated in tabular form for ease of application to the ANOVA.

Factor	Level	Level Mean	Level Effect	Total Effect
A Plastic Material	1	25.25	0.75	
	2	23.75	-0.75	1.50
B Fiberglass Additive	1	25.25	0.75	
	2	23.75	-0.75	1.50
C Binder Resin	1	26.5	2.00	
	2	22.5	-2.00	4.00

Step 3. Calculate the Sum of the Squares The next step is to calculate the sum of the squares for each of the following elements:

Sum of the Squares Total: SS_{Total} = [Distance of individual observations from the grand average]^2

Sum of the Squares for Factors/Interactions: SS_{Factor} = [Level effects at each level]^2 × [Number of observations at each level]

Sum of the Squares for Replications: SS_{Rep} = Sum of [Grand average − Average of each replication column]^2 × Number of test runs

Sum of the Squares for Error: $SS_{Error} = SS_{Total} - [SS_{Factors} + SS_{Interactions} + SS_{Replications}]$

Using the three levels of data from the parameter design DOE with the main effects, we obtain:

$$SS_A \text{ Plastic Material} = 9$$

$$SS_B \text{ Fiberglass Additive} = 9$$

$$SS_C \text{ Binder Resin} = 64$$

$$SS_{Total} = 96$$

$$SS_{Error} = 4$$

Step 4. Calculate the Mean Squares Next we calculate the mean squares (MS) for each factor, interaction, replication, and error. The mean squares are the sum of the squares divided by the degrees of freedom:

$$MS = \frac{Sum\ of\ the\ Squares}{Degrees\ of\ Freedom}$$

Using the data from our parameter design DOE for material, additive, and resin, calculate the mean squares.

$$MS_{Material} = \frac{9}{1} = 9$$

$$MS_{Additive} = \frac{9}{1} = 9$$

$$MS_{\text{Re sin}} = \frac{64}{1} = 64$$

$$MS_{Error} = \frac{14}{12} = 1.16$$

Step 5. Calculate the F_{Ratio} Calculate the F_{Ratio} for each treatment and interaction by dividing the mean square for the factor or interaction by the mean squares for error.

$$F_{Ratio} = \frac{Mean\ Squares\ for\ Factors\,/\,Interactions}{Mean\ Squares\ for\ Error}$$

$$F_{Material} = \frac{9}{1.16} = 7.71$$

$$F_{Additive} = \frac{9}{1.16} = 7.71$$

$$F_{\text{Re sin}} = \frac{64}{1.16} = 54.86$$

Step 6. Determine the critical value of F (F_{Crit}) Use the *F* Table in Appendix A for each factor and interaction at the 0.05 or 0.01 level of significance (95% or 99% confidence level) respectively. The *F* table columns represent the degrees of freedom in the numerator, and the rows represent the degrees of freedom in the denominator:

$$F_{Crit} = F\ (\alpha,\ df)$$

Using the data from the parameter design DOE for the main effects and interactions for diameter and location, we get the following:

Factor	F_{Ratio}	$F_{Critical}$ .95	.99	Significance
Material	7.71	4.75	9.33	Significant
Additive	7.75	4.75	9.33	Significant
Resin	55.17	4.75	9.33	Highly Significant

Step 7. Calculate the Pure Sum of the Squares Calculate the pure sum of the squares (SS′) as indicated in the following equation:

$$SS' = SS_{Factor} - [MS_{Error} \times df_{Factor}]$$

Using the data from the parameter design DOE for the main effects and interactions for diameter and location, we obtain these results:

$$SS'_{Material} = 9 - [1.16 \times 1] = 9 - 1.16 = 7.84$$

$$SS'_{Additive} = 9\ [1.16 \times 1] = 9\ 1.16 = 7.84$$

$$SS'_{Resin} = 64 - [1.16 \times 1] = 64 - 1.16 = 62.84$$

Step 8. Determine the Percent Contribution Calculate the percentage of the contribution to variability for each factor, interaction, and replication using the following equation:

$$\%\ Contribution = \frac{Pure\ Sum\ of\ the\ Squares}{Total\ Sum\ of\ the\ Squares} \times 100$$

To calculate the percent contribution for error ($\%\ CONT_{Error}$), use the following equation:

$$\%\ CONT_{Error} = 0 - \Sigma\ [\ \%\ \text{Contribution of all Factors/Interactions/Replicates}\ ]$$

Using the data from the parameter design DOE for the main effects and interactions for diameter and location, we obtain:

$$\%\ CONT_{Material} = \frac{7.84}{96} \times 100 = 8.16$$

$$\%\ CONT_{Additive} = \frac{7.84}{96} \times 100 = 8.16$$

$$\%\ CONT_{\text{Resin}} = \frac{62.84}{96} \times 100 = 65.46$$

$$\%\ CONT_{Error} = 0 - [8.16 + 8.16 + 65.46] = 0 - 81.78 = 18.22\%$$

With the Taguchi ANOVA, complete fact-based decision making becomes almost effortless. It is clear that all factors are making contributions to variation, but the degree of that contribution is clearly greater for resin. The resin factor is contributing 65.46% of the total variation in plastic sheet yield.

WHAT HAVE YOU LEARNED?

We have determined that a significant amount of information can be gained from fractional factorial experiments. Several key points contribute to our ability to fractionalize and use fractional factorial experiments:

- When there are several variables to be evaluated, the system being studied is most likely driven by some subset of main effects and possibly a few lower order interactions.
- The subset of significant factors identified in a fractional factorial experiment can be projected into a full factorial experiment and fully evaluated.
- Fractional factorial experiments can be used in a sequence of experiments used to refine and identify significant factors.
- Fractional factorial experiments can themselves be used to optimize systems where there is little or no direct concern about interactions.
- That fractional factorial experiments can be accomplished using a series of eight Steps.
  - Identify a problem or opportunity for improvement.
  - Perform a cause-and-effect analysis to select treatments.
  - Select treatments and treatment levels.
  - Select a fractional factorial experimental format.
  - Conduct the experiment and acquire data.
  - Determine the effects.
  - Graph the results.
  - Perform ANOVA.
- Taguchi Quality Engineering has made many significant contributions to the accomplishment of designed experiments.
- The Taguchi method of DOE uses orthogonal arrays that are forms of the fractional factorial arrays previously presented in this chapter. Taguchi indicates that there are no higher order interactions (three interactions or more) that are critical. Key two-way interactions can be considered, based on engineering and technical assessment, but they are assumed to be less significant than main effects. The Taguchi method of DOE can be accomplished in eight steps.
  - Identify an element of the system design for analysis.
  - Perform a cause-and-effect analysis.
  - Select treatments, levels, and values
  - Determine how experimental results will be expressed.
  - Select a designed experiment.
  - Conduct the experiment.
  - Perform data analysis.
  - Graph the results.

TERMINOLOGY

Fractional Factorial: An orthogonal array that has been reduced in size or fractionalized. This reduction provides for the acquisition of information on main effects and some interactions of your designed experiment.

Confounding: Occurs when you reduce the number of runs required in your experiment, by fractionalizing a full factorial design format. This causes certain interactions to be confounded with main effects and other interactions, that is, that some of the information acquired for main effects may contain the effects of interactions.

Alias: The interactions that are confounded with main effects or other interactions, that is, the main effect is an alias for the interaction ($A=BC$).

Robust Design: Robustness is synonymous with high quality, low-cost, and on-schedule processes. High quality, low cost, and on schedule is the optimum state for any system. Therefore, it is the goal of the design team to search for and find robust designs for products and processes. A robust design is the design of a product, process, or service that can withstand the noise of production and use.

Noise: Outer noises are factors that affect product and process performance from an external source. Outer noise factors are such things as ambient temperature, training, operating conditions, or anything that affects the system that is external to the product or process. Inner noises are the system conditions that cause variation from within the product or process. Inner noise factors are such things as tool wear, process procedures, component capability, different materials, and any other factor that is part of the system, product, or process that causes variability.

Quality Characteristics: The output or dependent variables of our designed experiment. A quality characteristic is usually some measure of quality, cost, or schedule.

On-Line Quality Engineering: Utilized as process and product improvement strategies. The purpose of on-line quality engineering is to control the process and products to ensure that they meet the robust design standards established during the design and development process and to improve those products and processes.

Off-line Quality Engineering: Performed early in the design and development process of a product. The intent of off-line quality engineering is to optimize the functionality of a system through a design strategy that will render a robust system.

System Design: Requires the design engineer and the design team to use their knowledge and experience to build a system that is technically able to meet the customer's needs, requirements ,and wants.

Parameter Design: In the parameter design phase we design the specifications of the system, that is, we improve quality without controlling or eliminating the sources of variance that occur in the customer or production environment.

Tolerance Design: Used when a further optimization of a system is required. If you have accomplished all you can by performing parameter design and the system still does not meet the requirements of the customer, you must now attempt to control the noise (environment).

EXERCISES

1. Fractionalize the following full factorial designed experiment using the design generators in Appendix A.

	Treatments														
Runs	A	B	C	D	AB	AC	AD	BC	BD	CD	ABC	ABD	ACD	BC D	ABC D
1	–	–	–	–	+	+	+	+	+	+	–	–	–	–	+
2	+	–	–	–	–	–	–	+	+	+	+	+	+	–	–
3	–	+	–	–	+	+	+	–	–	+	+	+	–	+	–
4	+	+	–	–	+	–	–	–	–	+	–	–	+	+	+
5	–	–	+	–	+	–	+	–	+	–	+	–	+	+	–
6	+	–	+	–	–	+	–	–	+	–	–	+	–	+	+
7	–	+	+	–	–	–	+	+	–	–	–	+	+	–	+
8	+	+	+	–	+	+	–	+	–	–	+	–	–	–	–
9	–	–	–	+	+	+	–	+	–	–	–	+	+	+	–
10	+	–	–	+	–	–	+	+	–	–	+	–	–	+	+
11	–	+	–	+	–	+	–	–	+	–	+	–	+	–	+
12	+	+	–	+	+	–	+	–	+	–	–	+	–	–	–
13	–	–	+	+	+	–	–	–	–	+	+	+	–	–	+
14	+	–	+	+	–	+	+	–	–	+	–	–	+	–	–
15	–	+	+	+	–	–	–	+	+	+	–	–	–	+	–
16	+	+	+	+	+	+	+	+	+	+	+	+	+	+	+

a. What are the design generators?

b. Draw the fractional factorial format.

2. What are the two- and three-level interactions confounded with the main effects of the fractional factorial design in Exercise 1? What are the aliases?

3. Draw a 2^{5-2} fractional factorial format.

a. In that fractional factorial design, in which column can the interactions of treatment factors *A* and *C* be evaluated?

b. How many main effects can be evaluated in this fractional factorial format if one column is used for interactions?

c. What other interaction is confounded into the column to evaluate interactions *A* and *C*?

d. Why should you be cautious in using this column to evaluate the interaction of *A* and *C*?

4. Calculate and graph the effects of the following fractional factorial experiment.

	Treatments			Data			
Runs	A	B	C	y^1	y^2	y^3	y^4
1	+	−	−	0.67	0.65	0.66	0.68
2	−	+	−	0.57	0.60	0.59	0.61
3	−	−	+	0.59	0.53	0.61	0.60
4	+	+	+	0.70	0.59	0.62	0.65

5. Perform and evaluate the ANOVA for the fractional factorial experiment in Exercise 4.

7

NONPARAMETRIC TESTS

Some nonparametric techniques are closely related to order statistics. Others rely on the probability of getting a specific combination out of a finite set of possible answers. Nonparametric tests generally are used when the standard assumptions of classical statistics are known not to be met. In particular, they are frequently used when the distribution in question is not even approximately normal or the data are not statistically independent (e.g., when there is sequential correlation: y_i depends on y_{i-1}). Unfortunately, as Box et al. [1] discuss, this apparent relaxation in requirements has led to serious misuse. For this reason, many statisticians have strong opinions about the subject of nonparametric methods. Noether [2] provides an elementary introduction to nonparametric methods, blended with classical ANOVA. In general, expect to use larger sample sizes with nonparametric tests than with parametric tests to get the same results.

RANDOM SEQUENCES

As an example of a nonparametric test, the following procedure may be used to check for randomness in sequences of data points [3]. The data must be expressible in the form of yes/no answers. For yes/no survey answers, use Y for a "yes" response and N for a "no" response, ignoring refusals to answer. This procedure can obviously be followed with any set of dichotomous responses. To define Y and N with a measurement that can vary continuously:

- Take a set of data points.
- Find the median.
- Call a point greater than the median a Y and a point less than the median an N; ignore points at the median itself.

- Express the results as a sequence of Y and N responses, separating them where they change.
- Let m equal the number of Y symbols and n equal the number of N symbols.
- Take enough data that m and n are both greater than 10. The test works better with a larger sample size, and the normal approximation also applies.

To see how to define Y and N with a measurement that can vary continuously, consider the following example, which shows the steps:

Step 1. Take enough data that there are at least 10 points which are not duplicates. Record it in the order taken. The test works better with a large sample size, and the normal approximation then applies. Table 7.1 shows the sample data set.

Step 2. Make a new copy of the data, placing it in ascending order, as in Table 7.2. Find the median. The example has 49 data points, so the 25th data point is the median; its value is 51.5.

Step 3. Count the number of points m greater than the median (51.5) and the number of points n less than the median.

In the example, there are $n = 22$ points below 51.5, 3 points at 51.5, and $m = 24$ points above 51.5.

Step 4. Go back to the original list of points, keeping them in the order taken, and place a Y next to those greater than the median (51.5) and an N next to those that were less than the median. Delete those points that are at the median. (See Table 7.3.)

Step 5. Count the number of runs (sets of identical letters).

In the example, there are R=9 runs.

Step 6. Approximate the mean and variance by

$$\mu_R = 1 + \frac{2mn}{m+n}$$

$$\sigma_R^2 = 1 + \frac{2mn(2mn - m - n)}{(m+n-1)(m+n)^2}$$

Step 7. Take the magnitude of $(R-\mu_R)/\sigma_R$ and compare it to $z_{\alpha/2}$=1.96 for 95% confidence. If the magnitude is smaller than 1.96, there is no evidence of a lack of randomness. In the example, μ_R=23.96, σ_R=3.35, and $|(21-23.96)/3.35| =$ $.88 < z_{\alpha/2}$=1.96, so there is no evidence for a lack of randomness.

The following example has the same m and n as the previous one (so μ_R and σ_R are also unchanged), but R=9. (See Table 7.4.)

51.8	51.6	57.3	51.2	55.1	42.3	55.4	51.6	49.1	48.7
51.4	51.7	43.3	50.4	41.4	50	46.9	46.4	56	44.6
56	54.7	60.7	45.6	50.2	54.9	52.1	47.5	39.9	65.4
52.5	55	57.6	51.5	49.4	48.1	55.6	50.6	42.1	48.1
51.5	62	52.7	68.5	49.7	46.7	53.2	51.5	63.1	

Table 7.1 Sequence of Measurements

39.9	41.4	42.1	42.3	43.3	44.6	45.6	46.4	46.7	46.9
47.5	48.1	48.1	48.7	49.1	49.4	49.7	50	50.2	50.4
50.6	51.2	51.4	51.5	51.5	*51.5*	51.6	51.6	51.7	51.8
52.1	52.5	52.7	53.2	54.7	54.9	55	55.1	55.4	55.6
56	56	57.3	57.6	60.7	62	63.1	65.4	68.5	

Table 7.2 Ranked Measurements

51.8	51.6	57.3	51.2	55.1	42.3	55.4	51.6	49.1	48.7
Y	Y	Y	N	Y	N	Y	Y	N	N

51.4	51.7	43.3	50.4	41.4	50	46.9	46.4	56	44.6
N	Y	N	N	N	N	N	N	Y	N

56	54.7	60.7	45.6	50.2	54.9	52.1	47.5	39.9	65.4
Y	Y	Y	N	N	Y	Y	N	N	Y

52.5	55	57.6	51.5	49.4	48.1	55.6	50.6	42.1	48.1
Y	Y	Y		N	N	Y	N	N	N

51.5	62	52.7	68.5	49.7	46.7	53.2	51.5	63.1	
	Y	Y	Y	N	N	Y		Y	

Table 7.3 Coded Measurements

YYYYYYY NNNNNNNNNNN YYYYY NNNNN YY NN YYYY NNNNNN YYYY

Table 7.4 Coded Measurements

Here, $|(9-23.96)/3.35|=4.47>z_{\alpha/2}=1.96$, so there *is* evidence for a lack of randomness. Although the large sample approximation was used, no assumption was made about what distribution Y or N actually followed (it was distribution-free), and the test did not compare parameters μ_Y to μ_N or σ_Y to σ_N (it was nonparametric). Trend tests can more efficiently be used to detect a gradual drift in one direction, for example, a gradual increase in sales as Christmas approaches.

WILCOXON RANK SUM TEST

The means of two continuous populations that have the same shape and spread, but which need not be normal, can be compared using the Wilcoxon rank sum test [2, 3]. This test is identical to the Mann-Whitney test, except for the counting procedure. In the Wilcoxon version, you just sum ranks, while in the Mann-Whitney version, you compare every x with every y. The Wilcoxon rank sum test is easy to implement on a spreadsheet. (See Table 7.5.) To do the Wilcoxon test:

Step 1. Call the smaller of the two samples x, and the larger y.

Step 2. Rank the combined data in increasing magnitude, keeping the x and y labels with them.

Step 3. If two or more of the x and/or y values are equal, give all of them the average rank.

Step 4. Let R_x be the sum of the ranks of the data labeled x; then $R_y=n_x(n_x+n_y+1)-R_x$.

Step 5a. Using a table of critical R values R^*_α,

Reject hypothesis H_0 if either R_x or R_y is *less* than or equal to R^*_α.

If R_x is *less* than or equal to R^*_α, then reject $\mu_x<\mu_y$.

This is NOT the same R^*_α as that for the other Wilcoxon test.

Rank	x	y
...	...	...
16	2	
17	2.1	
19		2.3
19	2.3	
19	2.3	
21		2.4
...	...	...

Table 7. 5 Ranking Ties

Step 5b. For n_x and n_y both greater than 8, you can use the normal approximation and z score test, but the approximations for the mean and variance are different.

Now use

$$\mu_R = \frac{n_x(n_x + n_y + 1)}{2} \quad \text{and} \quad \sigma_R^2 = \frac{n_x n_y(n_x + n_y + 1)}{12}$$

WILCOXON SIGNED RANK TEST

If the population distribution is continuous and symmetric, the Wilcoxon signed rank test can be used to test $\mu = \mu_0$ or to do the corresponding one-sided tests. It can be used to test $\mu_1 = \mu_2$ if the data are paired; this also allows the symmetry restriction to be dropped. The corresponding one-sided tests can be done.

To do the Wilcoxon signed rank test to test $\mu = \mu_0$:

Step 1. Rank the absolute differences $|X_i - \mu_0|$ in ascending order. Give ties the average of the ranks they would have if slightly different numbers. Give the ranks the signs of their differences. On a spreadsheet, you could just put them in different columns.

Step 2. Let R^+ be the sum of the ranks with a plus sign and R^- the magnitude of the sum of the ranks with a minus sign. Then R is the smaller of R^+ and R^-.

Step 3a. Look up critical values of R for $\alpha = .05$ and sample size n in Table 7.6. You can approximate this table by $R_c = (4.49n - 1.78)^2$ if you round off the result to the nearest integer. This can be handy for use with a pocket calculator.

n	Critical R	n	Critical R
7	2	21	58
8	3	22	65
9	5	23	73
10	8	24	81
11	10	25	89
12	13	26	98
13	17	27	107
14	21	28	116
15	25	29	126
16	29	30	137
17	34	31	147
18	40	32	159
19	46	33	170
20	52	34	182

Table 7.6 Critical values of R for $\alpha = .05$

$$\text{Reject } \mu > \mu_0 \text{ if } R^- < R^*_\alpha.$$

$$\text{Reject } \mu < \mu_0 \text{ if } R^+ < R^*_\alpha.$$

$$\text{Reject } \mu = \mu_0 \text{ if } R \leq. R^*_\alpha.$$

Step 3b. For $n > 50$ (with some texts saying only $n > 20$), you can use the normal approximation and z score test with

$$\mu_R = \frac{n(n+1)}{4} \qquad \text{and} \qquad \sigma_R^2 = \frac{n(n+1)(2n+1)}{24}$$

KRUSKAL-WALLIS TEST

The Kruskal-Wallis test is used for hypothesis testing when the data are nonparametric [3]. This is a nonparametric version of the completely randomized ANOVA that was developed before personal computers were available (1952) and was therefore designed for simplicity of application. It does not require that the samples be normally distributed. It uses ranks of the data, rather than actual values. Although it is more sensitive than many of the other nonparametric tests, it should not be used in preference to standard ANOVA on data for which both are equally valid. The Kruskal-Wallis Test is performed by the following steps, which are shown in the form of an example:

Step 1. Label the observations as seen in Table 7.7.

Step 2. Sort all the data in descending order without regard to the original label, as shown in Table 7.8.

Step 3. Make a third column with the rank of each number; that is, number each item in increasing size of the measured variable, as in Table 7.9. In case of ties, give all the tied items the same rank, the average of the ranks they would have had. For example, the sequence 31, 32, 32, 32, 33, 34, 35, 35, 36 would have ranks 1, 3, 3, 3, 5, 6, 7.5, 7.5. 9. The highest rank will equal the total number of items, n. In the example, $n=13$.

Step 4. Sort the three columns on the label column as in table 7.10.

Step 5. Square the value of each rank. Sum the squared values under each label; then divide by by the number of items with that label.

Notice that this whole process can be accomplished very simply with a spreadsheet, as illustrated in Table 7.11. If there are many ties in a nonparametric test, many authors recommend using a parametric test instead. A simpler alternative is given here after the final step.

A	B	C
32.40	32.10	33.00
32.50	31.00	31.50
33.60	34.70	34.20
32.60	33.30	32.90
		31.80

Table 7.7 Labeled Observations

Label	Data	Label	Data	Label	Data
B	31.00	A	32.40	C	33.00
C	31.50	A	32.50	B	33.30
C	31.80	A	32.60	A	33.60
B	32.10	C	32.90	C	34.20
				B	34.70

Table 7.8 Sorted data in Descending Order

Label	Data	Rank	Label	Data	Rank	Label	Data	Rank
B	31.00	1	A	32.40	5	C	33.00	9
C	31.50	2	A	32.50	6	B	33.30	10
C	31.80	3	A	32.60	7	A	33.60	11
B	32.10	4	C	32.90	8	C	34.20	12
						B	34.70	13

Table 7.9 Ranked Data

Label	Data	Rank	Label	Data	Rank	Label	Data	Rank
A	32.40	5	B	31.00	1	C	31.50	2
A	32.50	6	B	32.10	4	C	31.80	3
A	32.60	7	B	33.30	10	C	32.90	8
A	33.60	11	B	34.70	13	C	33.00	9
						C	34.20	12

Table 7.10 Sorted by Label

Label	Data	Rank	Label	Data	Rank	Label	Data	Rank
A	32.40	5	B	31.00	1	C	31.50	2
A	32.50	6	B	32.10	4	C	31.80	3
A	32.60	7	B	33.30	10	C	32.90	8
A	33.60	11	B	34.70	13	C	33.00	9
						C	34.20	12
$\sum R$	29.00		$\sum R$	28.00		$\sum R$	34.00	
$\left(\sum R\right)^2$	841.00		$\left(\sum R\right)^2$	784.00		$\left(\sum R\right)^2$	1156.00	
$\left(\sum R\right)^2/n$	210.25		$\left(\sum R\right)^2/n$	196.00		$\left(\sum R\right)^2/n$	231.20	

Table 7.11 Spreadsheet Calculations

Step 6. Calculate the Kruskal-Wallis test statistic H:

$$H = \frac{12}{n(n+1)} \sum_{i=1}^{k} \frac{R_i^2}{n_i} - 3(n+1)$$

where: R_i is the sum of the ranks in sample i

H is well approximated by a chi-square distribution with $k-1$ degrees of freedom. Since tables of the chi-square distribution are readily available, this is an additional attractive feature of the Kruskal-Wallis test. Because you will be working with only the ranks, the data can be scaled, for example, normalized, if convenient. As mentioned throughout this book, the original data should be from a randomized sampling procedure (even though you will later rank the data).

$$H = \frac{12}{13(13+1)}(210.25 + 196.00 + 231.20) - 13 + 1$$

$$H = (.066)(637.45) - 42 = 0.03$$

Step 7. Compare H, the calculated value of the Kruskal-Wallis statistic, to the critical value from a chi-square table. If the calculated value exceeds the expected value, the data are not homogeneous, and the differences in the data treatments are significant.

$$\chi^2_{df,\alpha} = \chi^2_{12,.05} = 21.03$$

$$H = 0.03 << 21.03$$

Since the calculated value of chi-square is much less than the expected value, we have failed to reject the null hypothesis. The test clearly shows that the difference in the samples is very insignificant.

If you have several ties in a Kruskal-Wallis Test, you can correct H to allow for this (though it is seldom necessary) by replacing H with H^*:

$$H^* = \frac{H}{1 - \left[\sum_j \left(t_j^3 - t_j\right)\right] \Big/ \left[n^3 - n\right]}$$

If you have data with tied ranks such as those indicated in Table 7.12, the above equation would be used to calculate the value of H^*.

Using this data with the equation for H^* confirms that the H statistic is essentially unchanged.

$$H^* = \frac{0.03}{1 - \left[\left(2^3 - 2\right) + \left(3^3 - 3\right)\right] \Big/ \left[13^3 - 13\right]}$$

$$H^* = \frac{0.03}{0.99} \approx 0.03$$

Label	Data	Rank	Label	Data	Rank	Label	Data	Rank
B	31.00	1	A	32.40	5	C	33.00	9
C	31.50	2.5	A	32.50	6	B	33.30	10
C	31.80	2.5	A	32.60	7	A	33.60	11
B	32.10	4	C	32.90	8	C	34.20	12
						B	34.70	13

Table 7.12 Data with Tied Ranks

FRIEDMAN TEST

Just as the Kruskal-Wallis test is a nonparametric version of the completely randomized ANOVA, the Friedman test is a nonparametric version of the randomized block ANOVA [3]. With the Kruskal-Wallis test, you divided the sample elements randomly among k treatments: n_1 into treatment 1, n_2 into treatment 2, etc. Here, you divide the total $N=nk$ units at random into n homogeneous blocks (homogeneous groups) of k units, such that each block has exactly one unit from each of the k treatments, and no unit is shared between blocks. Thus, one way that $\{A, B, C, D, a, b, c, d, 1, 2, 3, 4\}$ could be divided into $n=4$ blocks of $k=3$ treatments is:

$$[a, 1, B]\ [2, c, A]\ [b, C, 3]\ [4, d, D]$$

where the first treatment is given an uppercase letter, the second a lowercase letter, and the third a number. You assign ranks separately within each group, say

$$[1, 2, 3]\ [3, 1, 2]\ [2, 3, 1]\ [1, 2, 3]$$

You then add up the ranks for each treatment. The uppercase letters have $3+2+3+3=11$. The lowercase letters have $1+1+2+2=6$, and the numbers have $2+3+1+1=7$. The test statistic is

$$Q = \frac{12}{nk(k+1)} \sum_{i=1}^{k} R_i^2 - 3n(k+1)$$

which is to be compared to $\chi^2(\alpha_e, k-1)$, where

$$\alpha_e = \frac{\alpha}{k(k-1)}$$

is an "effective" α. In general, you will have to interpolate in a table to get this number (or use a special function call in a spreadsheet). If Q exceeds the table value, then you have evidence that the treatments do not have equal effect. Only a small number of treatments were used here for simplicity in explanation. The test works better for larger numbers of treatments.

	Data			Rank by Rows		
Lot	A	B	C	A	B	C
1	20	22	28	1	2	3
2	13	24	21	1	3	2
3	25	31	27	1	3	2
4	31	12	29	3	1	2
5	19	14	33	2	1	3
6	11	26	15	1	3	2
7	18	15	22	2	1	3
8	16	16	30	1.5	1.5	3
			Sum =	12.5	15.5	20
			Sum-sq =	156.25	240.25	400
				Chi-sq =	3.5625	

Table 7.13 Friedman's Test

Table 7.13 provides an example of how to use Friedman's test. Eight lots were prepared using three different treatments *A*, *B*, *C*. The resulting judge's scores are shown on the left in the table and the corresponding ranks on the right. Notice that each lot is judged and ranked separately.

The effective $\alpha_e = .05/(3)(4) = .0042$, so the estimated table value for $df = 3 - 1 = 2$ is 11, and the result does not show significance.

PAIRED COMPARISONS

This test is roughly the nonparametric analog to the Tukey test. Say you have just completed the Friedman test as just described, and you find that at least one of the treatments differs significantly from the others. To find out which one(s), compare rank sums in pairs. That is, compare $R_i - R_j$ with z_α for each pair of treatments, where z_α is the value of z corresponding to α in a table of the standard normal distribution.

RANK TRANSFORMATION

If data for an ANOVA are replaced by their ranks, a procedure called rank transformation, the *F* test can be used on the ranks [3]. Let N=number of ranks, a=number of samples, R_{ij}=rank of observation y_{ij}, $R_{i\bullet}$=total of the ranks in the ith treatment, and n_i=number of ranks in the ith treatment. Assign average ranks to ties and calculate

$$k = \frac{\left[\sum_{i=1}^{a} \frac{R_{i\bullet}^2}{n_i} - \frac{N(N+1)^2}{4} \right]}{S^2}$$

where

$$S^2 = \frac{\left[\sum_{i=1}^{a}\sum_{j=1}^{n_i} R_{ij}^2 - \frac{N(N+1)^2}{4}\right]}{N-1}$$

is the variance of the ranks.[3] Finally, use

$$F_O = \frac{K/(a-1)}{(N-1-k)/(N-a)}$$

as the test statistic for the F test on ranks. From the form of F_O, the degrees of freedom for the numerator should be $a-1$, and for the denominator, they should be $N-a$. Therefore, compare F_O to $F(a, a-1, N-a)$.

WHAT HAVE YOU LEARNED?

- Nonparametric tests have rules and restrictions just as classical tests do. When set up, they are usually easier for nontechnical personnel to apply.
- When you are concerned about the results of a classical test, try running the corresponding nonparametric/distribution-free test. It is generally considered more difficult to show significance with the latter. You will almost never have to take additional data, and agreement of the two results is excellent confirmation.

REFERENCES

1. George E. P Box, William G. Hunter, and J. Stuart Hunter, *Statistics for Experimenters*, New York: John Wiley & Sons, 1978.
2. Gottfried E. Noether, *Introduction to Statistics: A Nonparametric Approach*, 2nd ed., Boston: Houghton Mifflin, 1976.
3. William W. Hines and Douglas C. Montgomery, *Probability and Statistics in Engineering and Management Science*, 3rd ed., John Wiley & Sons, 1990.

8

FAULT ISOLATION AND FAILURE ANALYSIS

The previous discussions of quantitative comparisons all addressed the question of tuning a working process in order to improve some product or service. If suddenly the product or service changes, especially for no obvious reason, then comparison techniques may also be used to isolate the problem. They are not applied in exactly the same way, however, or they would attempt to optimize the overall process to the new conditions (e.g., try to "live with" a bad bearing, bad raw materials, a change in customer attitude due to advertising by a competitor, etc.), rather than helping you to track down the problem. This chapter may be considered a summary of troubleshooting tools.

There are several levels of comparison that may be used. The distinction is based primarily on the complexity of the problem. While personal preferences may vary, one useful pair of definitions considers *fault isolation* to be the localizing of a problem and *failure analysis* to be the determination of exactly what went wrong and how it happened. An important tool in accomplishing Fault Isolation and Failure Analysis (FIFA) is a structured approach. Figure 8.1 outlines an approach that has proven successful.

- **The Problem Statement:** This is very important. A properly stated problem statement will contribute significantly to the execution of FIFA. The problem statement should be clear, concise, and stated simply and directly. This is essentially your first screening, where you state what the problem is in technical, engineering, and business terms. Therefore, the problem statement should be developed by a cross-functional team of engineers, managers, technicians, and marketers. It should be readable by a nonexpert outsider.

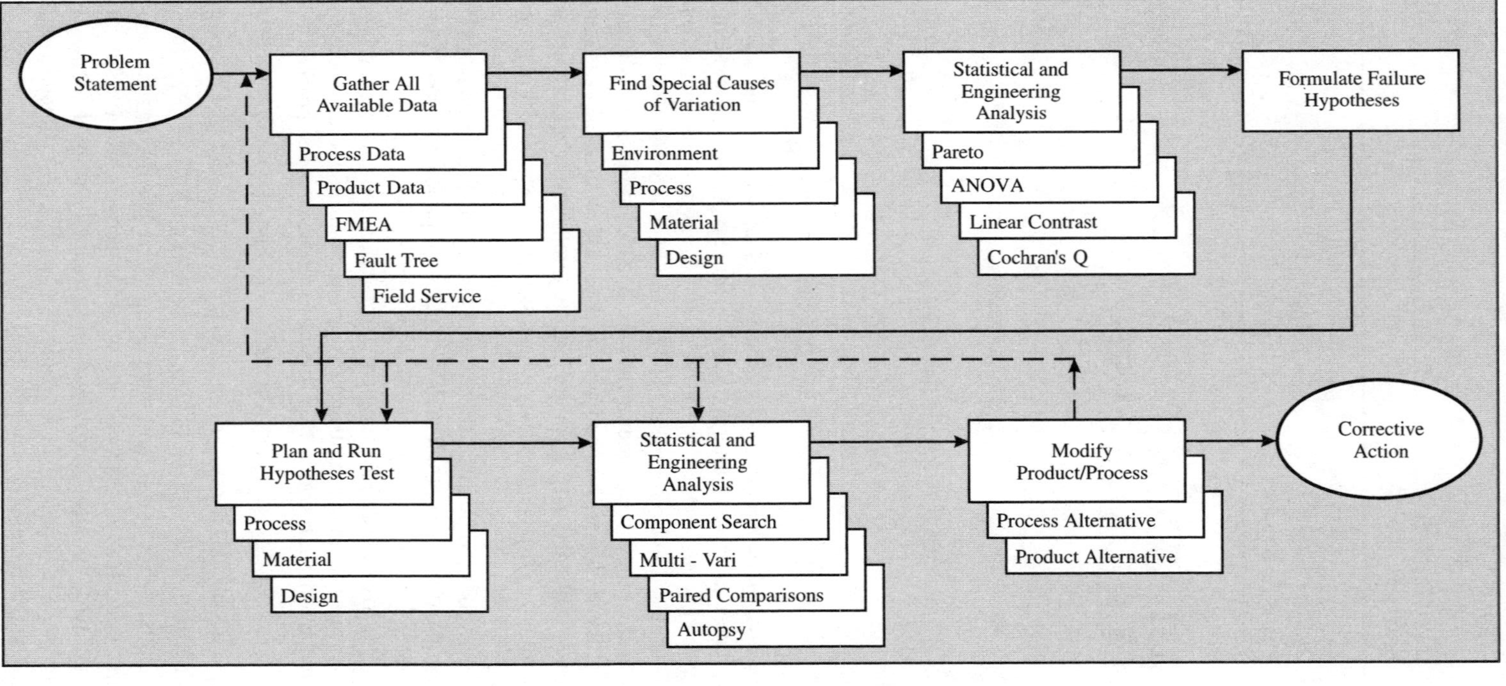

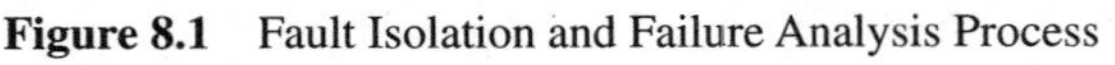

Figure 8.1 Fault Isolation and Failure Analysis Process

- Good Problem Statement: There has been a sudden increase in warranty repairs (63%) and returns (44%) in the last six-month period. This is driving our warranty costs up to $120,000/month and has increased customer complaints fourfold. Customers have noted that the unit "smelled hot" shortly before it failed, and Service says that most of the components replaced were somehow related to the power supply.
- Bad Problem Statement: Some of our customers appear to be displeased; let's find out why.

- **Gather All Available Data:** It is very important not to make assumptions at this point in the analysis. Trying to fit data to a preconceived notion of what is wrong will usually send you off on the wrong track. Review your process data, SPC charts, supplier data, failure modes and effects analysis (FMEA), field service data, and any other information you have concerning the product or service under analysis. The latter should include interviews with line managers and others about any changes, however apparently unrelated, that occurred during or just before the time that the problem began to appear. "Oh, yeah. That was around the time that the blue stuff started to ooze out of the ..."

  Some sources of this information:

  - Process data, product data, and field service data are available from within your activity, based on the history of the product or service.
  - FMEA is an engineering analysis that provides an indication of the possible failure modes and probabilities based on engineering analysis.
  - The fault tree is a combination of these where the probability of any given system failure is calculated using actual data.

- **Find Special Causes of Variation:** Special cause variation is that variation in processes, products, or services that is not explained by normal variation. This special cause variation can be attributed to some specific change or difference. The procedures here are used to isolate the special causes in your product or process.
- **Statistical and Engineering Analysis:** This is the fault isolation phase of the FIFA process. It includes using many of the statistical tools you have already become familiar with and a few that are specially designed for this purpose. Those special tools will be presented in this chapter.
- **Formulate Failure Hypotheses:** The hypothesis is formulated based on all the information you have gathered up to this point in the FIFA process. You should formulate up to three possible failure hypotheses to be tested together. The hypotheses should be tested together to determine whether they are having any interaction. If the top three causes of failure are not apparent, you should review the data you have gathered and consider performing further analyses, including gathering additional data.
- **Plan and Run Failure Hypotheses Test:** This is a designed experiment used to test your hypotheses and determine what your alternatives are.

- **Statistical and Engineering Analysis:** This is the failure analysis phase of the FIFA process. While it may use some of the some tools that were used in fault isolation, it looks for a more specific problem. It also tends to use more quantitative data than the fault isolation tests did. In a sense, fault isolation uses strategy, while failure analysis uses tactics.
- **Modify the Product or Process:** Based on the analysis you have completed, modify the product or process. At this point, you can implement the recommended changes and monitor the changes to determine their effectiveness. You may also return to another problem at this point.
- **Corrective Action:** You now know how to fix your problems and have verified that those fixes really work. You must make management decisions about whether the proposed changes are economically feasible and the possible impact of not making some of them [7]. You are now free to implement the corrective actions you have selected, secure in the knowledge that you have good technical and economic reasons for doing so.

Optimization versus FIFA

In Chapter 3, you made comparisons to look for changes and to detect any lack of homogeneity in your data. In failure analysis, this may be done to identify degraded components, but it may also be done in fault isolation to localize the source of a production problem by comparing good and bad units. Most of the same concepts may be used to locate sources of customer dissatisfaction in a service industry.

- In a destructive test, all the characteristics of good and bad units must obviously be measured first, and no repeat tests are possible. Of course, autopsies (Yes, this is really the technical term!) of the remains are usually possible.
- In a nondestructive test, it is possible to swap out a part from a good unit and a bad unit and see how the performance of each has changed or at least to replace suspect parts. This also allows you to check for "tolerance buildup" problems, in which allowed variations sum up to give a much tighter or much looser fit than expected, and to check for problems that are dependent on how a unit was assembled. For example, a cylinder that is just barely big enough with a piston that is just barely small enough may have problems as temperature changes. If a line worker hurries and jams delicate parts together, they may grind themselves free but leave residue in the system.
- If only failed units are available, then a remaining option is to compare types of damage and production errors over a number of units. For example, pin 14 of a certain integrated circuit may always fail "open" when a process control element ceases to function. This may only be a symptom of the real problem. As another example, if bearings in a rotor are always damaged no matter what the shaft tolerance errors happen to be, then this is quite different from having bearing damage only when the shaft is out of tolerance. The real situation is typically to have damage for most, but not all, of the items that display a certain characteristic, and this is where statistics comes into play.

As in your usual comparison tests, whether simple t tests or mixed-model ANOVA tests, the whole process reduces to common sense and careful bookkeeping. A number of simplified tests have been developed for use by people with no statistical training. These are usually comparisons that do not require the user to first test assumptions on the normality or homoscedasticity (equal variances) of the input data.

ANALYTICAL PROCEDURES FOR FAULT ISOLATION

Just as in medical work, you must first discover as many symptoms of a problem as you can before you look for the actual cause. To carry the medical analogy one step further, you may have to do an exploratory operation or even an autopsy. Often, there are several symptoms, and you may even have more than one cause of a breakdown in a complicated system. Breakdown in a single component can lead to damage in others by a domino effect. Flow of raw material carrying bits of metal or plastic from one machine into another is an obvious example. While the image of a detective sifting through clues and building up an explanation comes immediately to mind, a more specific procedure is needed.

The flow chart shown previously in Figure 8.1 is your general guide. There should be a logical flow from a specific statement of what is wrong, to how such a thing could happen, to prediction of what symptoms should appear for each hypothesis, to examination of data to see what hypotheses the data support (and don't). This same pattern should be followed for both fault isolation and failure analysis. Some of the data may have to be obtained from fresh experiments. Possibly the key point is that, as with any statistical procedure, you should start with a hypothesis and let the data prove you right or wrong. It is always dangerous to try to build a story around just the data you have. Obviously, in an emergency or other rush situation, these procedures will not be formal.

Most of the tests described here can be done either as parametric or nonparametric tests. As usual, the data must pass tests for normality and equality of variances for the parametric versions, but frequently the nonparametric versions will also demand these requirements. That is why both use sample statistics where possible, so a normal sampling distribution can be assumed. Notice, however, that failure analysis data tend to be qualitative, categorical, and/or nonnormal. Some people argue that parametric statistics should be used when sample sizes are large, and nonparametric should be used when sizes are small. They feel that nonparametric tests are more "forgiving."

Graphical Tests

z Score You can check for normality in your data by plotting it. Traditionally, this has been done by hand using "probability" graph paper. With computer software it is more commonly done by plotting the inverse, or z score, of the uniformly spaced values of the cumulative distribution function $F(y_i)=i/(n+1)$, or the more accurate

i	1	2	3	4	5	6	7	8	9	10
Col 2	42.15	44.57	45.64	46.45	46.87	48.14	48.38	49.44	50.61	52.14
Col 3	39.91	41.93	45.73	45.92	46.62	46.67	47.42	47.51	47.55	47.93

i	11	12	13	14	15	16	17	18	19	20
Col 2	52.47	52.53	53.16	54.86	55.03	55.6	57.62	61.97	65.43	68.54
Col 3	48.12	48.13	49.32	49.67	49.96	52.04	53.09	54.13	54.89	56.54

Table 8.1 Columns 2 and 3 of Table 2.3, Each Ranked

$F(y_i)=(i-3/8)/(n+1/4)$, for $i=1$ to n against rank ordered y values [2, 3]. This provides a means of seeing whether suspect data are skewed, squashed, stratified, contaminated, or otherwise nonnormal. The following steps show how to do this. As an example, the second and third columns of the standard data set in Chapter 2 are plotted on the same graph.

Step 1. Take the data and place it in ascending order (Table 8.1).

Step 2. Calculate the value of $z_i=F^{-1}[(i-3/8)/(n+1/4)]$ for each i (Table 8.2).

Step 3. Plot a graph with z_i on the horizontal axis and the corresponding data from columns 2 and 3 on the vertical axis (Figure 8.2).

Step 4. Interpret the graph. Both data sets are reasonably linear, given the sample size. There are no large kinks, steps, or other abnormalities. The tails do not bend wildly from a straight line through the central part of the data. In particular, the data do not approach a constant at either end, as they might if some process were limiting the values possible. For example, the output signal amplitude of a linear amplifier would "saturate" when it reached the power supply voltage. Both sets of data may be

rank	z score	Col 2	Col 3
1	−1.868	42.15	39.91
2	−1.403	44.57	41.93
3	−1.128	45.64	45.73
4	−0.919	46.45	45.92
5	−0.744	46.87	46.62
6	−0.589	48.14	46.67
7	−0.448	48.38	47.42
8	−0.315	49.44	47.51
9	−0.187	50.61	47.55
10	−0.062	52.14	47.93

rank	z score	Col 2	Col 3
11	0.062	52.47	48.12
12	0.187	52.53	48.13
13	0.315	53.16	49.32
14	0.448	54.86	49.67
15	0.589	55.03	49.96
16	0.744	55.6	52.04
17	0.919	57.62	53.09
18	1.128	61.97	54.13
19	1.403	65.43	54.89
20	1.868	68.54	56.54

Table 8.2 Worksheet for Plotting

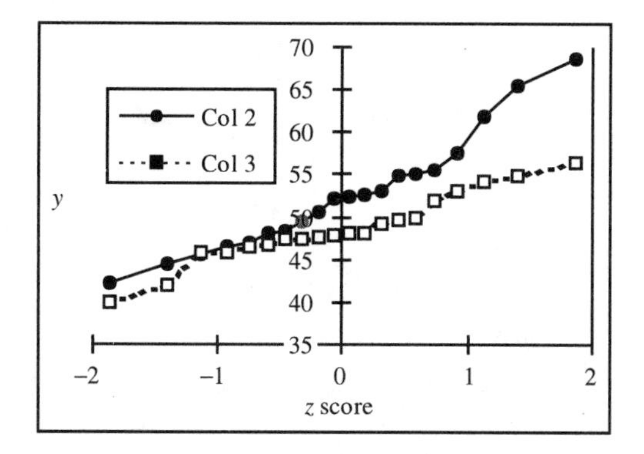

Figure 8.2 Graphical Check for Normality

considered normal. Note that if you draw a straight line through the data, its intercept with the vertical axis $z=0$ is the mean of the data. Subtracting the mean from the value of the line at $z=1$ yields the standard deviation of the data. This just solves for s in $z=(y-\bar{y})/s$ with $z=1$. It is the slope of the line. Since the intercepts for the two data sets are different, the means are different. Since the slopes are different, the standard deviations are different.

To identify special cause variation, you can use the fact that most scatter from measurements of a continuous variable will form a normal distribution. The linearity of this plot can serve as a detector of special cause variation. For example, if your plot bends significantly up or down at the ends, then your data suggests a special cause variation. You should expect the last point or two at each end of the curve to be a bit off the line. If there are several such points, so that the curve bends upward at both ends, this arises from a distribution that is skewed to the right. Figure 8.3(b) shows a uniform distribution, which is an extreme form of kurtosis when compared to the normal distribution in Figure 8.3(a). In this case, the curve bends down at the right, indicating a short tail to the right, because the points are less spread out into a tail than normal. On the left, the points are higher than normal, indicating a short tail to the left. Thus, the curve looks mesa-like or platykuric.

Pareto Although a Pareto chart is almost trivially simple, it really demonstrates the relative significance of the various failure symptoms, as seen in Figure 8.4, and is a very useful tool. Chapter 2 demonstrated how to make such a chart, so it won't be repeated here.

The term "symptom" will be used here repeatedly to emphasize the fact that observing the damage is only the first step in solving the problem. In a high percentage of typical cases, the symptoms are only indirectly related to the actual source of the problem (root cause). Identifying the source, however, requires an understanding of those symptoms.

Multi-vari Multi-vari was developed to look at variation in production lots [4, 5]. If you look at the maximum, minimum, and mean (or median) for a sequence of lots,

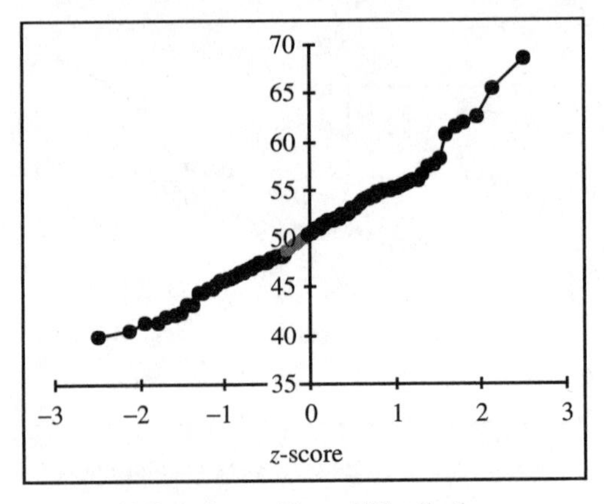

(a) Data from a Normal Distribution

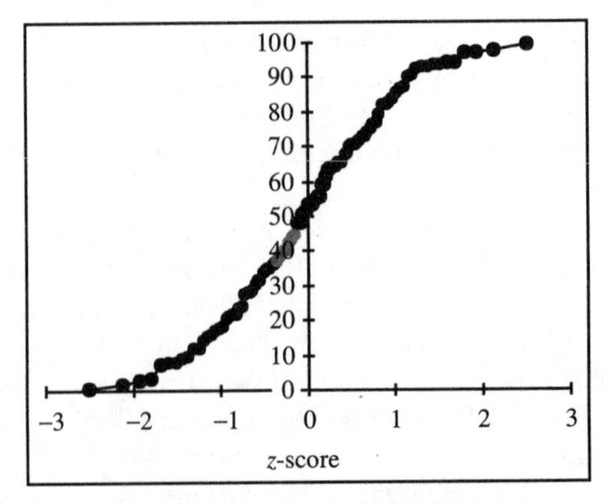

(b) Data from a Uniform Distribution

Figure 8.3 Data from Continuous Distributions

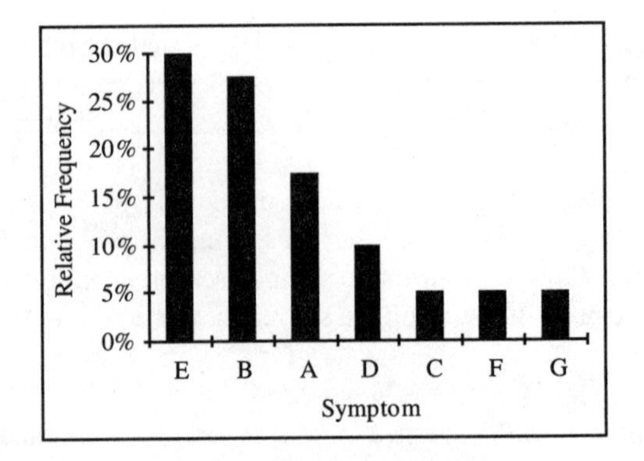

Figure 8.4 Pareto Diagram

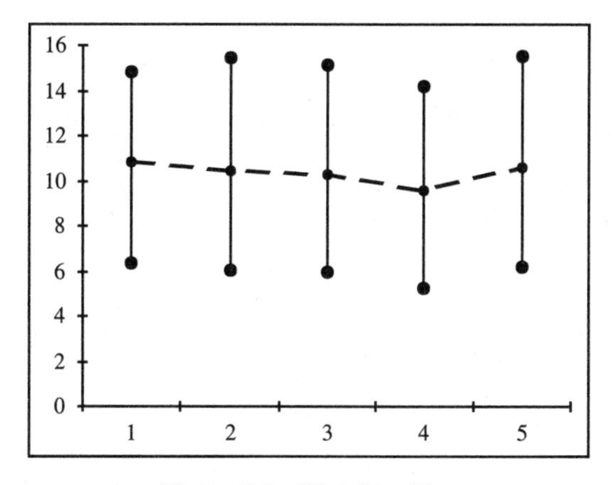

Figure 8.5 High-Low Plot

you can see the within-lot variation by noting the range, or length of the vertical bar, for each lot and the between-lot variation by comparing the means (or medians) of a sequence of lots. You can also see other sources of variation by looking at nonconsecutive sequences, say from different shifts. Multi-vari was developed using a high-low plot (Figure 8.5) because it is quick and easy, but a box-and-whisker plot (Figure 8.6), or a dot plot (Figure 8.7) can be used as well. The latter provide more details on distribution, which may also be useful. Chapter 2 discussed how to make these graphs.

The major criticism of multi-vari is that it presents you with symptoms of a problem that may have several possible causes. In that sense, it should be considered a first level of screening. It is better to work with multi-vari the other way around: You ask the question, "Could (part of) the problem be factor A?" The answer from multi-vari will then be either, "No, because there would be a much bigger variation in ..." or "Yes, it might be." After a few such checks, you are ready to do a quantitative screen such as a fractional factorial, or you may even be ready to do a full factorial analysis.

To make a multi-vari chart:

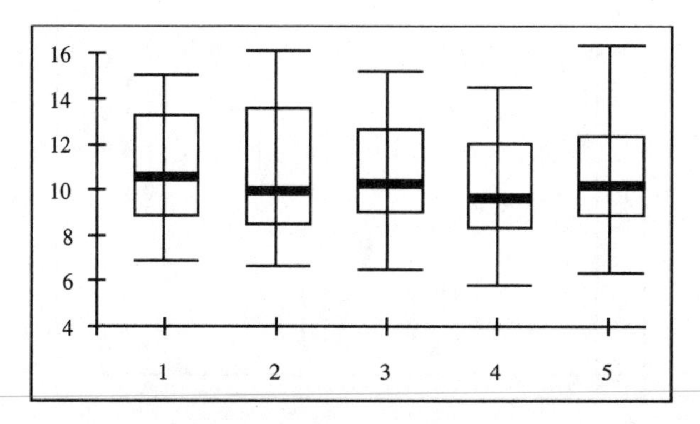

Figure 8.6 Box-and-Whisker Plot

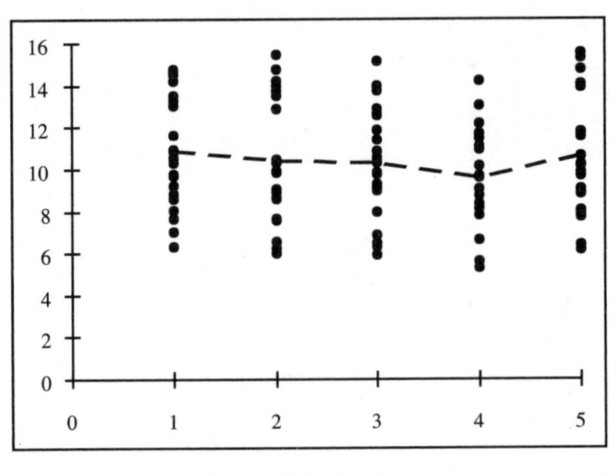

Figure 8.7 Dot Plot

Step 1. Plot the maximum, mean, and minimum for each of the first 5 samples.

Step 2. Skip about 25–30 samples, do step 1 for the next 5, skip 25–30 more, etc. Be sure to do at least 3–5 sequences. The skipping is important, because it allows any gradual buildup of variation to take place. Now you have enough data to look at within-sample variation, between-sample variation within a sequence, and between-sequence variation. The data are plotted in Figure 8.8. Notice that this is not a control chart technique, and it does not replace standard acceptance testing. What it does do is detect several types of variation, and it is done independently of other tests. The key concept is visual demonstration of types of variability, so you can look for patterns and unexpected results.

Step 3. To observe within-sample variation, compare only the ranges of the individual samples.

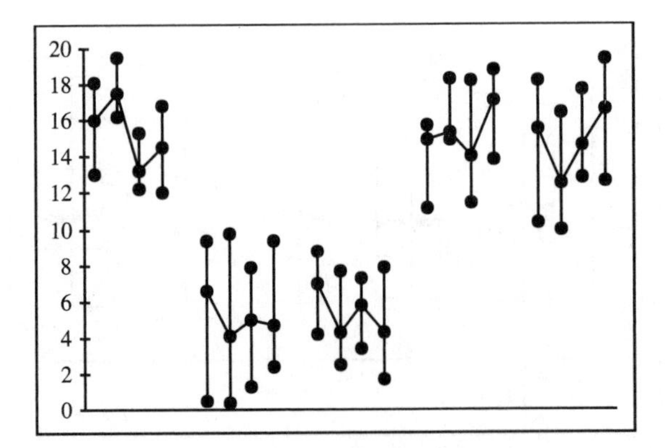

Figure 8.8 Multi-vari Chart

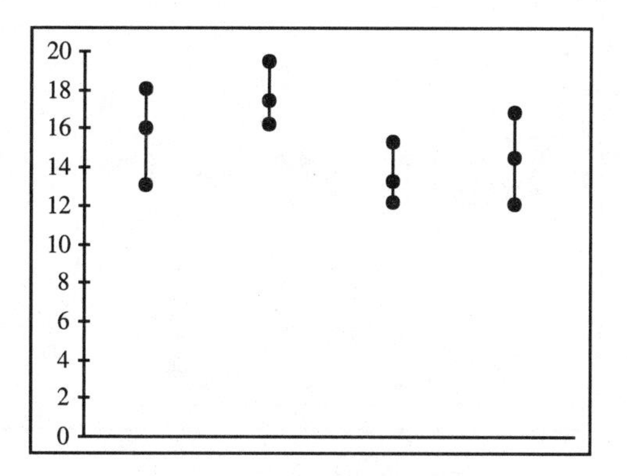

Figure 8.9 Within-sample Variation

Figure 8.9 shows the samples in the leftmost sequence in Figure 8.8. Note that the ranges are similar, even though the samples are shifted up and down relative to each other. Watch out for patterns in the variation of the range along the sequence. For example, a gradual increase in range may suggest a process that is going out of control.

Step 4. To observe between-sample variation within a sequence, compare only the medians of the samples in the sequence, as in Figure 8.10.

The maximum difference in the sample medians of the samples from Step 3 varies much more than the difference in ranges did. This might happen if your facility is in control, but your raw materials are varying in quality.

Step 5. To observe variation between sequences, compare groups of sequences like those in Figure 8.8.

If only the second sequence had been lower than the other four, this might have suggested a one-time-only "glitch." On the other hand, if further tests showed a regular pattern of two high sequences followed by two low sequences, this might suggest

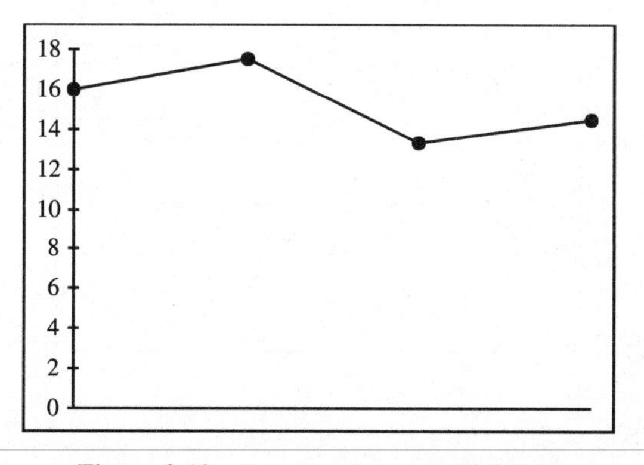

Figure 8.10 Between-sequence Variation

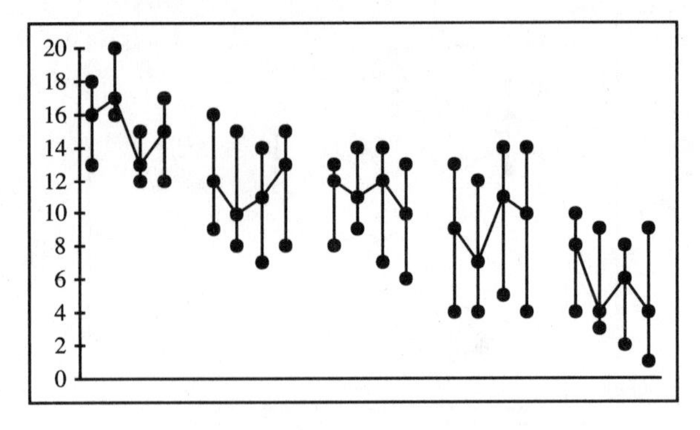

Figure 8.11 Long-term Drift

a procedural problem like confusion about what a setting should be. It might also indicate a difference in experience of people on different shifts. A staircase effect like that in Figure 8.11 suggests a probable long-term drift.

If you are looking for the cause of a sudden change in a process, use only new data; otherwise, the source of the change will not appear. If you had output from three different production lines, you could obviously plot clusters of data, as shown in Figure 8.12. The day shift is more consistent on all three lines, and line #2 always seems to be lowest of the three lines in maximum and minimum yields. Obviously, you still need ANOVA for statistical significance, but even if you only inquire into the reasons for the differences, it could be profitable. Multi-vari was developed by Seder [4].

Returning to the original lot testing example, if the lot ranges are relatively constant but there is a large, random variation in means as the day goes on, then you could be having problems with uniformity in your input materials. On the other hand, if the sequence of means exhibits a trend as the day goes on, then ambient temperature rise may be having an effect. If a sudden jump occurs between nonconsecutive lots, it could be for several reasons (different shift, different raw materials, etc.), but

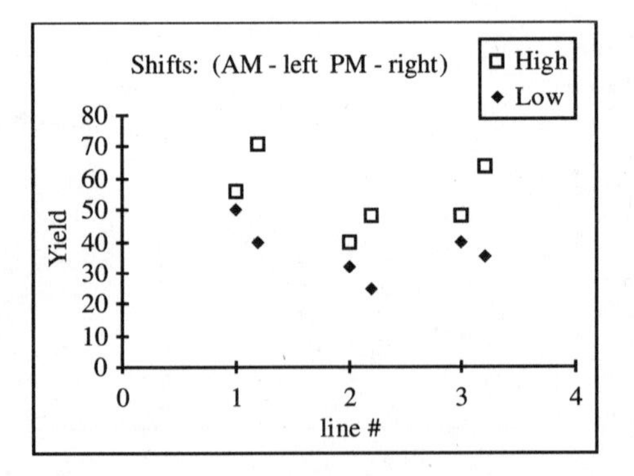

Figure 8.12 Performance Variation with Shift

a sudden change within a lot or sequence of supposedly identical lots should be investigated.

Quantitative Tests

Comparison of Failure Symptoms The following is a method of determining whether some symptoms of failure (charred insulation, heavy corrosion, etc.) are significantly more frequent statistically than others, given only their presence or absence in each item tested [8, 9]. Let 1 be the observation of a symptom, and 0 be the lack of that symptom.

Step 1. Determine whether each item shows a given symptom. This is a go/no-go test, so only a 1 or a 0 is a possible response.

Table 8.3 shows results of tests on a sample of size $m = 15$, each of which could have had any or all of the symptoms $A, B, \ldots, G$.

Step 2. Total s_i, the number of items that had symptom i, and place the result in the right column.

In the example, $S_E = 12$, etc.

Step 3. Total T_j, the number of symptoms for item j, and place the result in the bottom row.

In the example, item #6 exhibited 4 of the $N = 7$ possible symptoms, etc.

	1	2	3	4	5	6	7	8	9	10	11	12	13	14	15	Si
A	0	0	0	1	0	1	1	0	0	1	1	1	1	0	0	7
B	1	1	0	1	0	1	1	1	1	1	1	1	1	0	0	11
C	0	0	0	0	0	0	0	0	0	0	0	0	0	1	1	2
D	0	0	0	0	1	1	0	0	0	1	0	1	0	0	0	4
E	0	0	0	1	1	1	1	1	1	1	1	1	1	1	1	12
F	0	0	0	0	0	0	0	0	0	0	0	0	0	1	1	2
G	0	0	0	0	0	0	0	0	0	0	0	0	0	1	1	2
Tj	1	1	0	3	2	4	3	2	2	4	3	4	3	4	4	

$$m = 15$$

$$T = \sum_j T_j = 40$$

$$T2 = \sum_j T_j^2 = 130$$

$$S2 = \sum_i S_i^2 = 342$$

$$Q = 19.0$$

$$\chi^2(\alpha, m-1) = 23.7$$

Table 8.3 Test of Sample Homogeneity

Step 4. Use Cochran's [8] formula, which is a special case of Friedman's ANOVA, to calculate Q:

$$Q = (m-1)\frac{m\sum_{j=1}^{m} T_j^2 - T^2}{mT - \sum_{i=1}^{N} S_i^2}$$

Step 5. Compare Q to $\chi^2(\alpha, m-1)$.

The results for the example show that no item had an unusually high proportion of symptoms compared to the others, since Q is smaller than the critical value from the table. You could also have done a simple contingency table for this part of the test. The observed values would have been the S_i, and the expected values would each have been equal to T/m. Since you need to set up the array for the next step anyway, why not just use it?

Having established that there are no outliers, you can now compare the symptoms. The roles of S and T are reversed, and the same equation used, as seen in Table 8.4. Now, Q is much larger than $\chi^2(\alpha, m-1)$, so the frequency of occurrence of the various symptoms is not the same statistically. Symptoms B and E are much more frequent than the others. They should be investigated first.

To break down the results further, suppose that there were two different types of symptoms—mechanical (e.g., breakage) and chemical (e.g.)—and that four symptoms, A–D, were mechanical, while three, E–G, were chemical. To see whether the

	Symptom							
#	**A**	**B**	**C**	**D**	**E**	**F**	**G**	**Si**
1	0	1	0	0	0	0	0	1
2	0	1	0	0	0	0	0	1
3	0	0	0	0	0	0	0	0
4	1	1	0	0	1	0	0	3
5	0	0	0	1	1	0	0	2
6	1	1	0	1	1	0	0	4
7	1	1	0	0	1	0	0	3
8	0	1	0	0	1	0	0	2
9	0	1	0	0	1	0	0	2
10	1	1	0	1	1	0	0	4
11	1	1	0	0	1	0	0	3
12	1	1	0	1	1	0	0	4
13	1	1	0	0	1	0	0	3
14	0	0	1	0	1	1	1	4
15	0	0	1	0	1	1	1	4
Tj	7	11	2	4	12	2	2	

$m = 7$

$T = \sum_j T_j = 40$

$T2 = \sum_j T_j^2 = 342$

$S2 = \sum_i S_i^2 = 130$

$Q = 31.76$

$\chi^2(\alpha, m-1) = 12.59$

Table 8.4 Comparing Frequency of Occurrence

frequency of observation of the two types is significantly different, that is, to look at "between groups" variation, use the statistic

$$Q_{bg} = \frac{(m-1)(m_2U_1 - m_1U_2)^2}{m_1m_2\left(mT - \sum_{i=1}^{N} S_i^2\right)}$$

which is to be compared to $\chi^2(\alpha,1)$. To look at variation within each group, use the statistic

$$Q_{wg1} = \frac{m(m-1)\left(m_1\sum_{j=1}^{m_1} T_j^2 - U_1^2\right)^2}{m_1\left(mT - \sum_{i=1}^{N} S_i^2\right)}$$

and $\chi^2(\alpha,m_1-1)$ for the first group, and use

$$Q_{wg2} = \frac{m(m-1)\left(m_2\sum_{j=m_1+1}^{m} T_j^2 - U_2^2\right)^2}{m2\left(mT - \sum_{i=1}^{N} S_i^2\right)}$$

and $\chi^2(\alpha,m_2-1)$ for the second group.

For the example in Table 8.4, $Q_{bg}=.56$ and $\chi^2(.05,1)=3.84$, so the difference between the two groups is not significant. This is not surprising, since B dominates the first group, E dominates the second group, and they are about equal in size. Within the first group, $Q_{wg1}=32.97$ and $\chi^2(.05,3)=7.81$, so the large difference between B and the rest of the group is statistically significant. Similarly, $Q_{wg2}=18.67$ and $\chi^2(.05,2)=5.99$ indicate that E dominates the second group.

Locating Faulty Process Settings Process variables that are leading to unacceptable end-item performance (which may not be a breakage, so the technique can also be used for optimization) can be found by a process which in some ways resembles a factorial analysis.

Step 1. Identify two levels for each variable, one that you expect to provide good results (the high, or +, level), and one that you don't (the low, or − level). Selecting these levels will require some thought. If you are wrong, the test will tell you so.

Step 2. Screen 1: Make sure the answer lies somewhere within the chosen limits. To do this, make a set of runs with the variables all high (A^+B^+...) and a set with them all low (A^-B^-...). To do a nonparametric test at the .05 level, you could do 3 runs per set and demand that all high runs give results that were better than all those from low runs.

Step 3. Screen 2: Make a comparison of within-group versus between-group variation. This can be either a classical or nonparametric test. For the nonparametric test, use the difference in medians divided by the d_2 conversion factor (see Table A.1 in Appendix A) for range to standard deviation $s_{bg}=(\tilde{y}_{high}-\tilde{y}_{low})/d_2$ to get the between-group standard deviation for the numerator and $\bar{R}=(R_{high}+R_{low})/2$ to get the within-group standard deviation for the denominator from $s_{wg}=\bar{R}/d_2$. In the numerator, $d_2=1.128$, because two levels give 2 degrees of freedom. In the denominator, $d_2=1.693$, because there are 3 runs.

Step 4. Screen 3: Check individual variables. For each variable, you have a confidence band between two levels. Runs with only that variable reversed should fall within the band if the variable under investigation is having no effect on the process. In other words, keeping all else high and changing A to A^- should make the answer fall in the high band, while keeping all else low and changing A to A^+ should make the answer fall in the low band if A is having no effect.

This test has been called Variable Search and is attributed to Shainin by Bhote [1].

Fisher's Rank Randomization As you saw in Chapter 7, many nonparametric comparison techniques use ranks, rather than the data values themselves. Fisher's rank randomization procedure is a simple technique that forms the basis of many of the nonparametric methods [6]. It argues that if the numbers are really random, then the ranks will also be random. Since your comparison test is set up such that one or more combinations of ranks will appear if a significant change takes place, having that same combination show up at random can falsely show significance (a Type I error), just as it does with the actual data themselves.

For a process problem, levels of variables that might be causing the problem are set at what is expected to produce the best results (presumably, your existing settings), and then the same runs are made with those expected to produce the worst results. If the best and worst results differ by more than some agreed-upon amount, levels are swapped efficiently until the problem is isolated. This usually involves some sort of factorial analysis, even if it is done one factor at a time.

This technique can be used for ANOVA or factorial analysis (which is just a special case of ANOVA in which the treatments are selected with great symmetry). The null hypothesis of ANOVA is that none of the treatments have any effect, and the concern is that a random combination of data can falsely show significance (a Type I error). This is exactly the situation described earlier and can be handled in the same way. Only the order is needed in the example below; the rank itself may be omitted.

	A	B	Interaction					Avg
1	+	+	+	10.0	8.5	8.0	9.0	8.9
2	–	+	–	12.0	15.0	17.0	14.0	14.5
3	+	–	–	6.0	7.0	7.5	6.5	6.8
4	–	–	+	25.0	28.0	27.0	25.5	26.4

Table 8.5 Input Data from 2^2 Experiment

Step 1. Take data as you would for a factorial analysis, as shown in Table 8.5. In fact, you can use this test to cross-check an analysis that has you worried.

Step 2. Order the answers, keeping the "treatment" label with each data point, as shown in Table 8.6

Step 3. Find the Tukey-Duckworth [5] endcount by adding the number of like signs at one end of the column to the number of opposite signs at the other end. If the signs are not opposite at the other end, you can't use the test.

In *A*, there are 8 minuses at the top and 8 pluses at the bottom, for a total count of 16.

A	B	Interaction	Data
–	–	+	28.0
–	–	+	27.0
–	–	+	25.5
–	–	+	25.0
–	+	–	17.0
–	+	–	15.0
–	+	–	14.0
–	+	–	12.0
+	+	+	10.0
+	+	+	9.0
+	+	+	8.5
+	+	+	8.0
+	–	–	7.5
+	–	–	7.0
+	–	–	6.5
+	–	–	6.0

Table 8.6 Ordered Data

Variables	Confidence	Minimum Endcount
2	0.999	14
	0.99	11
	0.95	9
	0.9	8
3	0.999	16
	0.99	12
	0.95	10
	0.9	9
4	0.999	16
	0.99	13
	0.95	11
	0.9	10

Table 8.7 Critical Values of Endcount

Step 4. Use Table 8.7 to find the significance level.
A is significant at the 99.9% level.
In *B*, the same sign is at each end, so you don't show significance.
In the *AB* interaction, there is a total count of 4+4=8, so the *AB* interaction is significant at the 90% level.

Contingency Tables Say you have an excessive number of parts failures. You find that the times to failure for the manufacturer (*M*) and a second source (*S*) are as shown in Figure 8.13.

To compare the two parts:

Step 1. Set up Table 8.8 using the data from Figure 8.13.

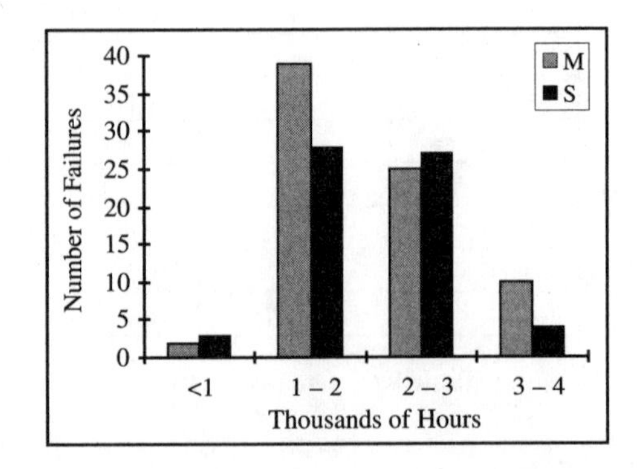

Figure 8.13 Graphical Comparison of Failure Rates

	Thousands of Hours				
Observed Failures	<1	1–2	2–3	3–4	Total
M	2	39	25	10	76
S	3	28	27	4	62
Total	5	67	52	14	138

Table 8.8 Numerical Comparison of Failure Rates

Expected Failures	<1	1–2	2–3	3–4	Total
M	2.75	36.90	28.64	7.71	76
S	2.25	30.10	23.36	6.29	62
Total	5	67	52	14	138

Table 8.9 Expected Distribution of Failures

Step 2. Calculate the expected number of failures for each cell.

Ideally, the proportion of failures in each time period should be nearly the same for the two sources, so the expected number of failures in less than 1000 hours should be

$$5(76)/138=2.75 \text{ for the part from the original manufacturer}$$

$$5(62)/138=2.25 \text{ for the second source}$$

as shown in Table 8.9.

Step 3. Call an observed value O and an expected value E. Calculate the quantity $(O-E)^2/E$ for each pair of corresponding cells. For example, the upper left cell is

$$(2-2.75)^2/2.75=.205$$

Step 4. Sum up all the numbers in Table 8.10.

$$\text{Sum}=.21+.12+....+.57+.83=3.27=\chi_o^2$$

Step 5. Find the critical minimum value that the sum must have in order to indicate a difference between the primary manufacturer (M) and second source (S).

	<1	1–2	2–3	3–4
M	0.21	0.12	0.46	0.68
S	0.25	0.15	0.57	0.83

Table 8.10 Chi-Square Terms

For chi-square with $df=(c-1)(r-1)=3$, $c=4$ columns, $r=2$ rows, and $\alpha=.1$, the critical value is 6.25. This is much greater than the sum calculated from the data, so there is no evidence that the part made by the second source is inferior to that from the manufacturer of the machine. Since there was no difference in the part itself, another cause for the high failure rate must be found. The next step might be to see whether certain production machines had more failures than others.

ANALYTICAL PROCEDURES FOR FAILURE ANALYSIS

If all of a group of systems were identical in every respect, we could simply change each component on one system until the system worked (preferably with some efficient system of narrowing down the problem quickly!), tolerance variation would not be a problem, and statistical analysis would not be necessary. Variation does exist in the real world and sometimes there is more than one problem, so versions of this concept have been developed that allow a statistical test.

A popular technique in a failure analysis is to select and compare the best n units and the worst n units (with best and worst based on whatever problem has been observed). A particular component (e.g., a carburetor) is swapped between good and bad units for every pair of units. If the bad units then work better and/or the good units no longer work well, then that component is one of the problems.

The following is a detailed procedure for finding a problem with an assembly (clock, motor, starter, meter, etc.) with removable components. It compares the medians of the "output" (quality of the function you are measuring) of good and bad units. Basically, it compares between-group variation to within-group variation, just as you would with any other comparison test. The two new ideas here are that: (1) the within-group variation is the result of variability due to repeated reassembly of the same unit (plus experimental error, of course); and (2) the between-group variation is due to a problem with the unit. If you discover that the problem is a change in raw materials from a supplier, for example, you may be back to your originally optimized process, and it may be unnecessary to test the rest of the variables. In this case, it would be a good idea to rank the proposed variables in order of decreasing likelihood of being the problem. On the other hand, if you were using this technique to shake out the wrinkles in a prototype design, you would probably want to check all the variables. Since you are only looking for a gross change and don't want to take a lot of data, you can get away with a rough check using nonparametrics instead of making your comparison with a classical t test. To do this, you use the range R instead of the standard deviation. Although the standard deviation is equal to the range divided by a conversion factor d_2 (for a normal distribution), this particular test does not require that conversion. In order to make sure that your approximations are not too rough, your first step is to make sure that there is enough difference in performance between bad and good units.

Step 1. To decide whether there is enough difference in the good and the bad, calculate the median and range of the good units and the bad units. If the performance of

	y1	y2	y3	Median	Range
Good	79	74	82	79	8
Bad	43	47	52	47	9

Table 8.11 Initial Data

any of the three bad units changes significantly, you have important information about improper tolerancing and the criticality of assembly procedures.

Data were taken on 3 good (old) units and 3 bad (recent) units, shown in Table 8.11.

Step 2. Let $D=\tilde{x}_g-\tilde{x}_b$ be the difference in medians of the data taken after each reassembly, and make sure $D>1.25\overline{R}_g$ and $D>1.25\overline{R}_b$. You must be able to disassemble and reassemble the good units at least three times, measuring the output each time, and still have $D>1.25\overline{R}_g$.

$$D=32 \qquad 1.25\overline{R}_g=10 \qquad 1.25\overline{R}_b=11.25$$

The data meet the requirements of the test.

Step 3. If none of the three bad units changes significantly, decide what the three most important components are—call them *A*, *B*, and *C*. Swap all the good *A* components with all the bad *A* components. Repeat the disassembly/test procedure.

Component *A* was swapped in all three pairs. The value of *A* was checked for a new run and found to be different; this was used for $A-$ in Table 8.12.

Step 4. If this totally fixes the problem, *A* was the bad component. However, if this only partially fixes the problem, there may be an interaction with another component. Replace all the bad *A* components, and go on to *B*. Do the same thing if *A* had no effect.

In the example, swapping $A+$ and $A-$ reversed the scores. *A* is the bad component.

Notice that the statistics are buried in the sample sizes of this test, so no "table-lookup" procedure is necessary. The median tests with *D* are somewhat of a side issue, except that the main test can't be done unless their conditions are met. Bhote[1] attributes development of the technical details of this test to Shainin, who calls it paired comparisons.

	y1	y2	y3	Median	Range
A +	45	51	52	51	7
A −	81	77	83	81	6

Table 8.12 Effect of *A*

A	2.9	3.0	2.8
B	2.5	2.2	2.3

Table 8.13 Test Data

Fisher's Rank Randomization

This test was shown previously for fault isolation, but it can also be used for failure analysis. For example, if three units produced by one machine (A) are compared with three units from a second (B) and all three of the A units are "better" than the B ones, then there is only a 5% chance that there is no difference in the machines, and the results occurred at random. The reasoning follows Fisher's randomization statistics: If there was no difference in the two machines, each arrangement of A and B outputs would occur with the same probability. On the other hand, the probability of all three from A being better than all three from B is one chance in 20:

$$P\{all\ new > all\ old\} = 1\Big/\binom{6}{3} = 3!3/6! = .05$$

To see details of this technique, consider the following comparison test of a sample of size 3 each from A and B .

Step 1. Take measurements y on 3 samples selected randomly from each machine, as noted in Table 8.13. These can actually be individual measurements, averages, or means, but keep the sizes equal! The units tested and the readings need not be paired in any fashion.

Step 2. Order the data, keeping the labels with each point, as in Table 8.14.

Step 3. Note whether the yields were always greater for one machine than for the other. If so, the result is significant. If not, it isn't.

Yield y was always greater for A than for B. As calculated previously, if all combinations of 3 A's and 3 B's have equal probability of occurring, the chance of this happening at random is 5%.

This technique can also be used for ANOVA or DOE (which is just a special case of ANOVA in which the treatments are selected with great symmetry). The null hypothesis of ANOVA is that none of the treatments have any effect, and the concern is that a random combination of data can falsely show significance (a Type I error). This is exactly the situation just described and can be handled in the

Label	B	B	B	A	A	A
y	2.2	2.3	2.5	2.8	2.9	3.0

Table 8.14 Combined and Ranked Data

	A	*B*	Interaction					Avg
1	+	+	+	10.0	8.5	8.0	9.0	8.9
2	–	+	–	12.0	15.0	17.0	14.0	14.5
3	+	–	–	6.0	7.0	7.5	6.5	6.8
4	–	–	+	25.0	28.0	27.0	25.5	26.4

Table 8.15 Data from 2^2 Experiment

same way. One application is to double-check your factorial analysis. To do the test:

Step 1. Use the same format that you would for a standard factorial analysis (see Table 8.15).

Step 2. Place all the data in descending order (keeping the label with each data point) in one long column, as in Table 8.16.

Step 3. Find Tukey's endcount[5] by adding the number of like signs at one end of the column to the number of opposite signs at the other end. If the signs are not opposite at the other end, you can't use the test. In *A*, there are 8 minuses at the top and 8 opposite signs (pluses) at the bottom, for a total count of 16.

Step 4. Use Table 8.17 to test for significance.

A	B	Interaction	Data
–	–	+	28.0
–	–	+	27.0
–	–	+	25.5
–	–	+	25.0
–	+	–	17.0
–	+	–	15.0
–	+	–	14.0
–	+	–	12.0
+	+	+	10.0
+	+	+	9.0
+	+	+	8.5
+	+	+	8.0
+	–	–	7.5
+	–	–	7.0
+	–	–	6.5
+	–	–	6.0

Table 8.16 Ordered Data

Variables	Confidence	Minimum Endcount
2	0.999	14
	0.99	11
	0.95	9
	0.9	8
3	0.999	16
	0.99	12
	0.95	10
	0.9	9
4	0.999	16
	0.99	13
	0.95	11
	0.9	10

Table 8.17 Table of Critical Values of Endcount

A is significant at the 99.9% level.

In *B*, the same sign is at each end, so you don't show significance.

In the *AB* interaction, there is a total count of $4+4=8$, so the *AB* interaction is significant at the 90% level.

Locating Bad Components by Autopsy

When you have assemblies that are potted or are too damaged to allow replacement of parts, you can do a visual comparison of the remains. To do this, you will need several bad units and at least as many good ones.

Step 1. Randomly select a sample of good units of size equal to the number of bad units you have available for test. Selecting both the good unit and the bad unit at random for pairing avoids confounding of effects.

Step 2. Observe one good unit and one bad unit, noting every characteristic in which they differ. Selecting both the good unit and the bad unit at random for pairing avoids dependence on order of test.

Step 3. Do the same for the next pair, etc., and count the number with each type of difference.

Step 4. If conflicting results occur for a given type, delete that type from the list. If several types remain, try to split them into groups according to a common subassembly or production process.

Cochran's *Q* test for significance can be performed. Bhote[1] attributes the technical procedure in this test to Shainin, who calls it paired comparisons.

WHAT HAVE YOU LEARNED?

- It is very important that you have a consistent means of screening the possible sources of error as you look more and more closely at the defective item or process. Your source of error must stay in the ballpark at every stage of the screening. At the same time, it is also very important that you be as creative and flexible in your thinking as you can at every stage of the screening. Some problems will not require a full, formal analysis, but you still should follow the pattern.
- If you start out with a hazy notion of what the *real* question is that you are trying to solve, you will almost certainly waste a lot of time, effort, and money. Along the same line of thought, gather as much information as you can up front so you don't have to cover the same ground again later. Talk to the people most closely involved; ask them what they've seen, especially little odd things that might not have been mentioned in a report or recorded on a data sheet. If you can, ask the designers what they think might be going wrong or areas that they think might be weak points.
- Pareto the symptoms—what broke, what wasn't up to specification, etc. The more complicated the problem, the more complicated the process, the more extensive the data available, the more important it is to make a Pareto chart. Don't assume that there is only one cause or one source of the symptoms you are seeing. If someone has made one mistake, they have just demonstrated to you that they can make another.
- Screen the symptoms you find. Pareto is a good screen, but there are others. Don't be misled by the fact that mostly new techniques were discussed here, and they were mostly nonparametric. You already saw the other techniques, for example, fractional factorial analysis, in earlier chapters. Nonparametrics were developed to provide quick and robust tests. This makes them perfect for FIFA. As you saw in the previous chapter, they also have their own implicit assumptions and requirements.
- With any comparison, you'd like to have as much control of the test conditions as you can. This means that you probably won't be able to use old data in a new comparison test, since you probably didn't control everything that should be. You probably also didn't measure everything you'd now like to know. That does not mean the old data are worthless.
- While you may be able to screen just once in your fault isolation, don't be surprised if you have to iterate in your failure analysis.

REFERENCES

1. Keki R. Bhote, "World Class Quality," AMACOM, 1991.
2. G. Blom, *Statistical Estimates and Transformed Beta-Variables*, New York: John Wiley & Sons, 1958.
3. John Mandel, *The Statistical Analysis of Experimental Data*, New York: Dover, 1984.

4. Leonard Seder, "Diagnostics with Diagrams, Part I," *Industrial Quality Control 6*(4), (January 1950), 11–19.
5. Alan F. Chow, "Variability Reduction: The Shainin Way," Hughes Aircraft Co. Manual.
6. William W. Hines and Douglas C. Montgomery, *Probability and Statistics in Engineering and Management Science*, 3rd ed., New York: John Wiley &Sons, 1990.
7. Harry Jackson and Normand L. Frigon, *Achieving the Competitive Edge*, New York: John Wiley & Sons, 1996.
8. W. G. Cochran, "The Comparison of Percentages in Matched Samples," *Biometrika 37*, (1950), 256–266.
9. Joseph L Fleiss, *Statistical Methods for Rates and Proportions*, 2nd ed., New York: John Wiley & Sons, 1981.

9

DOE APPLICATIONS TO INDUSTRIAL PROCESSES

APPLICATION OF DOE TO INDUSTRIAL PROCESSES

Design of Experiments (DOE) is used to design, develop, and improve industrial processes. This is a very effective, efficient selection and optimization method. DOE is routinely used to optimize industrial processes in several categories:

Machine	Methods	Material	Personnel	Environment	Measurement
Equipment selection	Planning	Material type	Training	Temperature	Instrumentation
Ware	Sequence	Supplier	Education	Humidity	Tolerances
Accuracy	Procedures	Material condition	Attitude	Vibration	Specifications
Settings	Techniques	Composition	Physical capabilities	Light	Technique
Speed	Welds	Raw materials	Experience	Pressure	Parameters
Manual	Processes	Composition	Shift	Contaminants	
Digital		Supplier	Location	Electrostatic discharge	
		Price			

Each of these categories is a treatment that can be used to optimize industrial processes by:

- Optimizing yield
- Reducing variability
- Reducing scrap rate
- Improving quality
- Decreasing operations cost
- Streamlining processes
- Robust process less process noise
- Reducing sensitivity to supplier variation
- Addressing customer concerns
- Reducing rework

We will review three case studies that apply DOE to practical industrial problems. Each of these case studies approaches the use of designed experiments in different ways for different purposes. Please note that these are not perfect academic examples of designed experiments where all of the data are always there and every experiment is successful. This is what you are likely to encounter in everyday business transactions.

Case Study 1—Brazing Process

This case study involves a process that was producing a high rate of in-process rework and end-of-production scrap of the final product. The process was not very well defined or documented, so the initial effort was to perform a process analysis and find the appropriate metrics for the process and product. That effort was then expanded to determine if there was any correlation between the several process metrics and the high rework and scrap rates. The focus of this analysis was to first evaluate the process and determine the gross cause(s) of in-process rework and scrap and then to determine what corrective action could be taken. The scrap rate was approximately 20% of the finished product. Manufacturing engineering indicated that this "was to be expected" with this type of system and doubted that anything could be done about it. An extensive amount of rework was being performed in the process. This was occurring with approximately 25% of the systems. The value of material being reworked was approximately $780,000 per quarter (and the scrap rate was valued at $300,000 per quarter).

Process Selection and Identification

To select the specific subprocess for analysis, we first reviewed all the process data available. The data were retrieved from the organization's database and from manual records covering a one-year period. The data were tabulated to provide the percentage failing for each process within the specific production area and to provide an evaluation of the proportion of total failures being caused by each of these

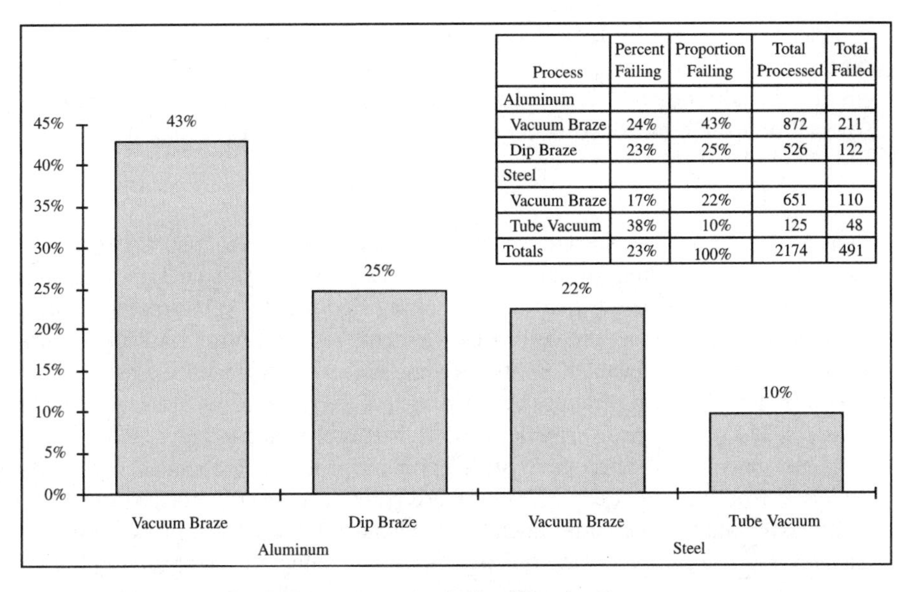

Process	Percent Failing	Proportion Failing	Total Processed	Total Failed
Aluminum				
Vacuum Braze	24%	43%	872	211
Dip Braze	23%	25%	526	122
Steel				
Vacuum Braze	17%	22%	651	110
Tube Vacuum	38%	10%	125	48
Totals	23%	100%	2174	491

Figure 9.1 Pareto Analysis of Brazing Processes

processes. As indicated in the Pareto Analysis in Figure 9.1, the Aluminum Vacuum Brazing Process is clearly the most significant contributor to process rework and scrap.

Then each of the items manufactured using this process was evaluated to further refine the failures to specific part numbers if possible. The analysis of failing part numbers is given in Figure 9.2. This analysis further refined the focus to the process and product for heat exchanger PN-370521 as the most significant contributor to the failures within the Aluminum Vacuum Brazing Processes.

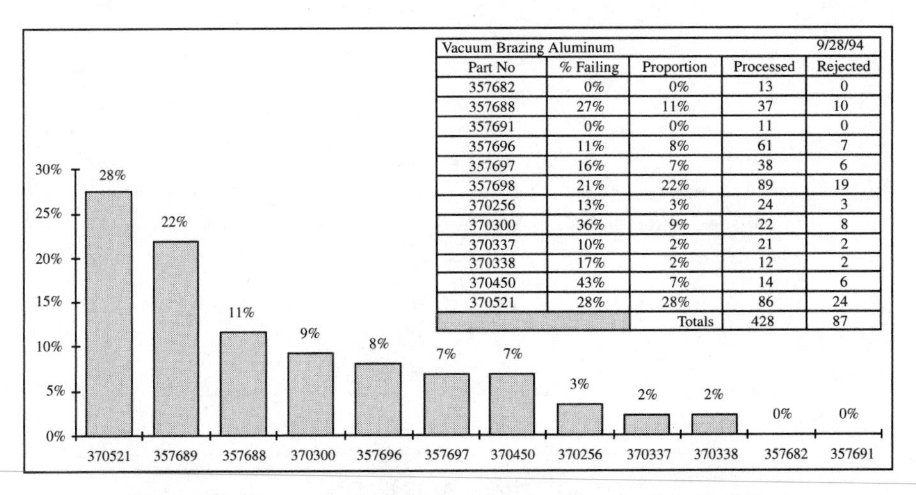

Vacuum Brazing Aluminum				9/28/94
Part No	% Failing	Proportion	Processed	Rejected
357682	0%	0%	13	0
357688	27%	11%	37	10
357691	0%	0%	11	0
357696	11%	8%	61	7
357697	16%	7%	38	6
357698	21%	22%	89	19
370256	13%	3%	24	3
370300	36%	9%	22	8
370337	10%	2%	21	2
370338	17%	2%	12	2
370450	43%	7%	14	6
370521	28%	28%	86	24
		Totals	428	87

Figure 9.2 Pareto Analysis by Part Number, First Six-Month Period

Please look carefully at the data provided with each of the graphs in Figures 9.1 and 9.2. In each case, using the percentage failing of the processes in Figure 9.1 or the Products in Figure 9.2 would have been incorrect. With the task of reducing overall scrap and rework costs, the correct metric is the proportion of overall production failures attributed to each process and product.

Process Analysis

An evaluation of the selected process analyzed the process input requirements, process data, each process element, and process outputs. This analysis indicated that three process elements may be contributing significantly to in-process failures, rework, and scrap. These are the process element 2—inputs from suppliers of parts and material, process element 3—the cleaning process, and process element 6—the variance in brazing temperatures between each of the two brazing ovens used in the process. The process analysis process flow chart is provided in Figure 9.3.

This analysis indicated there are significant problems associated with parts received from the suppliers. Many of these parts have a very poor process capability associated with their production as indicated in Figure 9.4. Parts are source inspected at the supplier; the 3,000 parts used to conduct the evaluation in Figure 9.4 (1,000 from each of three suppliers) were inspected on receipt at the production facility. There was significant disagreement between design engineering, supplier management, and manufacturing engineering concerning the effect of

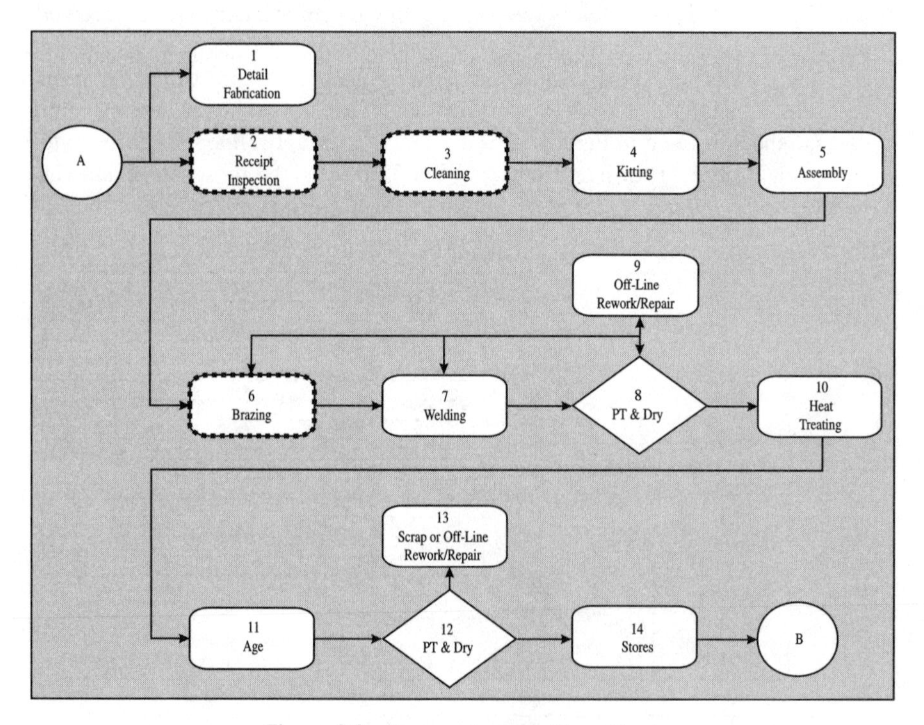

Figure 9.3 Process Analysis Flow Chart

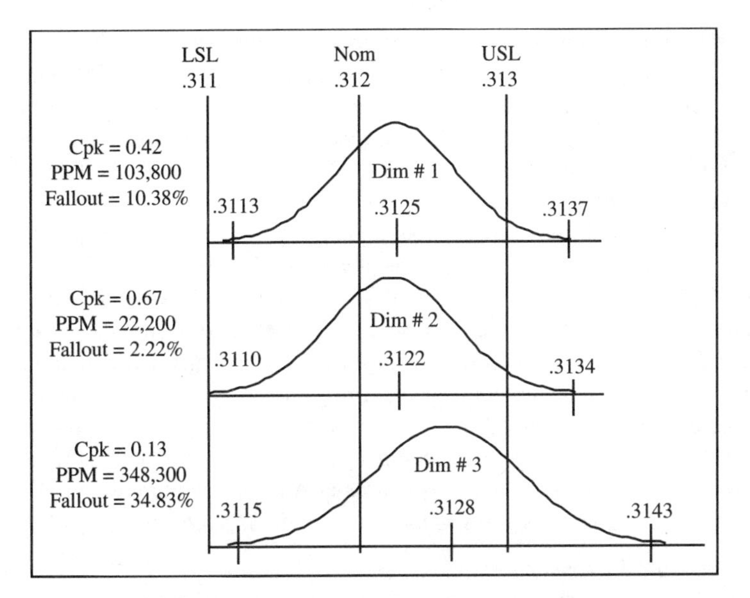

Figure 9.4 C_{pk} Resulting from Inspection of Parts

nonconforming parts on the overall process rework and scrap rates. Additionally, it was determined that the requirements for source inspection only looked at a percentage of the critical measurements. The data used to construct the graphs in Figure 9.4 were the result of a detailed inspection on receipt of all critical measures of the parts.

The cleaning process for parts cleaning is not maintained within the specification limits for the process. The process also fails to meet the standards for statistical process control. This is indicated by the graph in Figure 9.5. SPC at this

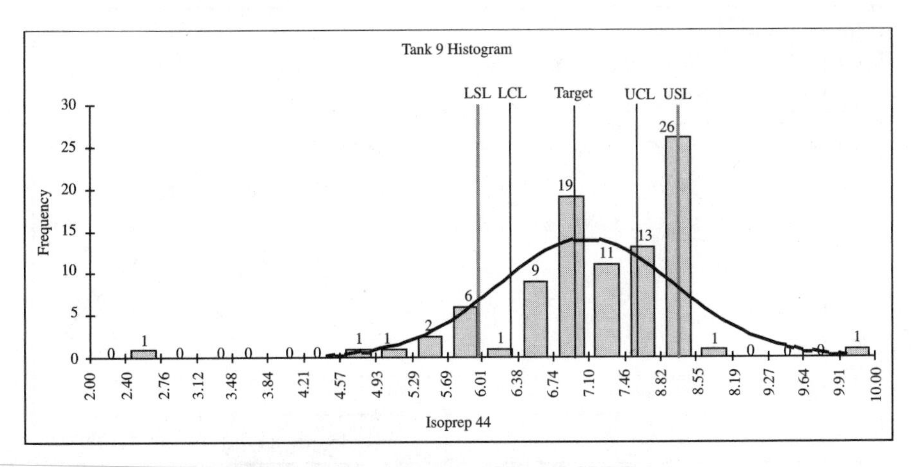

Figure 9.5 Histogram of Tank 9 SPC Data

workstation is operated by the chem. lab and is not available to the workstation operator as a resource. The operators not trained in the use of SPC cannot access the system at the workstation. The chemical analysis and adjustment of the chemicals in the cleaning tanks has been reduced from three times a week to once every two weeks as an economic move. Data are input to the SPC workstation when the chemicals in the tanks are analyzed and adjusted.

There are two different ovens used in the brazing process. They are of different size and manufacturer. One oven is the smaller of the two ovens with a capacity to braze 8 systems at a time. The second oven can braze 12 systems. There is a difference in the variance of the temperatures of these two ovens as indicated in the data in Figure 9.5. There is currently no SPC of these ovens. The data in Figure 9.6 were derived from the temperature recordings taken on paper disks for each oven during a single brazing cycle of 4 hours. Here also there was disagreement between the several groups in the company as to the effect of temperature variance on the brazing process. Both ovens maintained the temperature within the specifications for the brazing process as indicated by the product specifications.

The Designed Experiment

Based on this preliminary analysis, it was determined that the three treatments would be subjected to a designed experiment. The purpose of the designed experiment was twofold: First, to characterize these process treatments to determine what effect changes in the treatments had on the process, and second, to optimize the process yield, therefore reducing scrap and rework costs. The three treatments and levels selected were (1) conformance of parts coming into the process, (2) the cleaning process, and (3) the ovens.

The parts coming into the process were divided into two groups—parts received directly from the supplier that were not inspected at receipt, as was the normal process, and parts that were detail inspected at receipt, and only conforming parts

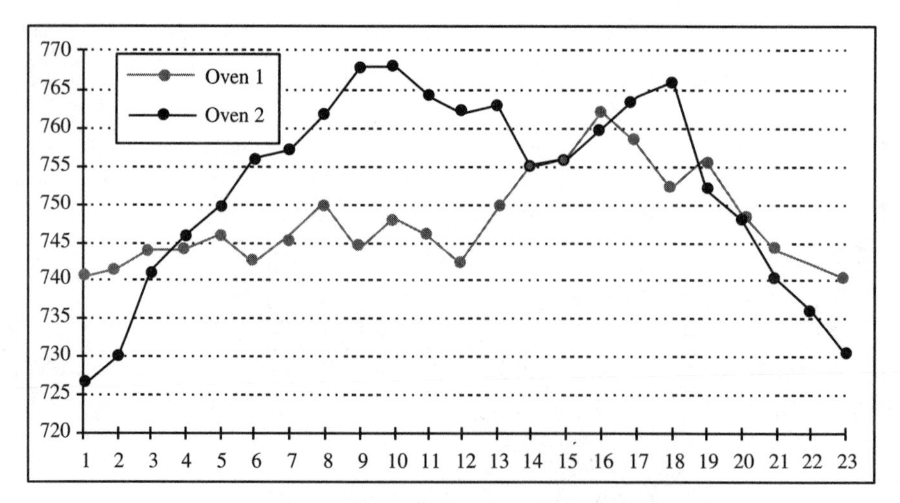

Figure 9.6 Oven Temperatures

<table>
<tr><td colspan="2" rowspan="2"></td><td colspan="6">Parts</td><td colspan="2" rowspan="2"></td></tr>
<tr><td colspan="3">A+</td><td colspan="3">A−</td></tr>
<tr><td rowspan="4">Chemical Cleaning</td><td rowspan="2">B+</td><td>A+</td><td>B+</td><td>C+</td><td>A−</td><td>B+</td><td>C+</td><td>C+</td><td rowspan="4">Oven Variance</td></tr>
<tr><td>A+</td><td>B+</td><td>C−</td><td>A−</td><td>B+</td><td>C−</td><td>C−</td></tr>
<tr><td rowspan="2">B−</td><td>A+</td><td>B−</td><td>C+</td><td>A−</td><td>B−</td><td>C+</td><td>C+</td></tr>
<tr><td>A+</td><td>B−</td><td>C−</td><td>A−</td><td>B−</td><td>C−</td><td>C−</td></tr>
</table>

Table 9.1 Test Matrix with Designated Test Cells

were used in the process. This would differentiate the parts and determine if the conformance to specification was contributing to failure and at what proportion.

The cleaning process was also divided into two specific groups—first with the cleaning tanks' chemical composition maintained with the current policy of testing and adjusting once every two weeks, and then with the tanks being tested and adjusted daily.

Finally, we want to determine "if the temperature variance has any effect on the process." The designed experiment would evaluate the two different ovens. These were placed out in a design matrix as indicated in Table 9.1.

- Parts were designated as Treatment A. The high level ($A+$) would be the parts accepted with the current practice of source inspection with no further inspection and sorting on receipt. The lower level ($A-$) would be parts detail inspected and sorted at receipt and only good parts inputted to the process.
- Chemical cleaning was designated as Treatment B. The high level ($B+$) would be the current practice of analysis and adjustment of the chemical cleaning baths weekly. The lower level ($B-$) would be the daily analysis and adjustment of the chemical baths.
- The brazing ovens were designated as Treatment C. The high level would be oven #1 ($C+$) with the greater temperature average and wider variance. The lower level would be oven #2 ($C-$) with the lower average temperature and smaller variance.

The designed experiment was to take place over a two-week period using the total output of 10 systems per day as the test sample. The daily output would be calculated as a scrap and rework rate and also as to the cost of scrap and rework. An engineering and scientific evaluation made by a team consisting of design engineering, manufacturing engineering, the chemical lab personnel and metallurgical specialists from the suppliers determined that, due to the gross nature of the failures, evaluating treatment interactions would be of no benefit at this time. Therefore, a 2^{3-1} fractional factorial designed experiment was selected. The selected fractional factorial orthogonal array is shown in Table 9.2.

The cleaning treatment would have been a problem due to the need to alternate from weekly to daily analysis and adjustment. This could have caused the neces-

	Treatments		
Runs	A	B	C
1	–	–	+
2	+	–	–
3	–	+	–
4	+	+	+

	Treatments		
Runs	Parts	Cleaning	Ovens
1	Source Inspection	Weekly	#1
2	Detailed Inspection	Weekly	#2
3	Source Inspection	Daily	#2
4	Detailed Inspection	Daily	#1

Table 9.2 2^{3-1} Fractional Factorial

sity to block this experiment into two blocks, weekly and daily analysis and adjustment. However, the assignment of the cleaning treatment to Treatment *B* provided for the weekly cleaning cycles to be run first (Runs 1 and 2) during the first two weeks of the experiments, and the daily analysis and adjustment to be run during the second two weeks. This prevented the need to block the experiment.

As indicated previously and in the process flow chart of Figure 9.3, the cost of scrap is directly related to yield; therefore, the scrap costs can be measured by the output variable yield directly. Rework, on the other hand, can occur at various stages in the process and can be reoccurring during the process. Therefore, it will be necessary to acquire a separate measurement for rework costs. This is accomplished during the same designed experiment. It is one of the strengths of a designed experiment that more than one output variable (Quality Characteristic or dependent variable) can be measured during the same designed experiment. For clarity and ease of analysis these two measures are placed on different test formats as indicated in Tables 9.3 and 9.4. They are, however, measured at the same time during the same designed experiment. In this way more than one quality characteristic can be measured and improved using the same designed experiment.

Using the design from Table 9.4, the next step was to perform the designed experiment and record the daily yield on the spreadsheet. The average yield was calculated and converted into a percentage. The resulting completed spreadsheet is contained in Table 9.5.

	Treatments			y_1	y_2	y_3	y_4	y_5
Runs	A	B	C	Yield	Yield	Yield	Yield	Yield
1	–	–	+					
2	+	–	–					
3	–	+	–					
4	+	+	+					

Table 9.3 Design Format for Yield

	Treatments			y_1	y_2	y_3	y_4	y_5
Runs	A	B	C	$ Rework	$ Rework	$ Rework	$ Rework	$ Rework
1	–	–	+					
2	+	–	–					
3	–	+	–					
4	+	+	+					

Table 9.4 Design Format for Rework Costs

	Treatments			y_1	y_2	y_3	y_4	y_5	$\bar{y}$
Runs	A	B	C	Yield	Yield	Yield	Yield	Yield	Yield %
1	–	–	+	7	8	8	7	9	0.78
2	+	–	–	9	8	8	9	9	0.86
3	–	+	–	8	7	8	9	8	0.80
4	+	+	+	10	9	10	10	10	0.98

Table 9.5 Design Format for Yield with Data Applied and Averages Calculated

Next, the effects were calculated using the methods and equations from Chapter 7. The calculations were performed directly on the same spreadsheet used to collect and summarize the data. The resulting calculations and effects appear in Table 9.6.

Using the data from Table 9.6, a graph of the effects was then made and is demonstrated in Figure 9.7. This graph clearly indicated that Treatment *A* (con-

	Treatments			y_1	y_2	y_3	y_4	y_5	$\bar{y}$
Runs	A	B	C	Yield	Yield	Yield	Yield	Yield	Yield %
1	–	–	+	7	8	8	7	9	0.78
2	+	–	–	9	8	8	9	9	0.86
3	–	+	–	8	7	8	9	8	0.80
4	+	+	+	10	9	10	10	10	0.98
$\Sigma+$	1.84	1.78	1.76						
$\Sigma-$	1.58	1.64	1.66						
$\Sigma+/n_+$	0.92	0.89	0.88						
$\Sigma-/n_-$	0.79	0.82	0.83						
Effect	0.13	0.07	0.05						

Table 9.6 Design Format for Yield with Effects Calculated

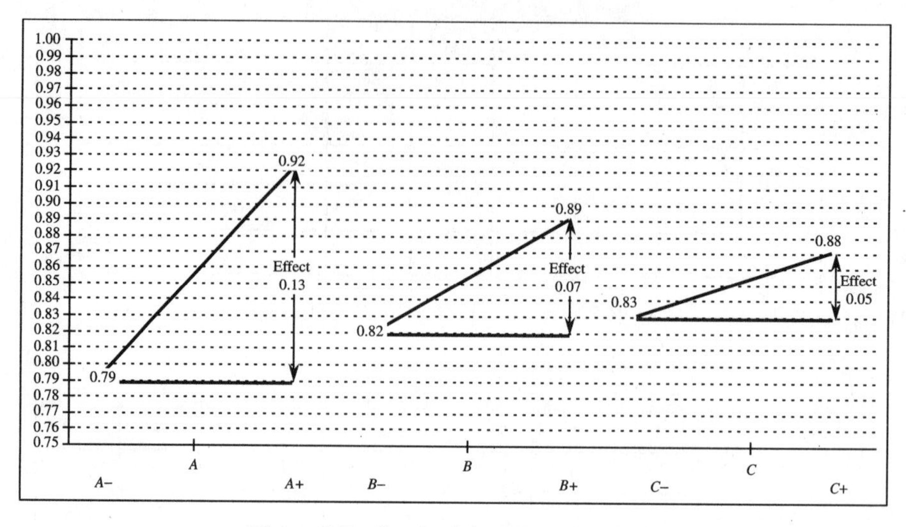

Figure 9.7 Graph of the Effects for Yield

formance of parts) had the most significant effect on the yield and therefore the scrap rate. A 13% improvement in the yield could be made by using only the 100% screened parts.

The significance of the analysis and the proportion of contribution was measured using ANOVA. The resulting ANOVA analysis and decision table are shown in Table 9.7. This ANOVA table contains an assessment of the criticality of the results at both the .99 (a=.01) and the .95 (a=.05) level. This method is sometimes employed to provide a clearer view of the level of criticality and help the analyst find the most significant treatments. In this case it could be said that Treatment A was highly critical (both at .99 and at .95) and that Treatment B was critical (only at .95). Additionally, the percent contribution also clearly indicates that Treatment A is the most significant, contributing 44% to the overall variance of the yield.

The next phase of the designed experiment was to calculate the results based on the cost of rework. This rework was accomplished off-line. This means that the repairs that occurred to the systems as they failed during the in process inspections

Source of Variation	Sum of the Squares	Degrees of Freedom	Mean Squares	F Ratio	F' Critical .99	F' Critical .95	% Contribution
A	8.40	1	8.40	19.53	8.53	4.49	44
B	2.40	1	2.40	5.58	8.53	4.49	13
C	1.25	1	1.25	2.91	8.53	4.49	7
Error	6.90	16	0.43				36
Total	18.95	19					

Table 9.7 Analysis of Variance for Yield

	Treatments			y_1	y_2	y_3	y_4	y_5	$\overline{y}$
Runs	A	B	C	$ Rework	$ Rework	$ Rework	$ Rework	$ Rework	$ Rework
1	–	–	+	11.70	11.50	12.20	11.50	10.00	11.38
2	+	–	–	10.00	8.70	7.20	9.50	9.70	9.02
3	–	+	–	7.20	6.90	7.10	6.80	7.50	7.10
4	+	+	+	5.20	4.30	5.70	4.00	5.50	4.94
$\Sigma+$	13.96	12.04	16.32						
$\Sigma-$	18.48	20.40	16.12						
$\Sigma+/n_+$	6.98	6.02	8.16						
$\Sigma-/n_-$	9.24	10.2	8.06						
Effect	–2.26	–4.18	0.1						

Table 9.8 Design Format for Cost of Rework with Effects Calculated

and tests, process elements 8 and 12, were charged separately from production. Therefore, the actual costs of rework could be obtained. Since a single system could (and frequently is) be in the repair cycle more than once, for different levels of repair, the cost of rework is a better measure of the level of rework than frequency. The approximate cost of rework was $11,800 per day. Here the objective was to minimize rework costs. The effects of this evaluation are given in Table 9.8.

Using the data from Table 9.8, a graph of the effects was then made and is demonstrated in Figure 9.8. In this graph the most significant effect is for Treatment *B*, the cleaning process element. Daily cleaning has a significant effect on rework costs.

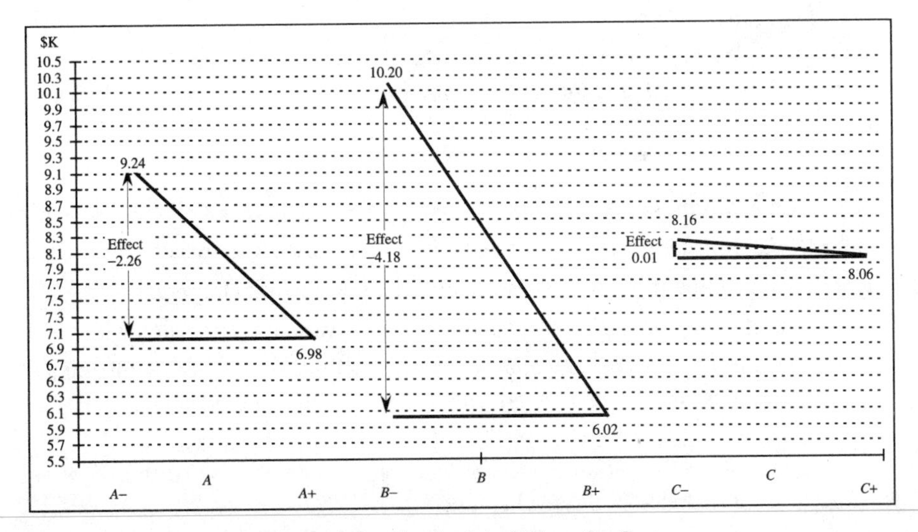

Figure 9.8 Graph of the Effects for Cost

Source of Variation	Sum of the Squares	Degrees of Freedom	Mean Squares	F Ratio	F' Critical .99	F' Critical .95	% Contribution
A	25.54	1	25.54	39.87	8.53	4.49	21
B	87.40	1	87.40	136.43	8.53	4.49	71
C	0.09	1	0.09	0.14	8.53	4.49	0
Error	10.25	16	0.64				8
Total	123.28	19					

Table 9.9 Analysis of Variance for Rework Costs

The conformance of parts also has an effect on rework costs. The variance in oven temperature (Treatment *C*) essentially has no effect on rework costs.

The ANOVA analysis and decision table for the cost of rework is given in Table 9.9. In this case it is clear that the effect of cleaning has the most significant effect on rework costs, constituting 71% of the total variance. The conformance of parts is also very significant and itself contributes 21% of the total variance. The small sum of the squares and percent contribution for the error (within) factor of the ANOVA table tells us that we have accounted for the major causes of variance in this experiment.

On completion of this designed experiment, further research determined that a change in the company's supplier management could correct the problem of nonconforming parts at the source (Supplier Source Inspection). This would add approximately 0.025% to the cost of the material. This is a fraction of the cost of scrap and rework. The total cost of this change would be approximately $750 per lot of 10,000 parts. The cost of sampling, analysis, and adjustment of the chemical cleaning tanks would increase the cost of cleaning by approximately $125 per day.

Conclusions

Based on the evaluations performed, the data collected, the analysis of the designed experiment, and the ANOVA, the following conclusions were drawn:

For the scrap rate associated with yield.

- Nonconforming parts adds significantly to the scrap rate and reduced yield.
- Nonconforming parts constituted 44% of the variance associated with yield.
- The cleaning process was also a contributor to rework and scrap.
- The cleaning process contributed 13% of the variance associated with yield.
- The variances of the temperatures in ovens #1 and #2 are not significant and their contribution to the rework and scrap associated with yield are negligible.
- This analysis accounted for 57% of the variance associated with the scrap rate. 36% of the variance was not accounted for and can be attributed to factors not included in this evaluation.

For the cost of rework associated with in-process failures.

- The most significant contributor to rework is the cleaning process.
- The cleaning process contributes 71% of the variance associated with rework costs.
- Nonconforming parts are also a significant contributor to the cost of rework.
- Nonconforming parts contribute 21% of the variability associated with rework costs.
- The temperature variance of the brazing ovens essentially has no effect on the cost of rework.
- This evaluation accounted for 92% of the total variance of process and product that contributed to rework costs.

Recommendations

Based on these conclusions it is recommended that the source selection policy be modified to include the critical parameters not currently evaluated during source selection. As an alternate measure or for an interim period, the company could go to receipt inspection of the materials. Using either method, a significant reduction in the scrap rate can be expected. The cleaning process SPC should be maintained daily and all cleaning tanks maintained within specification and SPC control. The following results were to be expected from these recommended changes.

For Scrap Rate: The current scrap rate is 20.00% quarterly. The scrap rate projected is 0.45%, less than 1%. This is an improvement from a yield of approximately 80% to a yield of 99.45%. In the following equation, A^+ is taken directly from Table 9.6. The optimized treatment level for Treatment A is 9.2. Likewise, the optimum treatment level for Treatment B^+, 8.9, is used.

$$\hat{y}_{Yield} = \frac{T}{n} + \left(A_+ - \frac{T}{n}\right) + \left(B_+ - \frac{T}{n}\right)$$

$$\hat{y}_{Yield} = \frac{171}{20} + \left(9.20 - \frac{171}{20}\right) + \left(8.9 - \frac{171}{20}\right)$$

$$\hat{y}_{Yield} = 8.55 + (9.20 - 8.55) + (8.9 - 8.55)$$

$$= 9.55$$

For Rework Costs: The current quarterly rework costs are $780,000 per quarter. This equates to $11,818 per day of manufacturing (66 work days per quarter). As indicated here, this rework cost can be reduced to $4,890 daily.

$$\hat{y}_{Yield} = \frac{T}{n} + \left(A_+ - \frac{T}{n}\right) + \left(B_+ - \frac{T}{n}\right)$$

$$\hat{y}_{Yield} = \frac{162.20}{20} + \left(6.98 - \frac{162.20}{20}\right) + \left(6.02 - \frac{162.20}{20}\right)$$

$$\hat{y}_{Yield} = 8.11 + (6.98 - 8.11) + (6.02 - 8.11)$$

$$4.89$$

Case Study 2

The M&F Industries Inc. is a manufacturing equipment supplier. The company consists of three divisions—a machine tool division, an electronics division, and a manufacturing equipment division. This designed experiment was conducted to improve the mean time between failure to meet benchmark requirements. The benchmark requirements for the M&F Industries RX05 series manufacturing equipment stated MTBF as a critical engineering design element of the new system. The benchmark requirement for the system MTBF is 3000 hours as stated in the QFD. The Electronic Control Division of M&F Industries was tasked to provide a digital control assembly that met this basic system requirement.

A team consisting of the M&F Industries RX05 program management, design engineers, service personnel, and analysts met to determine the approach to providing the needed assembly. The team first evaluated all the available process data for the existing assembly. From this data, they determined that the controller A12 board was the most significant contributor to MTBF for the digital control assembly. The customer service department's historical data on this board indicated that it currently exhibited an MTBF of 2750 operating hours in the customer's environment.

The team then performed a cause-and-effect analysis, using additional subject experts, to determine what factors contributed to the A12 board MTBF. The resulting analysis is demonstrated in Figure 9.9. Based on this cause-and-effect analysis

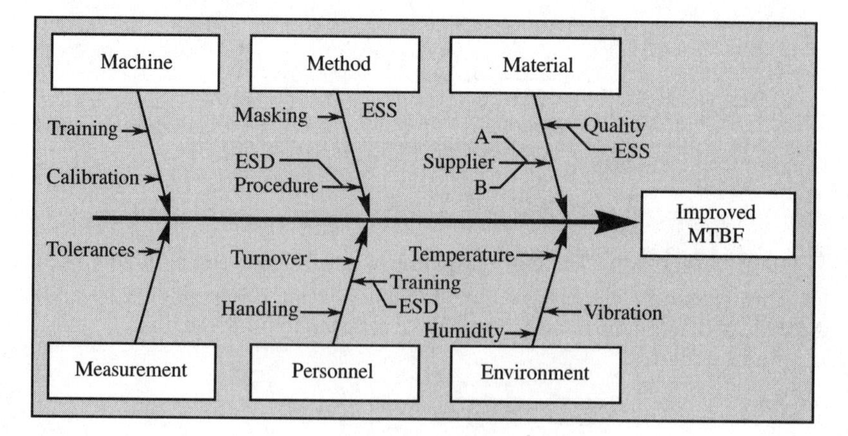

Figure 9.9 Cause-and-Effect Analysis

and the knowledge of the subject specialists, the team selected three factors or treatments that could contribute most significantly to MTBF:

Board Supplier: M&F Industries is using two board suppliers for existing digital controllers. The boards had the same performance characteristics, but the suppliers produced them using different methods and technologies. Both boards had previously met all M&F Industries supplier quality requirements and were one-to-one interchangeable. These treatments will be Supplier A1 and Supplier A2.

Vibration Protection: The vibration induced on the assembly by the equipment itself and by the customer's environment was a concern to the subject specialists and the system engineers. The existing controller assembly and its associated boards are assembled to the system without any special vibration damping. The treatments will be called Vibration Protection (Damper) B1 and No Vibration Protection B2.

Electrostatic Discharge (ESD): Sensitivity to ESD had not been very well defined in the past. M&F Industries is currently using ESD standards that are acceptable in the industry. There was very little information on the effects of ESD, and no known correlation between ESD and customer service calls for the existing systems. The engineers and subject experts wanted to determine if, in fact, ESD had any effect on system service life. Improved ESD protection standards, devices, and training are available. The treatments selected for this evaluation are Existing ESD Standards C1 and Improved ESD Standards C2.

The team determined that the experimental results would be expressed (quality characteristic, output variable, dependent variable) as hours MTBF. The experimental process would be incorporated at the end of the ESS cycle as an accelerated life cycle test and the systems would be tested to failure or to the equivalent of 3750 hours (125% of requirement). There was a consensus among the team members that no interactions would be significant. Therefore, the designed experiment considered the following factors:

Treatment A: Supplier
Treatment B: Vibration Protection
Treatment C: ESD Protection
Experimental Results: Hours MTBF

The resulting design matrix is demonstrated in Table 9.10. To evaluate this experiment completely for main effects and interactions would require a 2^3 full factorial experiment or eight test runs. The expense and time requirements for accelerated life cycle testing are significant and are of concern to the team and management. Therefore, a Taguchi L_4 orthogonal array was selected as meeting these needs and the requirements for orthogonality.

The team next prepared a worksheet for the experiment to manage the data more effectively and to estimate the data management requirements. The experiment was then run, using the data from four samples after initial ESS was completed. The resulting data table is shown in Table 9.11.

		A Supplier			
		A1	A2		
B Damping	B1	A1B1C1	A2B1C1	C1	C ESD Protection
		A1B1C2	A2B1C2	C2	
	B2	A1B2C1	A2B2C1	C1	
		A1B2C2	A2B2C2	C2	

Table 9.10 Test Matrix

Column / Run	A	B	C	Results 1	2	3	4	$\bar{y}$
1	+	+	+	5365	5791	4902	5523	5395.25
2	+	–	–	4615	5110	4790	4988	4875.75
3	–	+	–	5112	4976	5378	5500	5241.50
4	–	–	+	5045	4798	5110	4811	4941.00
Level 1	5135.5	5318.38	5168.13					
Level 2	5091.25	4908.38	5058.63	Grand Average = 5113				

Table 9.11 MTBF Data Table

The next step in evaluating the Taguchi DOE for improving MTBF is to determine the level effects and total effects. To accomplish this, the team used the spreadsheet in Table 9.12.

Figure 9.10 provides a graphical representation of the level effects and total effects calculated from the Taguchi DOE. To determine the level of significance of

Treatment	Name	Level	Level Mean	Level Effect	Total Effect
A	SUPP	+	5135.5	22	44
		–	5091.25	–22	
B	VIB	+	5318.38	205	410
		–	4908.38	–205	
C	ESD	+	5168.13	55	110
		–	5058.63	–55	

Table 9.12 MTBF Level and Total Effect

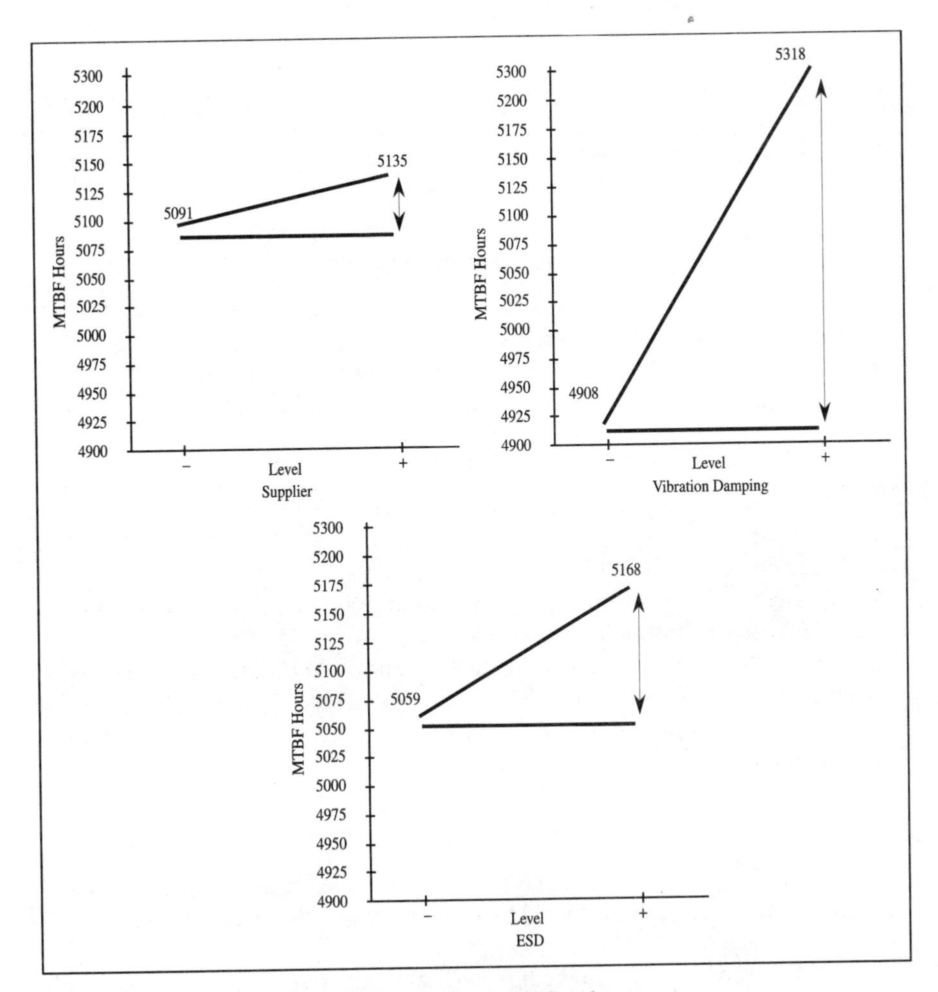

Figure 9.10 DOE Graph

the effects and the percent contribution the team then performed a Taguchi-Style ANOVA.

The team then placed all this data into the Taguchi ANOVA Table in Table 9.13 to evaluate the results. The team evaluated the results of the analysis and was able to make the following "fact-based decisions."

- Treatment *B* is making a significant contribution to the MTBF of the A12 Board in this experiment.
- Treatments *A* and *C* are not significant.
- Treatment *B* is contributing 43.49% of the total variation in MTBF.
- Due to the relatively large *SS*, *MS*, and percent contribution for error, there may be significant contributors to MTBF that were not considered in this DOE or that are attributable to common cause variation.

Source of Variation	*df*	*SS*	*MS*	*F*	*F'*	*SS'*	% CONT
A	1	7832	7832	0.117	4.75	2241.33	0
B	1	672400	672400	10.022	4.75	666809.10	44
C	1	47961	47961	0.715	4.75	42370.08	3
Error	12	805092	67091				53
Total	15	1533285					

Table 9.13 MTBF DOE Taguchi ANOVA Table

Based on this information, the team determined that optimization of the MTBF would occur with all treatments set at the high level (+), based on the "greater is better" principle. Therefore, the estimate of the results of the optimized function is:

$$\hat{Y}=5113.4+(5135.5-5113.4)+(5318.4-5113.4)+(5168.1-5113.4)=$$

$$5113.4+22.1+205.0+54.75=\mathbf{5395.3}$$

Based on the results of this designed experiment, the team determined that the QFD design and manufacturing matrix goal of 5000 hours could be met by the A12 controller board with an optimized MTBF of 5395. The team also determined that there were other opportunities to improve MTBF based on treatments, interactions, environment, and other factors not included in this study. The team recommended damping as a vibration protection as the most significant contribution to MTBF.

Case Study 3

The purpose of this designed experiment is to characterize the compost panel production process and optimize yield based on three quality characteristics.

- **Pull Strength:** Pull Strength is a variable measurement that will be made using test equipment from which quantitative measurements will be made. The target pull strength is 4 Ft LB.
- **Color:** Color is a qualitative measurement based on the closeness of the color of the panels compared to a standard. This will be a pass/fail measurement.
- **Staining:** Staining is a qualitative measurement based on the absence of any staining on the panel. This is also a pass/fail measurement.

This is a new process and the process and procedures have just recently been developed. Many of the process input and settings are those recommended by suppliers of raw material or recommended by the customer. The effects of these inputs has not been characterized by the company and there is no knowledge of the effect of sleeting different options may have on the product. Additionally, it is unknown if this process can meet the new requirement for a minimum pull strength of 4 Ft LB.

These concerns, the input options, and the quality characteristics were the subject of several meetings between the company, its suppliers, and the customer requiring the product. The engineering and technical knowledge of the process and the product has indicated that there are six input variables that will affect the quality characteristics selected. These treatments are listed:

A. Materials
SWG
Rogers

B. Temperature
130
150

C. Bonding Chemical C11A9
12%
7%

D. Binder Chemical S620
30
50

E. Stay Time
25 Seconds
50 Seconds

F. Special Cleaning
Yes New Method
No Standard Method

These treatments and levels will require a six-treatment designed experiment. Because of resource considerations, time, and the cost of sample coupons, the best design to use will be a 2^{6-3} fractional factorial designed experiment. In addition, the experiment must be blocked to minimize the changes in the application of Bonding Chemical C11A9 and Binder Chemical S620. This consideration will require the experimental design to be blocked into four blocks. There will be two experiments conducted as screening experiments and a confirming run conducted on the optimized process.

Experiment #1 Screening experiment to characterize the six variables at two levels each and determine if any of the selected input variables and options had a significant effect on the quality characteristics. Depending on the results of this run, the process may attempt to be optimized.

Experiment #2 A confirming run of the optimized design that also will be integrated with the first production run of compost panels. This run will also be used to determine if there is any interaction between the critical input treatments.

Designed Experiment #1

This is the 2^{6-3} fractional factorial experiment. The 2^{6-3} fractional factorial format in Table 9.14 was taken from Table A.22 in Appendix A.

Table 9.15 provides the runs, treatments, and levels of the experiment as they are actually going to be run. The design format was done this way to provide assistance to the manufacturing engineers and technicians on the shop floor when running the experiment. This view also makes it very clear how the experiment will need to be blocked to minimize the changes required in the C11A9 and S620 mixtures.

	Treatments					
Runs	A	B	C	D	E	F
1	−	−	−	+	+	+
2	+	−	−	−	−	+
3	−	+	−	−	+	−
4	+	+	−	+	−	−
5	−	−	+	+	−	−
6	+	−	+	−	+	−
7	−	+	+	−	−	+
8	+	+	+	+	+	+

Table 9.14 2^{6-3} Fractional Factorial Format

The attached randomized block design (Table 9.16) for Experiment #1 will be run in four blocks to minimize the changes required in the C11A9 and S620 solutions. Each of these blocks will be run separately as indicated. The data will then be combined in the proper run order on the data sheets for each characteristic measured.

As indicated in Figure 9.11, 7 of 40 required data elements for this analysis are missing (No Test or No Data). This constitutes 17.5% of the data for the analysis. Test run number 5 has a very significant shift in the data from test elements y_1, y_2, and y_3 with an average of 7.5 peel strength to runs y_4 and y_5 with an average peel strength of 6.16. These data elements are nonhomogenous, indicating that they come from distinctly different populations. This indicated that an unknown environmental or test control factor caused this shift of data. These anomalies are demonstrated in Figure 9.11.

	Treatments					
Runs	Material	Temperature	C11A9	S620	Stay Time	Cleaning
1	Rogers	130	7%	30	25	Yes
2	SW/G	130	7%	40	50	Yes
3	Rogers	150	7%	40	25	No
4	SW/G	150	7%	30	50	No
5	Rogers	130	12%	30	50	No
6	SW/G	130	12%	40	25	No
7	Rogers	150	12%	40	50	Yes
8	SW/G	150	12%	30	25	Yes

Table 9.15 2^{6-3} Fractional Factorial Format with Treatments

	Treatments					
Runs	Material	Temperature	C11A9	S620	Stay Time	Cleaning
Block 1						
4	SW/G	150	7%	30	50	No
1	Rogers	130	7%	30	25	Yes
Block 2						
3	Rogers	150	7%	40	25	No
2	SW/G	130	7%	40	50	Yes
Block 3						
8	SW/G	150	12%	30	25	Yes
5	Rogers	130	12%	30	50	No
Block 4						
7	Rogers	150	12%	40	50	Yes
6	SW/G	130	12%	40	25	No

Table 9.16 2^{6-3} Fractional Factorial Format with Treatments

The basis for performing designed experiments and for the test plan we established for this designed experiment is orthogonality. That is that the test runs are balanced, providing data indicating the variability around the data generated by the selected factors. This also requires that the data be of equal sample size to allow each factor an equal opportunity to vary. The missing data from this experiment caused the test data to be unevenly distributed. Only the data for columns containing data elements y_1, y_2, and y_3 can therefore be used.

This caused a significant problem for the team conducting the experiment. These were very expensive test runs and the company could not afford to repeat the complete experiment. It was determined by the industrial engineers that there were two pull test machines used to perform the experiment, and that one of the machines was faulty. The faulty machine had been used for the missing data, causing the destruction of the samples, and also was used in run #5 for samples y_1, y_2, and y_3.

	A	B	C	D	E	F	y-1	y-2	y-3	y-4	y-5	$\bar{y}$
1	–	–	–	+	–	–	4.36	3.96	4.2	4.8		4.33
2	+	–	–	–	+	–	4.2	[illegible]	4.6	4.12	4.8	4.46
3	–	+	–	–	+	+	6.24	5.78	5.52			5.85
4	+	+	–	+	–	+	4.58	4.8	5.04	5.4	4.44	4.85
5	–	–	+			+	8	7.5	7	5.76	4.56	6.56
6	+	–	+	–	–	+	4.56	4.8	4.56	4.8		4.68
7	–	+	+	–	–	–	[illegible]	[illegible]	5.88	4.8		5.25
8	+	+	+	+	+	–	4.68	4.56	5.04			4.76

Missing Data

Suspect Data Run

Missing Data

Figure 9.11 First Test Results

> Author's Note: This is a very good example of what can happen during a designed experiment that is not properly supervised and controlled. The second pull test machine was not part of the experimental design and had not been calibrated. It was used as an expedient to "get things moving a little faster." Plan and Control your experiment.

It was critical to attempt to salvage at least part of this experiment. The team met and did a very careful review of the situation. It was determined that the data that were taken from the one calibrated pull test machine were "good." The data from the other machine were special cause variation and must be removed from the experiment. That left three samples from repeats Y_1, Y_2, and Y_3 with the exception of run #5 that was usable. There was no way around the necessity of rerunning test run #5. This was not a good situation and certainly not a perfectly designed experiment. This situation was mitigated by the fact that a full factorial experiment was to be run as a confirming experiment and to determine if there were any interactions affecting the process.

Test run #5 was rerun and carefully monitored to ensure that the conditions of the rerun experiment were the same as the original experiment. The results of the designed experiment with rerun #5 are given in Table 9.17.

Using the data from Figure 9.17, a graph of the effects of the six treatments were then made. These graphs are presented in Figures 9.12 and 9.13.

Reviewing the effects of the designed experiment and the resulting graphs, peel strength would be optimized using $A-$, $B+$, $C+$, $D-$, $E+$, and $F+$. Apply-

	Treatment						Data			
Run	A	B	C	D	E	F	y-1	y-2	y-3	$\bar{y}$
1	–	–	–	+	–	–	4.36	3.96	4.20	4.17
2	+	–	–	–	+	–	4.20	4.56	4.60	4.45
3	–	+	–	–	+	+	6.24	5.78	5.52	5.85
4	+	+	–	+	–	+	4.58	4.80	5.04	4.81
5	–	–	+	+	+	+	6.21	5.98	6.10	6.10
6	+	–	+	–	–	+	4.56	4.80	4.56	4.64
7	–	+	+	–	–	–	4.56	5.76	5.88	5.40
8	+	+	+	+	+	–	4.68	4.56	5.04	4.76
$\Sigma+$	18.66	20.81	20.90	19.84	21.16	21.39				
$\Sigma-$	21.52	19.36	19.28	20.34	19.02	18.79				
$\Sigma+/n_+$	4.67	5.20	5.22	4.96	5.29	5.35				
$\Sigma-/n_-$	5.38	4.84	4.82	5.09	4.76	4.70				
Effect	-0.71	0.36	0.40	-0.13	0.53	0.65				

Table 9.17 Completed Designed Experiment with Data and Effects

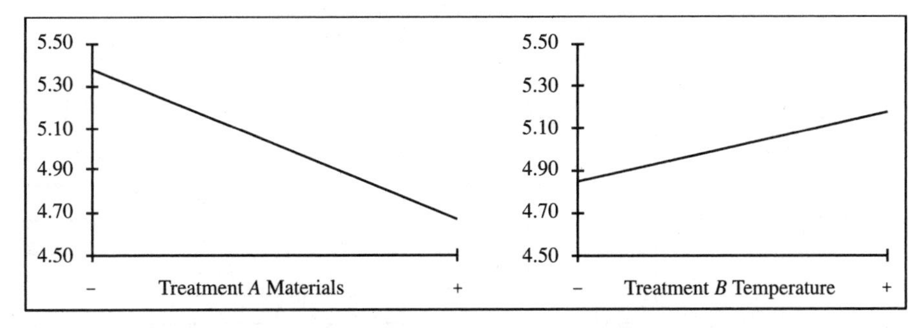

Figure 9-12 DOE Graph of Treatment Effects *A* and *B*

ing the information from this experiment to analysis of variance results in ANOVA Table 9.18, the level of criticality was selected at .05.

The review of the ANOVA table indicated that Treatments *A*, *E*, and *F* are significant. These three treatments combined contribute 63% of the total variance of pull strength. The additional quality characteristics of color and staining were all Pass; therefore, no evaluation was relevant. The team discussed the results of this experiment and the resulting ANOVA. There was some concern that 29% of the variation was unaccounted for and that this variance could be caused by an interaction between the critical parameters that needed to be controlled.

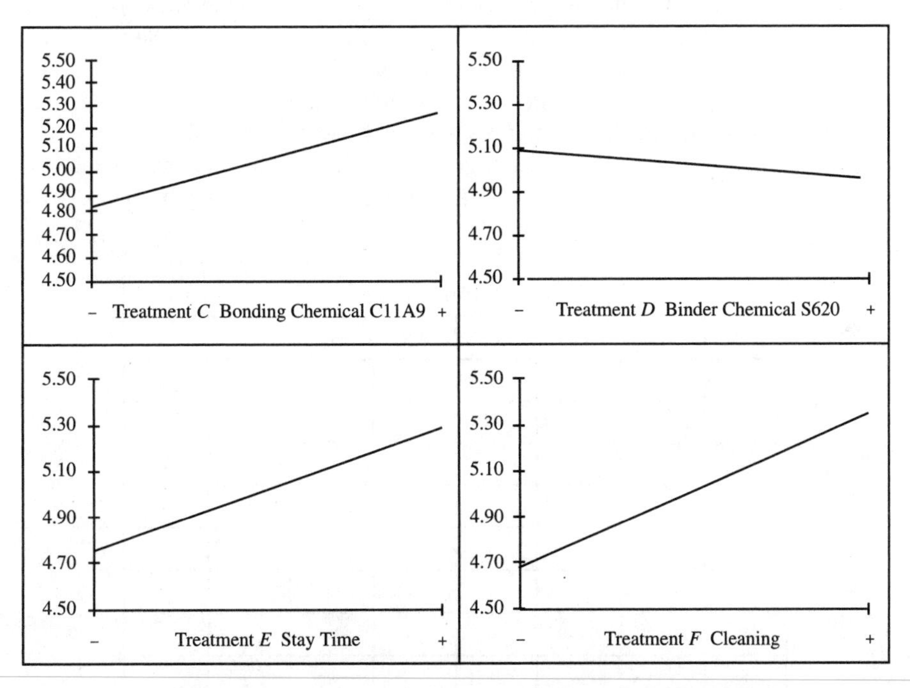

Figure 9.13 DOE Graph of Treatment Effects *C*, *D*, *E*, and *F*

Source of Variation	Sum of the Squares	Degrees of Freedom	Mean Squares	F Ratio	F' Critical .05	% Contribution
A	3.06	1	3.06	10.20	4.450	26
B	0.79	1	0.79	2.63	4.450	7
C	0.10	1	0.10	0.33	4.450	1
D	0.10	1	0.10	0.33	4.450	1
E	1.71	1	1.71	5.70	4.450	15
F	2.54	1	2.54	8.47	4.450	22
Error	3.35	17	0.20			29
Total	11.65	23				

Table 9.18 ANOVA Table

It was therefore determined that a full factorial experiment would be conducted, as planned and budgeted, as the confirming run for the process optimization (for Critical Factors *A*, *E*, and *F*) and also to determine if there were any interactions. Since there were three significant factors, the full factorial designed experiment would be a 2^3 design format as in Table 9.19. This design format can be found in Table A.4 in Appendix A.

For the purpose of clarity and understanding, the factors were redesigned as treatments *A*, *B*, and *C* as indicated in Table 9.20. *A* was the Materials, *B* the Stay Time, and *C* the Cleaning process with and without the new methods.

The experiment was then run, with the results indicated in Table 9.21. This appears to confirm the optimum process settings as $A-$, the Rogers material, $B+$ 50 Seconds stay time in the chemical vats, and $C+$, the new cleaning methods. There also may be an interaction (as suspected between *B* and *C*).

Using the data from Table 9.22, graphs of the main effects and interactions were then made. Figure 9.14 provides the calculations for plotting the interactions.

	Treatments						
Runs	A	B	C	AB	AC	BC	ABC
1	–	–	–	+	+	+	–
2	+	–	–	–	–	+	+
3	–	+	–	–	+	–	+
4	+	+	–	+	–	–	–
5	–	–	+	+	–	–	+
6	+	–	+	–	+	–	–
7	–	+	+	–	–	+	–
8	+	+	+	+	+	+	+

Table 9.19 2^3 Design Format

	Treatments						
	Temperature	Stay Time	Cleaning	Interactions			
Runs	A	B	C	AB	AC	BC	ABC
1	90.00	25.00	Standard	+	+	+	–
2	120.00	25.00	Standard	–	–	+	+
3	90.00	50.00	Standard	–	+	–	+
4	120.00	50.00	Standard	+	–	–	–
5	90.00	25.00	New Method	+	–	–	+
6	120.00	25.00	New Method	–	+	–	–
7	90.00	30.00	New Method	–	–	+	–
8	120.00	30.00	New Method	+	+	+	+

Table 9.20 Redesigned Treatments and Design Format

	Treatments							Test Results				
	Main Effects			Interactions				Pull Strength				
Runs	A	B	C	AB	AC	BC	ABC	y_1	y_2	y_3	y_4	$\bar{y}$
1	–	–	–	+	+	+	–	4.55	5.78	5.58	5.67	5.40
2	+	–	–	–	–	+	+	4.59	4.79	4.90	5.05	4.83
3	–	+	–	–	+	–	+	4.76	5.10	4.90	4.80	4.89
4	+	+	–	+	–	–	–	4.35	4.41	4.70	4.50	4.49
5	–	–	+	+	–	–	+	4.65	4.72	4.76	4.91	4.76
6	+	–	+	–	+	–	–	4.20	4.32	4.10	4.25	4.22
7	–	+	+	–	–	+	–	5.98	6.10	6.20	5.90	6.05
8	+	+	+	+	+	+	+	4.66	4.59	5.04	4.77	4.77
$\Sigma+$	18.31	20.19	19.79	19.41	19.27	21.04	19.25					
$\Sigma-$	21.09	19.21	19.61	19.99	20.13	18.36	20.15					
$\Sigma+/n_+$	4.58	5.05	4.95	4.85	4.82	5.26	4.81					
$\Sigma-/n_-$	5.27	4.80	4.90	5.00	5.03	4.59	5.04					
Effect	-0.70	0.25	0.04	-0.14	-0.22	0.67	-0.23					

Table 9.21 Completed Designed Experiment with Data and Effects

	B–	B+		C–	C+		C–	C+
A–	5.08	5.47	A–	5.14	5.40	B–	5.11	4.49
A+	4.53	4.63	A+	4.66	4.49	B+	4.69	5.41

Table 9.22 Calculations for Plotting Interactions

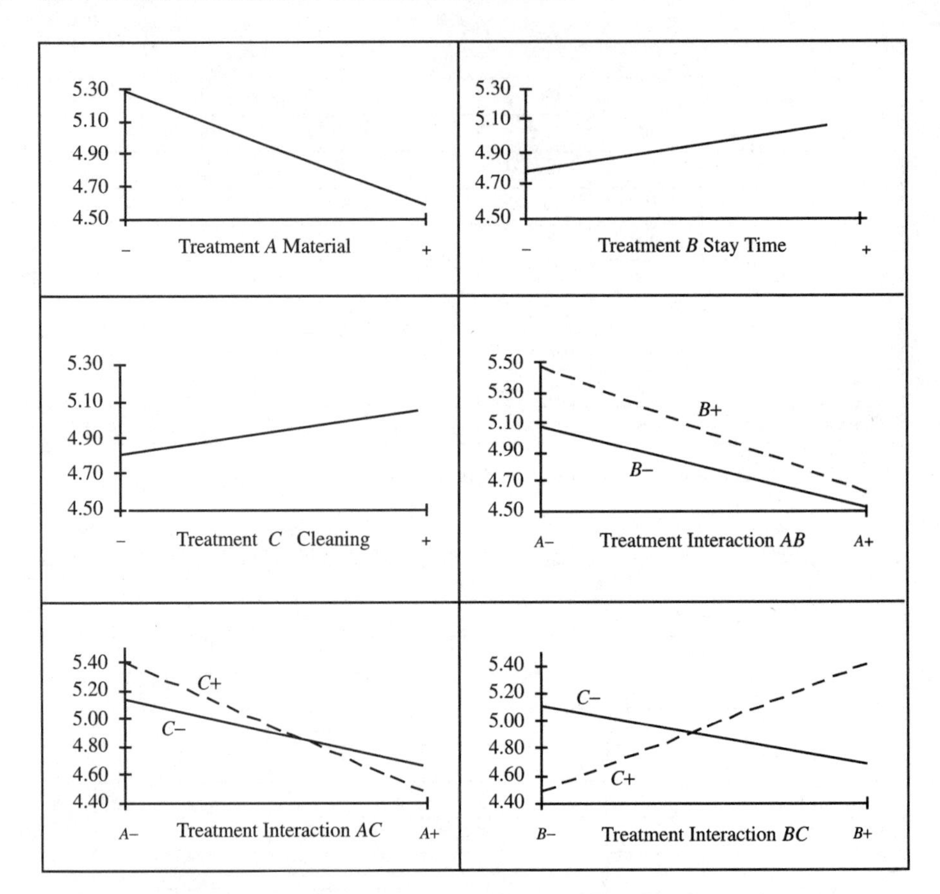

Figure 9.14 DOE Graph of Main Effects and Two-Way Interactions

This equation is provided just as a reminder on how to calculate interactions.

$$Interaction_{A-B-} = \frac{\sum y_{A-B-}}{n_{A-B-}} = \frac{5.40 + 4.76}{2} = 5.08$$

Reviewing the effects of the designed experiment and the resulting graphs indicates that the process would be optimized with $A-$, $B+$, and $C+$. This confirms the information from the fractional factorial. In addition, the suspected interactions between stay time and the cleaning process were confirmed. It was a surprise that there was any interaction between material and stay time; that had not been suspected by the engineers and material specialists from the suppliers. Applying the information from this experiment to ANOVA resulted in Table 9.23. Since this was the final confirming run, the team decoded to set the criticality level at .01.

Source of Variation	Sum of the Squares	Degrees of Freedom	Mean Squares	F Ratio	F' Critical .01	% Contribution
A	3.880	1	3.880	70.545	7.820	37
B	0.490	1	0.490	8.909	7.820	5
C	0.020	1	0.020	0.364	7.820	0
AB	0.630	1	0.630	11.455	7.820	6
AC	0.370	1	0.370	6.727	7.820	4
BC	3.590	1	3.590	65.273	7.820	35
ABC	0.660	1	0.660	12.000	7.820	6
Error	0.730	24	0.030			7
Total	10.370	31				

Table 9.23 ANOVA Table

The review of the ANOVA table indicated that Treatments *A*, *B*, *AB*, and *BC* were significant. These treatments and treatment combinations clearly account for the majority of the % contribution to overall variance. Additionally, only 7% of the variability of this process remains unaccounted for. This is very good. The company can move forward with this process optimization with confidence that their steps will have the desired effect. It is clear from this series of experiments that:

- The process can be established and optimized using the treatments and levels as $A-, B+, C+$.
- The processes are capable of producing compost material with a pull strength of 4 or greater.
- The process is affected by the interaction of Treatments *BC*. These can be monitored and controlled by controlling the significant factor *B*.
- There is very little variance that is not accounted for (7%). The most important treatments have been included in this series of experiments by the team.

The team can now sit back and enjoy its success! Right? No. Process and products must be constantly improved and refined to maintain a product or service as competitive in the global marketplace. The team must now use all the tools at its disposal to control the process, measure it, and implement continuous measurable improvement.

10

OTHER DOE APPLICATIONS

Traditionally Design of Experiments (DOE) has been applied to industrial processes and products. As more and more of business becomes nonindustrial in nature, there has been a growing trend to the application of fact-based decision making. DOE is a valuable tool in making these fact-based decisions. As with all business processes, their optimization and the selection of quality characteristics (output variables) fall into the same three basic categories of quality, cost, and schedule. One of the difficulties often encountered in nonindustrial DOE is determining how to measure an administrative/service process or product. This at times does not appear as clear as in industrial processes. The following are some typical measures of these processes and products.

Quality	Cost	Schedule
Acceptance of services	Product costs	Timelines
Repeat customers	Production costs	Throughput
Responses received	Wages	Work hours
Rejections	Material costs	Down time
Work that is repeated	Interest costs (cost of money)	Delay time

The general scenario for performing a DOE for these nonindustrial processes and products is the same as for industrial processes. You must still know and understand very clearly your customer, your products, and your processes. Therefore, much of the preparatory work of getting started is the same. There is one very large difference, however; in an industrial process you are measuring things, whereas in a nonindustrial process you are measuring people.

This difference will have a significant effect on your designed experiment. This will become very obvious during two phases of the process analysis, process or product analysis, and data acquisition. Here you will encounter people's natural tendency to be protective (even defensive) of themselves, their product, or their process. You must remember that this is "fact-based decision making" and proceed to perform your designed experiment. The goals of performing DOE for nonindustrial processes are much the same as for any process or product; some are just stated differently:

- Optimize output
- Reduce variability
- Reduce work that needs to be redone
- Improve customer satisfaction
- Streamline processes
- Reduce sensitivity to supplier variation
- Address customer concerns
- Decrease operations cost

We will review three case studies that apply DOE to practical service, administrative, and R&D problems. Each of these case studies approaches the use of designed experiments in different ways for different purposes. Here again, just as with industrial processes, these are not perfect academic examples of designed experiments where all data are always there and every experiment is successful. This is what you can expect to encounter in performing a DOE in the business environment.

Case Study 1

A large company was having a problem with receivables. The average age of receivables due was 200 days after delivery of material. The company currently had a value of $67,000K in excess of 60 days old. Typically these were invoices for large assemblies that were in the $10,000 range and higher. These assemblies were used in the manufacture of automated equipment. At the time the study began, the company had $130 million that was 30 days or older after receipt by the customer. The cost of money for this delay was significant. Additionally, this delay was causing a cash flow problem at the company.

The first action taken was to review the 30 most recent invoices and attempt to determine the cause of the delay. It was determined that there were very few errors in the actual invoices or receipts of material. These shipments were typically of one or two very large pieces of equipment and little could be wrong with count or other things that were typically wrong with receipts. Although large, the assemblies were very reliable and there were no receipt inspection rejections by the customers. When the companies were contacted and asked for the cause of the significant delays of payment, they all objected very strongly and indicated they paid promptly after receipt of an invoice. The data were viewed in two different ways—as the days after delivery of the material and as the days after invoice. These data are presented in Table 10.1.

Invoice No.	Customers A DAD		B DAD		C DAD		D DAD		E DAD	
	DAD	DAI	DAD	DAI	DAD	DAI	DAD	DAI	DAD	DAI
1	186	41	183	40	193	49	189	44	201	56
2	172	27	169	26	179	35	175	30	187	42
3	190	45	187	44	197	53	193	48	205	60
4	222	77	219	76	229	85	225	80	237	92
5	221	76	218	75	228	84	224	79	236	91
6	240	95	237	94	247	103	243	98	255	110
7	169	24	166	23	176	32	172	27	184	39
8	198	53	195	52	205	61	201	56	213	68
9	186	41	183	40	193	49	189	44	201	56
10	185	40	182	39	192	48	188	43	200	55
11	198	53	195	52	205	61	201	56	213	68
12	201	56	198	55	208	64	204	59	216	71
13	221	76	218	75	228	84	224	79	236	91
14	196	51	193	50	203	59	199	54	211	66
15	184	39	181	38	191	47	187	42	199	54
16	197	52	194	51	204	60	200	55	212	67
17	184	39	181	38	191	47	187	42	199	54
18	186	41	183	40	193	49	189	44	201	56
19	176	31	173	30	183	39	179	34	191	46
20	199	54	196	53	206	62	202	57	214	69
Average	195.55	50.55	192.55	49.55	202.55	58.55	198.55	53.55	210.55	65.55
DAD: Days After delivery				DAI: Days After Invoice						

Table 10.1 Invoice Payment Delay Data

One company policy was to invoice for payment after the customer accepted the material. Therefore, there could be some delay expected between shipment, receipt, and invoice. The data indicated that this was averaging a 181-day delay within the company before an invoice (request for payment) was sent to the customer. One overseas company had actually sent the company a letter requesting to be billed so they could clear their books before the beginning of a new fiscal year. This made it apparent that the problem statement had changed from " What is the cause of the aging of receivables" to "How can we optimize the process of billing."

The first step that was performed was to do a complete process analysis and flow chart on the invoicing process. This process analysis flowcharting was completed and the results appear in Figure 10.1. This process flow chart was estab-

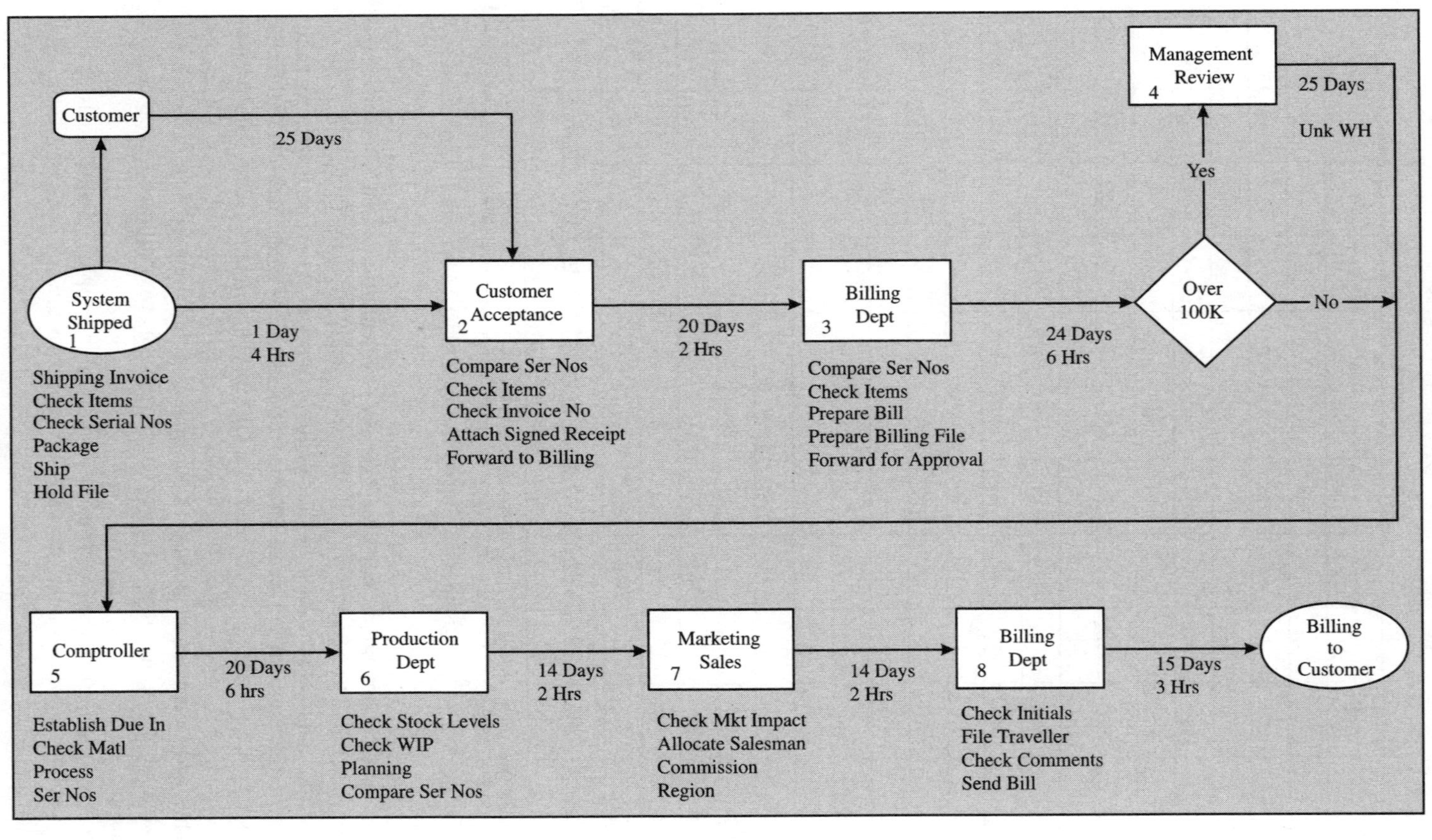

Figure 10.1 Process Analysis Flow Chart

lished by walking through the actual billing process, as there was no procedure specifying the process. The only specific policy requirement was that all bills for over $100K be sent to the management for review and initialing.

The times associated with each process step are the number of days from signout at one activity through signout at the next. The number of work hours required was estimated when talking to the people in the process. It was also learned that at no time in the process had a bill been rejected, found incorrect, or been returned for any reason. It was also noted during the process analysis that an automated system was on every desk in the company. However, the billing signature and approval cycle were completely manual.

As is true in many companies, this process had evolved over time and, as an administrative process, was not subject to the same control, documentation, or procedures as a production activity. It was quickly determined that the invoices and bills were absolutely required by the Comptroller, Production Department, and Marketing department to keep their records and to do planning. There was really no reason for any of these process elements to have the original bills or to see the bills before they went out. Therefore, process elements 5, 6, and 7 could be eliminated with a gross savings of 48 days. The company management indicated that they still wanted notification "before" billing of over $100K went to a customer. This was very important to them in maintaining a close relationship with those customers. They agreed to being notified as the bill was sent, rather than holding up billing on $100K for up to 25 days as the invoice and bill were processed through the corporate offices.

There were several other options available that may further reduce billing time that were less clear as to their effect on aging of receivables. These options were:

- Bill directly on the invoice.
- Automate the billing and invoicing system.
- Provide follow-up to the customers at 30 and 45 days by telephone or in writing.
- Contract out the billing department to a professional billing activity.

These options lent themselves to evaluation using a designed experiment. The first option was to bill directly on the invoice or with the invoice at time of shipping. This is how billing is normally done in many companies. We all have received bills enclosed with shipments. It traditionally had not been done that way in this company and there was concern that it would not have a significant effect and might cause losses and delays with the customers' routines.

The second option was to completely automate the billing system which is now partially automated. This would provide the company with the option of informing management on-line concerning billings over $100K. In addition, the other process elements could be provided automatic automated copies of the billings and invoices. The real change to this option is that the actual bills can go to the customers directly from the billing department at time of invoice without waiting for the hard copy of the invoice.

The third option is follow-up. This company had never followed up with a single customer by telephone or in writing. If a billing was very late only management would bring it up in a call to the customer's management. There was a lot of sensitivity to this in the company; the owners (a closely held company) did not want to be viewed as harassing their customers. But they did need to reduce the aging of receivables. Follow-up was to be a "polite letter" at 45 and 60 days, and the other option was a phone call from management at 45 and 60 days.

Another option was to completely contract out the invoice and billing process. This would call for shipping and receiving to send the invoices to a professional billing company and allow them to do all of the activities concerning billing and receivables. This is somewhat of a new trend and the actual effect on the age of receivables resulting from this is unknown. This resulted in the following treatments and levels for the designed experiment.

- A. Billing
  - −1. Directly on the invoice with the shipment
  - 1. Billing from the billing department mailed separately from the shipment.
- B. Automation
  - −1. Automate the complete billing process with all billing generated automatically on shipment.
  - 1. Maintain the current system in which the generation of billing is automated but the bills and invoices are transmitted and routed in hard copy.
- C. Follow-up
  - −1. Follow up by letter only at 45 and 60 days.
  - 1. Follow up by telephone at 45 and 60 days.
- D. Contract
  - −1. Contract out the billing and follow-up
  - 1. Keep the billing and follow-up in house.

These treatments and levels are a 2^{4-1} fractional factorial experiment. The experimental format and the layout of the levels are given in Table 10.2. It was also determined that the trial would take place over a six-month period.

Using the design from Table 10.2, the next step was to perform the designed experiment and record the monthly average for the six-month period on the spreadsheet. The average yield was calculated in days since billing. The resulting completed spreadsheet is contained in Table 10.3. We have already reduced the age of deliverables by 74 days in the first phase of the analysis. Here we are looking to improve further on that gain.

Using the data from Table 10.3 a graph of the effects was then made and is demonstrated in Figure 10.2. It is clear from these graphs that Treatments *A* and *C*

	Treatments			
Runs	A	B	C	D
1	−1	−1	−1	−1
2	1	−1	−1	1
3	−1	1	−1	1
4	1	1	−1	−1
5	−1	−1	1	1
6	1	−1	1	−1
7	−1	1	1	−1
8	1	1	1	1

	Treatments			
Runs	A	B	C	D
1	Invoice	Complete	Letter	Contract
2	Separate	Complete	Letter	In House
3	Invoice	Partial	Letter	In House
4	Separate	Partial	Letter	Contract
5	Invoice	Complete	Telephone	In House
6	Separate	Complete	Telephone	Contract
7	Invoice	Partial	Telephone	Contract
8	Separate	Partial	Telephone	In House

Table 10.2 2^{4-1} Fractional Factorial Experiment

are making a difference in the aging of receivables. It is equally obvious that treatments *B* and *D* are having little effect.

Author's Note: This experimental format uses the −1 and 1 notation for the levels of the treatments. It is no more difficult to work with than the + and − notation or the 1 and 2 notation. The analytical methodology remains the same. This format is very common and you will see it as you use the software packages that are available and in some textbooks.

	Treatments				Data						
Runs	A	B	C	D	y_1	y_2	y_3	y_4	y_5	y_6	$\bar{y}$
1	−1	−1	−1	−1	49	46	56	59	47	44	50
2	1	−1	−1	1	79	84	86	78	86	91	84
3	−1	1	−1	1	51	55	64	53	63	61	58
4	1	1	−1	−1	93	96	81	79	80	88	86
5	−1	−1	1	1	47	46	44	51	40	49	46
6	1	−1	1	−1	59	61	69	62	54	66	62
7	−1	1	1	−1	46	52	49	55	59	42	51
8	1	1	1	1	62	61	64	68	69	60	64
$\Sigma+$	296.00	258.50	278.17	252.00							
$\Sigma-$	204.67	242.17	222.50	248.67							
$\Sigma+/n_+$	74.00	64.63	69.54	63.00							
$\Sigma-/n_-$	51.17	60.54	55.63	62.17							
Effect	22.83	4.08	13.92	0.83							

Table 10.3 2^{4-1} Fractional Factorial Experiment

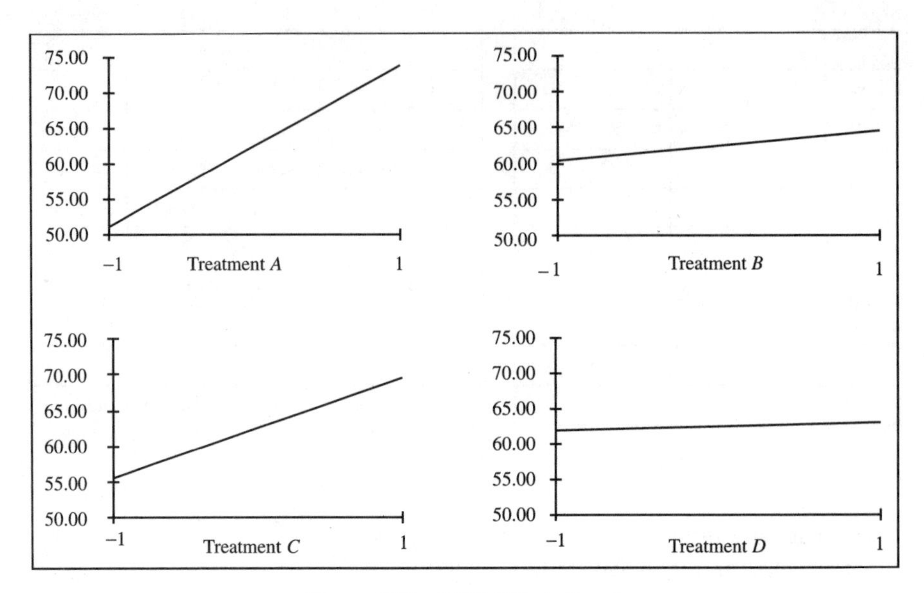

Figure 10.2 DOE Effects Graph

Well, how big is big? was the next question asked by the company owners. What am I going to gain from making these changes? First we needed to determine the statistical significance of the analysis using ANOVA. The resulting ANOVA appears as Table 10.4.

Based on the process analysis, the designed experiment, and the analysis of variance, the following determinations were made.

- Process elements 5, 6, and 7 can be accomplished off-line with information copies of the invoices or bills. This will reduce the average time to process an invoice and bill by 43 days.
- The executive review of all billing over $100K can be accomplished by PC. At the time the invoices are prepared and submitted to billing, they will be sent vial e-mail to the management staff. This will save 25 days of internal process time.

Sources of Variation	Sum of the Squares	Degrees of Freedom	Mean Squares	*F* Ratio	*F′* Critical .025	% Contribution
A	6256	1	6256	130.33	5.42	58
B	200	1	200	4.17	5.42	2
C	2324	1	2324	48.42	5.42	21
D	9	1	9	0.19	5.42	0
Error	2055	43	48			19
Total	10844	47				

Table 10.4 ANOVA Table

- Optimizing the remaining process will occur by making two changes indicated in the process analysis:
  - Billing directly to the customer on the invoice at time of shipment. (Treatment $A -1$)
  - Following up by telephone 45 and 60 days after shipment and invoice. (Treatment $B -1$)

The projected time to actually receive a payment (Receivables) was then calculated as follows:

$$\hat{y} = \frac{T}{n} + \left(A_{-1} - \frac{T}{n}\right) + \left(C_{-1} - \frac{T}{n}\right)$$

$$\hat{y} = 63 + (51.17 - 63) + (55.63 - 63)$$

$$y = 43.77 \approx 44 \text{ days}$$

Using the information from the process analysis and the designed experiment, the receivables age can be reduced from the current average of 200 days to 44 days. This is a significant reduction in the cost of money and an improvement in the company cash flow.

Case Study 2

The Rowan family of Cadiz, Alaska has been in the fishing business for three generations. They operate two medium-sized commercial fishing vessels and fish the Bering Sea and northern Pacific. Recently the fishing industry in the area has fallen on hard times. This is partially due to the depletion of fish in Alaskan waters, environmental changes, and competitive fishing by others. Some changes must be made in this business if it is to survive and thrive.

Although the boats are current and updated as far as equipment is concerned, a new fish-finding sonar is available that is much more accurate and has a greater range than the existing systems. It is, however, very expensive. There is also the option to start fishing outside Alaskan waters and venturing into the northern Pacific. There is also a difference in the types of fish that can be targeted by the boats. That provides us with three opportunities for a designed experiment.

- New Sonar System
- Fishing Area
- Fish

How to measure the results of this experiment is somewhat problematic. Different fish sell for different prices. The costs of fishing differ between equipment and fishing area. There is also a time differential: When fishing in local waters, the catch is returned and sold daily, whereas fishing in waters outside the immediate area will cause the catches to be sold weekly. It was decided that the measure would increase earnings before interest and taxes (EBIT). The break-even point for this small business is $25K per week. This is a typical financial measurement that takes into account expenses and costs. Since there may be some interaction between the new sonar, and the fish, a full factorial experiment was used. The following were to be the treatments and treatment levels.

A. Sonar	B. Fishing Area	C. Fish
+ New System	+ Local Alaskan Waters	+ Mid-Ocean Fish
− Current System	− Northern Pacific	− Bottom-Dwelling Fish

This resulted in a 2^3 full factorial experimental format as indicated in Table 10.5. The experiment was to take place over a four-week period with the EBIT being summarized weekly. This was the period of time the new sonar would be provided to the fishermen to use and make a decision concerning purchase. This will provide for four replicates to use in the analysis.

These experiments were then accomplished as indicated in Table 10.6.

The next step was to perform the experiment and take the data. As you can see from Table 10.7, there was a significant week-to-week between the EBIT in the fishing industry.

	Treatments						
	Main Effects			Interactions			
Runs	A	B	C	AB	AC	BC	ABC
1	−	−	−	+	+	+	−
2	+	−	−	−	−	+	+
3	−	+	−	−	+	−	+
4	+	+	−	+	−	−	−
5	−	−	+	+	−	−	+
6	+	−	+	−	+	−	−
7	−	+	+	−	−	+	−
8	+	+	+	+	+	+	+

Table 10.5 2^3 Full Factorial Experimental Format

	Treatments and Levels		
Runs	Sonar	Fishing Area	Fish Type
1	Current	North Pacific	Type B
2	New	North Pacific	Type B
3	Current	Local	Type B
4	New	Local	Type B
5	Current	North Pacific	Type A
6	New	North Pacific	Type A
7	Current	Local	Type A
8	New	Local	Type A

Table 10.6 Experimental Format for Executing the Experiment

Directly from the data from this format you can determine that in some cases the average EBIT was less then the break-even point. Those occurred in runs 2 and 5. Next the effect of the experimental design was calculated and graphed as indicated in Table 10.8 and Figure 10.3.

The next analysis that was performed was the ANOVA. The results of the ANOVA appear in Table 10.9.

There are three sets of information for us to review—the designed experiment and its associated effects, the graph of those effects, and the analysis of variance

	Treatments							Test Resutls				
	Main Effects			Interactions				Pull Strength				
Runs	A	B	C	AB	AC	BC	ABC	y_1	y_2	y_3	y_4	$\bar{y}$
1	−	−	−	+	+	+	−	22.10	27.30	26.50	31.20	26.78
2	+	−	−	−	−	+	+	19.20	21.60	18.70	16.50	19.00
3	−	+	−	−	+	−	+	33.60	37.20	31.00	29.40	32.80
4	+	+	−	+	−	−	−	44.50	49.50	32.00	36.70	40.68
5	−	−	+	+	−	−	+	27.50	19.60	18.40	20.10	21.40
6	+	−	+	−	+	−	−	47.60	44.30	32.50	40.00	41.10
7	−	+	+	−	−	+	−	56.20	33.50	42.10	36.80	42.15
8	+	+	+	+	+	+	+	47.50	32.00	38.20	42.50	40.05

Table 10.7 Experimental Format with Data Applied

	Treatments							Test Resutls				
	Main Effects			Interactions				Pull Strength				
Runs	A	B	C	AB	AC	BC	ABC	y_1	y_2	y_3	y_4	$\bar{y}$
1	–	–	–	+	+	+	–	22.10	27.30	26.50	31.20	26.78
2	+	–	–	–	–	+	+	19.20	21.60	18.70	16.50	19.00
3	–	+	–	–	+	–	+	33.60	37.20	31.00	29.40	32.80
4	+	+	–	+	–	–	–	44.50	49.50	32.00	36.70	40.68
5	–	–	+	+	–	–	+	27.50	19.60	18.40	20.10	21.40
6	+	–	+	–	+	–	–	47.60	44.30	32.50	40.00	41.10
7	–	+	+	–	–	+	–	56.20	33.50	42.10	36.80	42.15
8	+	+	+	+	+	+	+	47.50	32.00	38.20	42.50	40.05
$\Sigma +$	140.83	155.68	144.70	128.90	140.73	127.98	113.25					
$\Sigma -$	123.13	108.28	119.25	135.05	123.23	135.98	150.70					
$\Sigma +/n_+$	35.21	38.92	36.18	32.23	35.18	31.99	28.31					
$\Sigma -/n_-$	30.78	27.07	29.81	33.76	30.81	33.99	37.68					
Effect	4.43	11.85	6.36	-1.54	4.38	-2.00	-9.36					

Table 10.8 Experimental Format with Effects Calculated

(ANOVA). Based on the review of this information, several key determinations can be made.

The analysis of the DOE effects and the graphs indicate that the EBIT from fishing can be optimized as $A+$, $B+$, and $C+$. That is with the new sonar system an increase of \$4.43K over the current sonar system. The local fishing area is more profitable, yielding \$11.85K more than fishing in the north Pacific. And fish type A is 6.36K better than fishing for fish type B.

The ANOVA table is very interesting and makes some decisions very clear. First, the .01 level was selected for this analysis because the decision was very critical and the future profitability of the company depended on a correct decision. Therefore, the most stringent decision level was selected. The ANOVA indicated that main effects *B* and *C* were critical at this level. It also indicated that these two treatments made up for 43% of the potential variability in EBIT. None of the two level interactions were found to be critical.

However, the three-level interaction for *ABC* is critical. This is highly unusual, but it does occur. What this means is that controlling the three-level interaction can have a significant effect on EBIT. This means that although treatment *A* was not found to be critical as a stand-alone treatment, it must also be controlled in order to control the interaction of *ABC*. You control an interaction by controlling the main effect. Therefore, the result of optimizing the EBIT for fishing is:

- Treatment *A* as the current sonar system.
- Treatment *B* as the local Alaskan fishing waters.
- Treatment *C* as Fish type *A*.

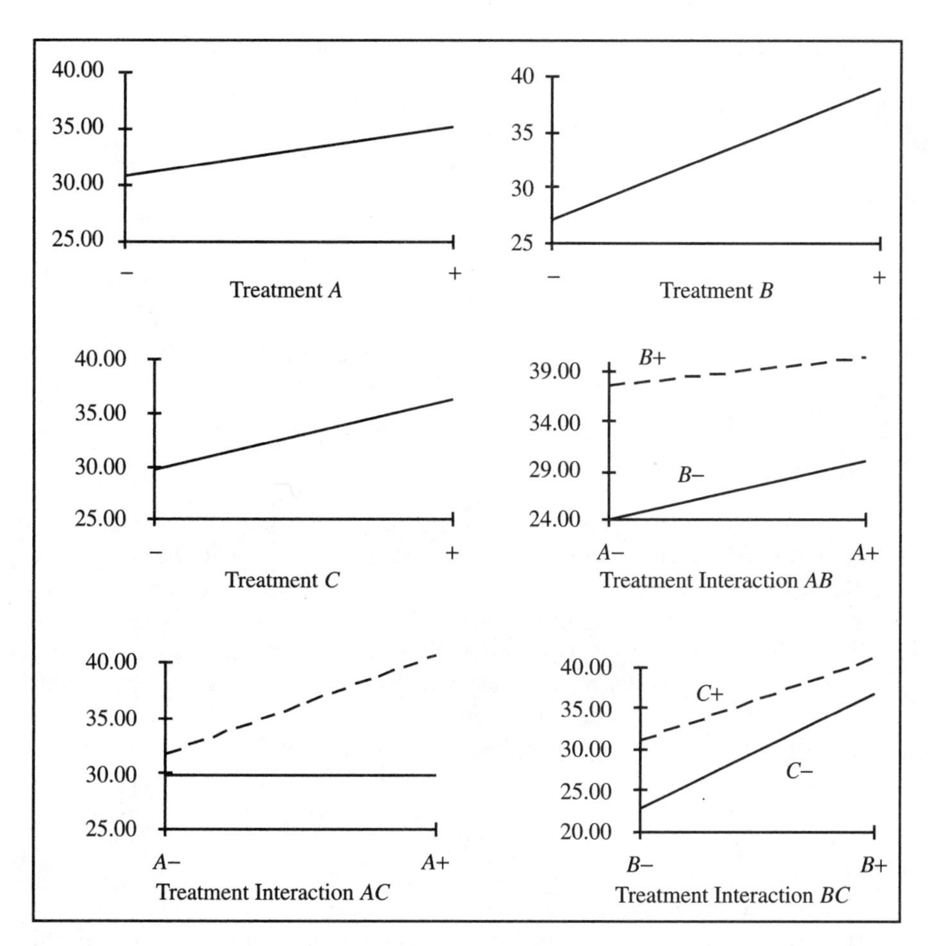

Figure 10.3 Graph of the Effects

Source of Variation	Sum of the Squares	Degrees of Freedom	Mean Squares	F Ratio	F' Critical .01	% Contribution
A	156.650	1	156.650	4.287	7.820	5
B	1123.380	1	1123.380	30.744	7.820	33
C	323.850	1	323.850	8.863	7.820	10
AB	18.910	1	18.910	0.518	7.820	1
AC	153.120	1	153.120	4.190	7.820	5
BC	183.820	1	183.820	5.031	7.820	5
ABC	555.150	1	555.150	15.193	7.820	16
Error	876.960	24	36.540			26
Total	3391.840	31				

Table 10.9 Completed ANOVA Table

This results in an estimate of the optimized function as follows:

$$\hat{y}_{\text{EBIT}} = \frac{T}{n} + \left(A_+ - \frac{T}{n}\right) + \left(B_+ - \frac{T}{n}\right) + \left(C_+ - \frac{T}{n}\right)$$

$$\hat{y} = 32.99 + (35.21 - 32.99) + (38.92 - 32.99) + (36.18 - 32.99)$$

$$y = 44.33K \approx 44\text{K EBIT}$$

The treatment for sonar (A) was found to be insignificant during the analysis of variance. It was included in the estimate for expected EBIT because it is a factor in controlling the interaction of treatments ABC.

Case Study 3

This case study concerns the cost of workman's compensation. The company is self-insured. They give a sum of money to a third party, who holds the money in an escrow fund and manages the workman's compensation program. When an employee is injured, a report is initiated by his/her supervisor. This report is logged by the safety office and then submitted to the "third party" for processing.

The third party evaluates the injury report to determine the potential liability and this is approved by a senior adjuster. At that time, the company provides the funding to the third party. When the medical charges are paid, they are charged against this liability amount established by the adjuster. The company was concerned that their worker compensation costs are excessive. In addition, their employees take legal action in more than 50% of the work-related disability compensation cases.

The company is working to reduce worker compensation liability, reduce the time its employees are off work for their injuries, and improve its safety record. The company would also like to improve workman's compensation so the employees would be less likely to feel they need to take legal action to collect (justified) workman's compensation. The normal process for workman's compensation is indicated in Figure 10.4.

A review of this flow chart and the data that are available from the reports indicated that the following input variables were contributing to workman's compensation. Since there were so many variables associated with departments, job classifications, and wage or salary levels, the measure was determined to be cost (Dollars). The analysis was conducted to characterize the cost of disability compensation using the following factors:

- Job Classification
  - White collar salaries
  - White collar wage
  - Blue collar salaries
  - Blue collar wage

- Process Being Performed
  - Production
  - Administrative
- Time of Injury
  - Morning
  - Afternoon
- Type of Injury
  - Back
  - Repetitive motion
  - Cuts
  - Burns
  - Other

Using existing data, a histogram of the cost of these factors was made. That histogram is shown in Figure 10.5.

A review of this histogram revealed some surprising facts. It had been assumed that the back injury to production workers was contributing most significantly to

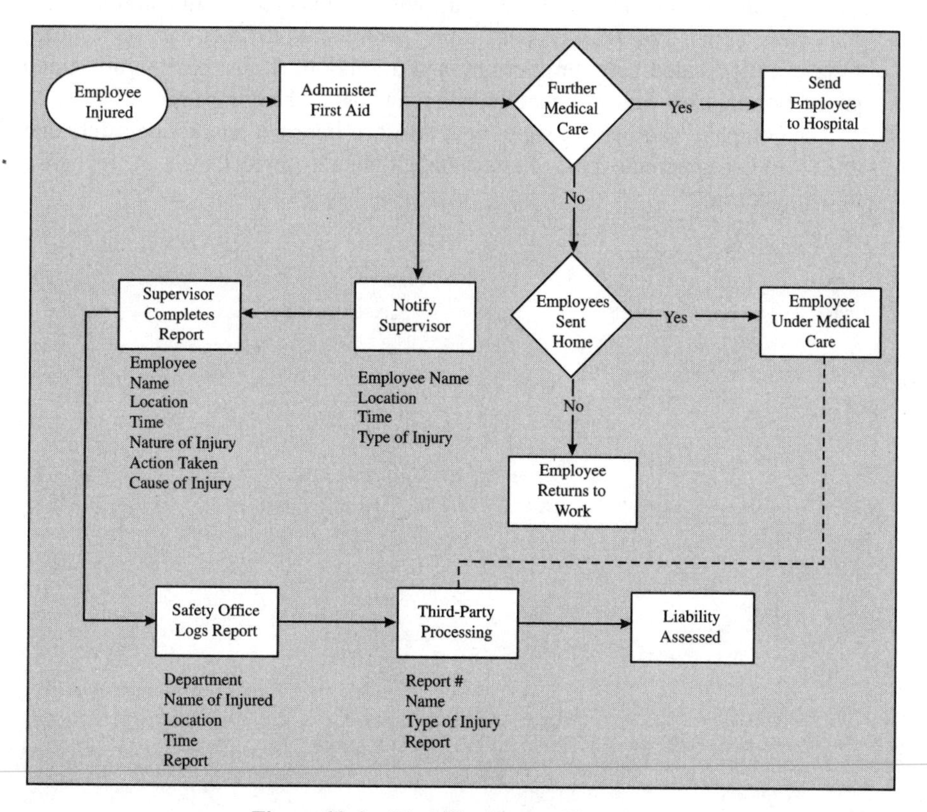

Figure 10.4 Disability Process Flow Chart

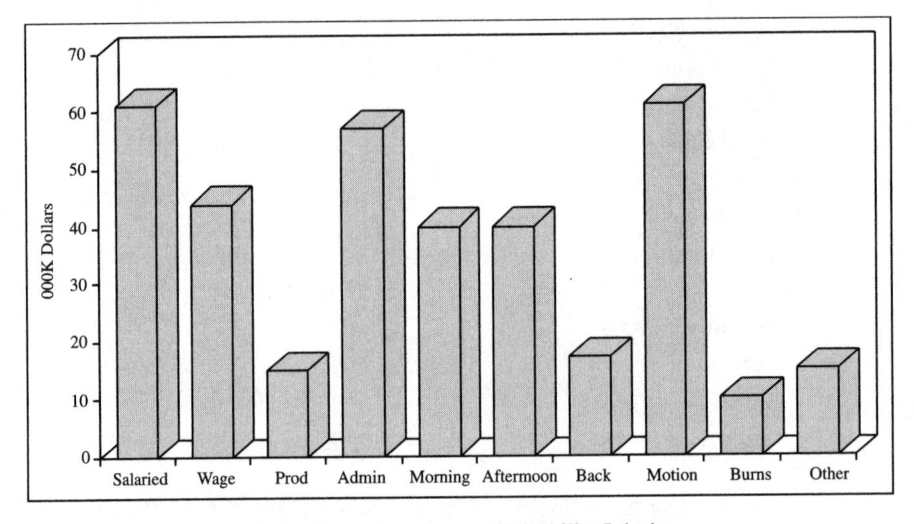

Figure 10.5 Histogram of Disability Injuries

the cost of disability. The histogram indicates that the most costly injuries to the company are salaried and wage hour white collar workers for motion injuries. This same trend appears for workers performing administrative functions. The injuries were evenly divided between morning and afternoon. Burns, back injuries, and production injuries were comparatively low in cost to the company.

The company had made a significant effort to improve production safety and that effort had apparently paid off. Now the problem to approach was the repetitive injury syndrome.

STATISTICAL TABLES

Subgroup Size	Factors for Standard Deviation	$\overline{X}$ & R Charts Factors For			$\overline{X}$ & S Charts Factors for			$\overline{X}$ & Rm Charts Factors for		
		$\overline{X}$ - Chart Control Limits	R - Chart Control Limits LCL	R - Chart Control Limits UCL	$\overline{X}$ - Chart Control Limits	S - Chart Control Limits LCL	S - Chart Control Limits UCL	$\overline{X}$ - Chart Control Limits	Rm - Chart Control Limits LCL	Rm - Chart Control Limits UCL
	d_2	A_2	D_3	D_4	A_3	B_3	B_4	E_2	D_3	D_4
2	1.128	1.880	0	3.267	2.659	0	3.267	2.260	0	3.267
3	1.693	1.023	0	2.574	1.954	0	2.568	1.772	0	2.574
4	2.059	0.729	0	2.282	1.628	0	2.266	1.457	0	2.282
5	2.326	0.577	0	2.114	1.427	0	2.089	1.290	0	2.114
6	2.534	0.483	0	2.004	1.287	0.030	1.970	1.184	0	2.004
7	2.704	0.419	0.076	1.924	1.182	0.118	1.882	1.011	0.076	1.924
8	2.847	0.373	0.136	1.864	1.099	0.185	1.815	1.054	0.136	1.864
9	2.970	0.337	0.184	1.816	1.032	0.239	1.761	1.010	0.184	1.816
10	3.078	0.308	0.223	1.777	0.975	0.284	1.716	0.975	0.223	1.777
11	3.173	0.285	0.256	1.744	0.927	0.321	1.679	0.946	0.256	1.744
12	3.258	0.266	0.283	1.717	0.886	0.354	1.646	0.921	0.283	1.717
13	3.336	0.249	0.307	1.693	0.850	0.382	1.618	0.899	0.307	1.693
14	3.407	0.235	0.328	1.672	0.817	0.406	1.594	0.881	0.328	1.672
15	3.472	0.223	0.347	1.653	0.789	0.428	1.572	0.864	0.347	1.653
16	3.532	0.212	0.363	1.637	0.763	0.448	1.552	0.849	0.363	1.637
17	3.588	0.203	0.378	1.622	0.739	0.466	1.534	0.836	0.378	1.622
18	3.640	0.194	0.391	1.608	0.718	0.482	1.518	0.824	0.391	1.608
19	3.689	0.187	0.403	1.597	0.698	0.497	1.503	0.813	0.403	1.597
20	3.735	0.180	0.415	1.585	0.680	0.510	1.490	0.803	0.415	1.585
21	3.778	0.173	0.425	1.575	0.663	0.523	1.477	0.794	0.425	1.575
22	3.819	0.167	0.434	1.566	0.647	0.534	1.466	0.785	0.434	1.566
23	3.858	0.162	0.443	1.577	0.633	0.545	1.455	0.778	0.443	1.557
24	3.895	0.157	0.451	1.548	0.619	0.555	1.445	0.770	0.451	1.548
25	3.931	0.153	0.459	1.541	0.606	0.565	1.435	0.763	0.459	1.541

Table A.1 Factors for Control Charts and Standard Deviation Formulas

df	*df* NUMERATOR										
DENOM	1−α	1	2	3	4	5	6	7	8	9	10
1	0.90	39.90	49.50	53.60	55.80	57.20	58.20	58.90	59.40	59.90	60.20
	0.95	161.00	200.00	216.00	225.00	230.00	234.00	237.00	239.00	241.00	242.00
2	0.90	8.53	9.00	9.16	9.24	9.29	9.33	9.35	9.37	9.38	9.39
	0.95	18.50	19.00	19.20	19.20	19.30	19.30	19.40	19.40	19.40	19.40
	0.99	98.50	99.00	99.20	99.20	99.30	99.30	99.40	99.40	99.40	99.40
3	0.90	5.54	5.46	5.39	5.34	5.31	5.28	5.27	5.25	5.24	5.23
	0.95	10.10	9.55	9.28	9.12	9.10	8.94	8.89	8.85	8.81	8.79
	0.99	34.10	30.80	29.50	28.70	28.20	27.90	27.70	27.50	27.30	27.20
4	0.90	4.54	4.32	4.19	4.11	4.05	4.01	3.98	3.95	3.94	3.92
	0.95	7.71	6.94	6.59	6.39	6.26	6.16	6.09	6.04	6.00	5.96
	0.99	21.20	18.00	16.70	16.00	15.50	15.20	15.00	14.80	14.70	14.50
5	0.90	4.06	3.78	3.62	3.52	3.45	3.40	3.37	3.34	3.32	3.30
	0.95	6.61	5.79	5.41	5.19	5.05	4.95	4.88	4.82	4.77	4.74
	0.99	16.30	13.30	12.10	11.40	11.00	10.70	10.50	10.30	10.20	10.10
6	0.90	3.78	3.46	3.29	3.18	3.11	3.05	3.01	2.98	2.96	2.94
	0.95	5.99	5.14	4.76	4.53	4.39	4.28	4.21	4.15	4.10	4.06
	0.99	13.70	10.90	9.78	9.15	8.75	8.47	8.26	8.10	7.98	7.87
7	0.90	3.59	3.26	3.07	2.96	2.88	2.83	2.78	2.75	2.72	2.70
	0.95	5.59	4.74	4.35	4.12	3.97	3.87	3.79	3.73	3.68	3.64
	0.99	12.20	9.55	8.45	7.85	7.46	7.19	6.99	6.84	6.72	6.62
8	0.90	3.46	3.11	2.92	2.81	2.73	2.67	2.62	2.59	2.56	2.54
	0.95	5.32	4.46	4.07	3.84	3.69	3.58	3.50	3.44	3.39	3.35
	0.99	11.30	8.65	7.59	8.01	6.63	6.37	6.18	6.03	5.91	5.81
9	0.90	3.36	3.01	2.81	2.69	2.61	2.55	2.51	2.47	2.44	2.42
	0.95	5.12	4.26	3.86	3.63	3.48	3.37	3.29	3.23	3.18	3.14
	0.99	10.60	8.02	6.99	6.42	6.06	5.80	5.61	5.47	5.35	5.26
10	0.90	3.28	2.92	2.73	2.61	2.52	2.46	2.41	2.38	2.35	2.32
	0.95	4.96	4.10	3.71	3.48	3.33	3.22	3.14	3.07	3.02	2.98
	0.99	10.00	7.56	6.55	5.99	5.64	5.39	5.20	5.06	4.94	4.85
11	0.90	3.23	2.86	2.66	2.54	2.45	2.39	2.34	2.3	2.27	2.25
	0.95	4.84	3.98	3.59	3.36	3.20	3.09	3.01	2.95	2.90	2.85
	0.99	9.65	7.21	6.22	5.67	5.32	5.07	4.89	4.75	4.63	5.54
12	0.90	3.18	2.81	2.61	2.48	2.39	2.33	2.26	2.24	2.21	2.19
	0.95	4.75	3.89	3.49	3.26	3.11	3.00	2.91	2.85	2.80	2.75
	0.99	9.33	6.93	5.95	5.41	5.06	4.82	4.64	4.50	4.39	4.30
15	0.90	3.07	2.70	2.49	2.36	2.27	2.21	2.16	2.12	2.09	2.06
	0.95	4.54	3.68	3.29	3.06	2.90	2.79	2.71	2.64	2.59	2.54
	0.99	8.68	6.36	5.42	4.89	4.56	4.32	4.14	4.00	3.89	3.80
20	0.90	2.97	2.59	2.38	2.25	2.16	2.09	2.04	2.00	1.96	1.94
	0.95	4.35	3.49	3.10	2.87	2.71	2.60	2.51	2.45	2.39	2.35
	0.99	8.10	5.85	4.94	4.43	4.10	3.87	3.70	3.56	3.46	3.37
30	0.90	2.88	2.49	2.28	2.14	2.05	1.98	1.93	1.88	1.85	1.82
	0.95	4.17	3.32	2.92	2.69	2.53	2.42	2.33	2.27	2.21	2.16
	0.99	7.56	5.39	4.51	4.02	3.70	3.47	3.30	3.17	3.07	2.98
40	0.90	2.84	2.44	2.23	2.09	2.00	1.93	1.87	1.83	1.79	1.76
	0.95	4.08	3.23	2.84	2.61	2.45	2.34	2.35	2.18	2.12	2.08
	0.99	7.31	5.18	4.31	3.83	3.51	3.29	3.12	2.99	2.89	2.80
60	0.90	2.79	2.39	2.18	2.04	1.95	1.87	1.82	1.77	1.74	1.71
	0.95	4.00	3.15	2.76	2.53	2.37	2.25	2.17	2.10	2.04	1.99
	0.99	7.08	4.98	4.13	3.65	3.34	3.12	2.95	2.82	2.72	2.63
120	0.90	2.75	2.35	2.13	1.99	1.90	1.82	1.77	1.72	1.68	1.65
	0.95	3.92	3.07	2.68	2.45	2.29	2.17	2.09	2.02	1.96	1.91
	0.99	6.85	4.79	3.95	3.48	3.17	2.96	2.79	2.66	2.56	2.47

Table A.2 *F* Test

(*continues*)

df	*df* NUMERATOR									
DENOM	1−α	12	15	20	30	40	50	60	100	120
1	0.90	60.70	61.20	61.70	62.30	62.50	62.70	62.80	63.00	63.10
	0.95	244.00	246.00	248.00	250.00	251.00	252.00	252.00	253.00	253.00
2	0.90	9.41	9.42	9.44	9.46	9.47	9.47	9.47	9.48	9.48
	0.95	19.40	19.40	19.40	19.50	19.50	19.50	19.50	9.48	19.50
	0.99	99.40	99.40	99.40	99.50	99.50	99.50	99.50	99.50	99.50
3	0.90	5.22	5.20	5.18	5.17	5.16	5.15	5.15	5.14	5.14
	0.95	8.74	8.70	8.66	8.62	8.59	8.58	8.57	8.55	8.55
	0.99	27.10	26.90	26.70	26.50	26.40	26.40	26.30	26.20	26.20
4	0.90	3.90	3.87	3.84	3.82	3.80	3.80	3.79	3.78	3.78
	0.95	5.91	5.86	5.80	5.75	5.72	5.70	5.69	5.56	5.66
	0.99	14.40	14.20	14.00	13.80	13.70	13.70	13.70	13.60	13.60
5	0.90	3.27	3.24	3.21	3.17	3.16	3.15	3.14	3.13	3.12
	0.95	4.68	4.62	4.56	4.50	4.46	4.44	4.43	4.41	4.40
	0.99	9.89	9.72	9.55	9.38	9.29	9.24	9.20	9.13	9.11
6	0.90	2.90	2.87	2.84	2.80	2.78	2.77	2.76	2.75	2.74
	0.95	4.00	3.94	3.87	3.81	3.77	3.75	3.74	3.71	3.70
	0.99	7.72	7.56	7.40	7.23	7.14	7.09	7.06	6.99	6.97
7	0.90	2.67	2.63	2.59	2.56	2.54	2.52	2.51	2.50	2.49
	0.95	3.57	3.51	3.44	3.38	3.34	3.32	3.30	3.27	3.27
	0.99	6.47	6.31	6.16	5.99	5.91	5.86	5.82	5.75	5.74
8	0.90	2.50	2.46	2.42	2.38	2.36	2.35	2.34	2.32	2.32
	0.95	3.28	3.22	3.15	3.08	3.04	3.02	3.01	2.97	2.97
	0.99	5.67	5.52	5.36	5.20	5.12	5.07	5.03	4.96	4.95
9	0.90	2.38	2.34	2.30	2.25	2.23	2.22	2.21	2.19	2.18
	0.95	3.07	3.01	2.94	2.86	2.83	2.80	2.79	2.76	2.75
	0.99	5.11	4.96	4.81	4.65	4.57	5.52	4.48	4.42	4.40
10	0.90	2.28	2.24	2.20	2.16	2.13	2.12	2.11	2.09	2.08
	0.95	2.91	2.85	2.77	2.70	2.66	2.64	2.62	2.59	2.58
	0.99	4.71	4.56	4.41	4.25	4.17	4.12	4.08	4.01	4.00
15	0.90	2.02	1.97	1.92	1.87	1.85	1.83	1.82	1.79	1.79
	0.95	2.48	2.40	2.33	2.25	2.20	2.18	2.16	2.12	2.11
	0.99	3.67	3.52	3.37	3.21	3.13	3.08	3.05	2.98	2.96
20	0.90	1.89	1.84	1.79	1.74	1.71	1.69	1.68	1.65	1.64
	0.95	2.28	2.20	2.12	2.04	1.99	1.97	1.95	1.91	1.90
	0.99	3.23	3.09	2.94	2.78	2.69	2.64	2.61	2.54	2.52
30	0.90	1.77	1.72	1.67	1.61	1.57	1.55	1.54	1.51	1.50
	0.95	2.09	2.01	1.93	1.84	1.79	1.76	1.74	1.70	1.68
	0.99	2.84	2.70	2.55	2.39	2.30	2.25	2.21	2.13	2.11
40	0.90	1.71	1.66	1.61	1.54	1.51	1.48	1.47	1.43	1.42
	0.95	2.00	1.92	1.84	1.74	1.69	1.66	1.64	1.59	1.58
	0.99	2.66	2.52	2.37	2.20	2.11	2.06	2.02	1.94	1.92
60	0.90	1.66	1.60	1.54	1.48	1.44	1.41	1.40	1.36	1.35
	0.95	1.92	1.84	1.75	1.65	1.59	1.56	1.53	1.48	1.47
	0.99	2.50	2.35	2.20	2.03	1.94	1.88	1.84	1.75	1.73
120	0.90	1.60	1.55	1.48	1.41	1.37	1.34	1.32	1.27	1.26
	0.95	1.83	1.75	1.66	1.55	1.50	1.46	1.43	1.37	1.35
	0.99	2.34	2.19	2.03	1.86	1.76	1.70	1.66	1.56	1.53

Table A.2 *F* Test (*Continued*)

	Treatments		
Runs	A	B	AB
1	–	–	+
2	+	–	–
3	–	+	–
4	+	+	+

Table A.3 2^2 Full Factorial Design Format

	Treatments						
Runs	A	B	C	AB	AC	BC	ABC
1	–	–	–	+	+	+	–
2	+	–	–	–	–	+	+
3	–	+	–	–	+	–	+
4	+	+	–	+	–	–	–
5	–	–	+	+	–	–	+
6	+	–	+	–	+	–	–
7	–	+	+	–	–	+	–
8	+	+	+	+	+	+	+

Table A.4 2^3 Full Factorial Design Formats

	Treatments														
Runs	A	B	C	D	AB	AC	AD	BC	BD	CD	ABC	ABD	ACD	BCD	ABCD
1	−	−	−	−	+	+	+	+	+	+	−	−	−	−	+
2	+	−	−	−	−	−	−	+	+	+	+	+	+	−	−
3	−	+	−	−	+	+	+	−	−	+	+	+	−	+	−
4	+	+	−	−	+	−	−	−	−	+	−	−	+	+	+
5	−	−	+	−	+	−	+	−	+	−	+	−	+	+	−
6	+	−	+	−	−	+	−	−	+	−	−	+	−	+	+
7	−	+	+	−	−	−	+	+	−	−	−	+	+	−	+
8	+	+	+	−	+	+	−	+	−	−	+	−	−	−	−
9	−	−	−	+	+	+	−	+	−	−	−	+	+	+	−
10	+	−	−	+	−	−	+	+	−	−	+	−	−	+	+
11	−	+	−	+	−	+	−	−	+	−	+	−	+	−	+
12	+	+	−	+	+	−	+	−	+	−	−	+	−	−	−
13	−	−	+	+	+	−	−	−	−	+	+	+	−	−	+
14	+	−	+	+	−	+	+	−	−	+	−	−	+	−	−
15	−	+	+	+	−	−	−	+	+	+	−	−	−	+	−
16	+	+	+	+	+	+	+	+	+	+	+	+	+	+	+

Table A.5 2^4 Full Factorial Design Formats

	Treatments														
Runs	A	B	C	D	E	AB	AC	AD	AE	BC	BD	BE	CD	CE	DE
1	−	−	−	−	−	+	+	+	+	+	+	+	+	+	+
2	+	−	−	−	−	−	−	−	−	+	+	+	+	+	+
3	−	+	−	−	−	−	+	+	+	−	−	−	+	+	+
4	+	+	−	−	−	+	−	−	−	−	−	−	+	+	+
5	−	−	+	−	−	+	−	+	+	−	+	+	−	−	+
6	+	−	+	−	−	−	+	−	−	−	+	+	−	−	+
7	−	+	+	−	−	−	−	+	+	+	−	−	−	−	+
8	+	+	+	−	−	+	+	−	−	+	−	−	−	−	+
9	−	−	−	+	−	+	+	−	+	+	−	+	−	+	−
10	+	−	−	+	−	−	−	+	−	+	−	+	−	+	−
11	−	+	−	+	−	−	+	−	+	−	+	−	−	+	−
12	+	+	−	+	−	+	−	+	−	−	+	−	−	+	−
13	−	−	+	+	−	+	−	−	+	−	−	+	+	−	−
14	+	−	+	+	−	−	+	+	−	−	−	+	+	−	−
15	−	+	+	+	−	−	−	−	+	+	+	−	+	−	−
16	+	+	+	+	−	+	+	+	−	+	+	−	+	−	−
17	−	−	−	−	+	+	+	+	−	+	+	−	+	−	−
18	+	−	−	−	+	−	−	−	+	+	+	−	+	−	−
19	−	+	−	−	+	−	+	+	−	−	−	+	+	−	−
20	+	+	−	−	+	+	−	−	+	−	−	+	+	−	−
21	−	−	+	−	+	+	−	+	−	−	+	−	−	+	−
22	+	−	+	−	+	−	+	−	+	−	+	−	−	+	−
23	−	+	+	−	+	−	−	+	−	+	−	+	−	+	−
24	+	+	+	−	+	+	+	−	+	+	−	+	−	+	−
25	−	−	−	+	+	+	+	+	−	+	−	−	−	−	+
26	+	−	−	+	+	−	−	+	+	+	−	−	−	−	+
27	−	+	−	+	+	−	+	−	−	−	+	+	−	−	+
28	+	+	−	+	+	+	−	+	+	−	+	+	−	−	+
29	−	−	+	+	+	+	−	−	−	−	−	−	+	+	+
30	+	−	+	+	+	−	+	+	+	−	−	−	+	+	+
31	−	+	+	+	+	−	−	−	−	+	+	+	+	+	+
32	+	+	+	+	+	+	+	+	+	+	+	+	+	+	+

Table A.6 2^5 Full Factorial Design Formats

(*continues*)

	Treatments															
Runs	ABC	ABD	ABE	ACD	ACE	ADE	BCD	BCE	BDE	CDE	ABCD	ABCE	ABDE	ACDE	BCDE	ABCDE
1	−	−	−	−	−	−	−	−	−	−	+	+	+	+	+	−
2	+	+	+	+	+	+	−	−	−	−	−	−	−	−	+	+
3	+	+	+	−	−	−	+	+	+	−	−	−	−	+	−	+
4	−	−	−	+	+	+	+	+	+	−	+	+	+	−	−	−
5	+	−	−	+	+	−	+	+	−	+	−	−	+	−	−	+
6	−	+	+	−	−	+	+	+	−	+	+	+	−	+	−	−
7	−	+	+	+	+	−	−	−	+	+	+	+	−	−	+	−
8	+	−	−	−	−	+	−	−	+	+	−	−	+	+	+	+
9	−	+	−	+	−	+	+	−	+	+	−	+	−	−	−	+
10	+	−	+	−	+	−	+	−	+	+	+	−	+	+	−	−
11	+	−	+	+	−	+	−	+	−	+	+	−	+	−	+	−
12	−	+	−	−	+	−	−	+	−	+	−	+	−	+	+	+
13	+	+	−	−	+	+	−	+	+	−	+	−	−	+	+	−
14	−	−	+	+	−	−	−	+	+	−	−	+	+	−	+	+
15	−	−	+	−	+	+	+	−	−	−	−	+	+	+	−	+
16	+	+	−	+	−	−	+	−	−	−	+	−	−	−	−	−
17	−	−	+	−	+	+	−	+	+	+	+	−	−	−	−	+
18	+	+	−	+	−	−	−	+	+	+	−	+	+	+	−	−
19	+	+	−	−	+	+	+	−	−	+	−	+	+	−	+	−
20	−	−	+	+	−	−	+	−	−	+	+	−	−	+	+	+
21	+	−	+	+	−	+	+	−	+	−	−	+	−	+	+	−
22	−	+	−	−	+	−	+	−	+	−	+	−	+	−	+	+
23	−	+	−	+	−	+	−	+	−	−	+	−	+	+	−	+
24	+	−	+	−	+	−	−	+	−	−	−	+	−	−	−	−
25	−	+	+	+	+	+	+	+	−	−	−	−	+	+	+	−
26	+	−	−	−	−	+	+	+	−	−	+	+	−	−	+	+
27	+	−	−	+	+	−	−	−	+	−	+	+	−	+	−	+
28	−	+	+	−	−	+	−	−	+	−	−	−	+	−	−	−
29	+	+	+	−	−	−	−	−	−	+	+	+	+	−	−	+
30	−	−	−	+	+	+	−	−	−	+	−	−	−	+	−	−
31	−	−	−	−	−	−	+	+	+	+	−	−	−	−	+	−
32	+	+	+	+	+	+	+	+	+	+	+	+	+	+	+	+

Table A.6 2^5 Full Factorial Design Formats (*Continued*)

	Treatments									
RUN	A	B	C	D	E	F	AB	AC	AD	AE
1	–	–	–	–	–	–	+	+	+	+
2	+	–	–	–	–	–	–	–	–	–
3	–	+	–	–	–	–	–	+	+	+
4	+	+	–	–	–	–	+	–	–	–
5	–	–	+	–	–	–	+	–	+	+
6	+	–	+	–	–	–	–	+	–	–
7	–	+	+	–	–	–	–	–	+	+
8	+	+	+	–	–	–	+	+	–	–
9	–	–	–	+	–	–	+	+	–	+
10	+	–	–	+	–	–	–	–	+	–
11	–	+	–	+	–	–	–	+	–	+
12	+	+	–	+	–	–	+	–	+	–
13	–	–	+	+	–	–	+	–	–	+
14	+	–	+	+	–	–	–	+	+	–
15	–	+	+	+	–	–	–	–	–	+
16	+	+	+	+	–	–	+	+	+	–
17	–	–	–	–	+	–	+	+	+	–
18	+	–	–	–	+	–	–	–	–	+
19	–	+	–	–	+	–	–	+	+	–
20	+	+	–	–	+	–	+	–	–	+
21	–	–	+	–	+	–	+	–	+	–
22	+	–	+	–	+	–	–	+	–	+
23	–	+	+	–	+	–	–	–	+	–
24	+	+	+	–	+	–	+	+	–	+
25	–	–	–	+	+	–	+	+	–	–
26	+	–	–	+	+	–	–	–	+	+
27	–	+	–	+	+	–	–	+	–	–
28	+	+	–	+	+	–	+	–	+	+
29	–	–	+	+	+	–	+	–	–	–
30	+	–	+	+	+	–	–	+	+	+
31	–	+	+	+	+	–	–	–	–	–
32	+	+	+	+	+	+	+	+	+	+

Table A.7 2^6 Full Factorial

(*continues*)

	Treatments									
RUN	A	B	C	D	E	F	AB	AC	AD	AE
33	–	–	–	–	–	+	+	+	+	+
34	+	–	–	–	–	+	–	–	–	–
35	–	+	–	–	–	+	–	+	+	+
36	+	+	–	–	–	+	+	–	–	–
37	–	–	+	–	–	+	+	–	+	+
38	+	–	+	–	–	+	–	+	–	–
39	–	+	+	–	–	+	–	–	+	+
40	+	+	+	–	–	+	+	+	–	–
41	–	–	–	+	–	+	+	+	–	+
42	+	–	–	+	–	+	–	–	+	–
43	–	+	–	+	–	+	–	+	–	+
44	+	+	–	+	–	+	+	–	+	–
45	–	–	+	+	–	+	+	–	–	+
46	+	–	+	+	–	+	–	+	+	–
47	–	+	+	+	–	+	–	–	–	+
48	+	+	+	+	–	+	+	+	+	–
49	–	–	–	–	+	+	+	+	+	–
50	+	–	–	–	+	+	–	–	–	+
51	–	+	–	–	+	+	–	+	+	–
52	+	+	–	–	+	+	+	–	–	+
53	–	–	+	–	+	+	+	–	–	–
54	+	–	+	–	+	+	–	+	–	+
55	–	+	+	–	+	+	–	–	+	–
56	+	+	+	–	+	+	+	+	–	+
57	–	–	–	+	+	+	+	+	–	–
58	+	–	–	+	+	+	–	–	+	+
59	–	+	–	+	+	+	–	+	–	–
60	+	+	–	+	+	+	+	–	+	+
61	–	–	+	+	+	+	+	–	–	–
62	+	–	+	+	+	+	–	+	+	+
63	–	+	+	+	+	+	–	–	–	–
64	+	+	+	+	+	+	+	+	+	+

Table A.7 2^6 Full Factorial (*Continued*)

	Treatments										
RUN	AF	BC	BD	BE	BF	CD	CE	CF	DE	DF	EF
1	+	+	+	+	+	+	+	+	+	+	+
2	−	+	+	+	+	+	+	+	+	+	+
3	+	−	−	−	−	+	+	+	+	+	+
4	−	−	−	−	−	+	+	+	+	+	+
5	+	−	+	+	+	−	−	−	+	+	+
6	−	−	+	+	+	−	−	−	+	+	+
7	+	+	−	−	−	−	−	−	+	+	+
8	−	+	−	−	−	−	−	−	+	+	+
9	+	+	−	+	+	−	+	+	−	−	+
10	−	+	−	+	+	−	+	+	−	−	+
11	+	−	+	−	−	−	+	+	−	−	+
12	−	−	+	−	−	−	+	+	−	−	+
13	+	−	−	+	+	+	−	−	−	−	+
14	−	−	−	+	+	+	−	−	−	−	+
15	+	+	+	−	−	+	−	−	−	−	+
16	−	+	+	−	−	+	−	−	−	−	+
17	+	+	+	−	+	+	−	+	−	+	−
18	−	+	+	−	+	+	−	+	−	+	−
19	+	−	−	+	−	+	−	+	−	+	−
20	−	−	−	+	−	+	−	+	−	+	−
21	+	−	+	−	+	−	+	−	−	+	−
22	−	−	+	−	+	−	+	−	−	+	−
23	+	+	−	+	−	−	+	−	−	+	−
24	−	+	−	+	−	−	+	−	−	+	−
25	+	+	−	−	+	−	−	+	+	−	−
26	−	+	−	−	+	−	−	+	+	−	−
27	+	−	+	+	−	−	−	+	+	−	−
28	−	−	+	+	−	−	−	+	+	−	−
29	+	−	−	−	+	+	+	−	+	−	−
30	−	−	−	−	+	+	+	−	+	−	−
31	+	+	+	+	−	+	+	−	+	−	−
32	+	+	+	+	+	+	+	+	+	+	−

Table A.7 2^6 Full Factorial (*Continued*)

	Treatments										
RUN	AF	BC	BD	BE	BF	CD	CE	CF	DE	DF	EF
33	−	+	+	+	−	+	+	−	+	−	−
34	+	+	+	+	−	+	+	−	+	−	−
35	−	−	−	−	+	+	+	−	+	−	−
36	+	−	−	−	+	+	+	−	+	−	−
37	−	−	+	+	−	−	−	+	+	−	−
38	+	−	+	+	−	−	−	+	+	−	−
39	−	+	−	−	+	−	−	+	+	−	−
40	+	+	−	−	+	−	−	+	+	−	−
41	−	+	−	+	−	−	+	−	−	+	−
42	+	+	−	+	−	−	+	−	−	+	−
43	−	−	+	−	+	−	+	−	−	+	−
44	+	−	+	−	+	−	+	−	−	+	−
45	−	−	−	+	−	+	−	+	−	+	−
46	+	−	−	+	−	+	−	+	−	+	−
47	−	+	+	−	+	+	−	+	−	+	−
48	+	+	+	−	+	+	−	+	−	+	−
49	−	+	+	−	−	+	−	−	−	−	+
50	+	+	+	−	−	+	−	−	−	−	+
51	−	−	−	+	+	+	−	−	−	−	+
52	+	−	−	+	+	+	−	−	−	−	+
53	−	−	+	−	−	−	+	+	−	−	+
54	+	−	+	−	−	−	+	+	−	−	+
55	−	+	−	+	+	−	+	+	−	−	+
56	+	+	−	+	+	−	+	+	−	−	+
57	−	+	−	−	−	−	−	−	+	+	+
58	+	+	−	−	−	−	−	−	+	+	+
59	−	−	+	+	+	−	−	−	+	+	+
60	+	−	+	+	+	−	−	−	+	+	+
61	−	−	−	−	−	+	+	+	+	+	+
62	+	−	−	−	−	+	+	+	+	+	+
63	−	+	+	+	+	+	+	+	+	+	+
64	+	+	+	+	+	+	+	+	+	+	+

Table A.7 2^6 Full Factorial (*Continued*)

	Treatments									
RUN	ABC	ABD	ABE	ABF	ACD	ACE	ACF	ADE	ADF	AEF
1	−	−	−	−	−	−	−	−	−	−
2	+	+	+	+	+	+	+	+	+	+
3	+	+	+	+	−	−	−	−	−	−
4	−	−	−	−	+	+	+	+	+	+
5	+	−	−	−	+	+	+	−	−	−
6	−	+	+	+	−	−	−	+	+	+
7	−	+	+	+	+	+	+	−	−	−
8	+	−	−	−	−	−	−	+	+	+
9	−	+	−	−	+	−	−	+	+	−
10	+	−	+	+	−	+	+	−	−	+
11	+	−	+	+	+	−	−	+	+	−
12	−	+	−	−	−	+	+	−	−	+
13	+	+	−	−	−	+	+	+	+	−
14	−	−	+	+	+	−	−	−	−	+
15	−	−	+	+	−	+	+	+	+	−
16	+	+	−	−	+	−	−	−	−	+
17	−	−	+	−	−	+	−	+	−	+
18	+	+	−	+	+	−	+	−	+	−
19	+	+	−	+	−	+	−	+	−	+
20	−	−	+	−	+	−	+	−	+	−
21	+	−	+	−	+	−	+	+	−	+
22	−	+	−	+	−	+	−	−	+	−
23	−	+	−	+	+	−	+	+	−	+
24	+	−	+	−	−	+	−	−	+	−
25	−	+	+	−	+	+	−	−	+	+
26	+	−	−	+	−	−	+	+	−	−
27	+	−	−	+	+	+	−	−	+	+
28	−	+	+	−	−	−	+	+	−	−
29	+	+	+	−	−	−	+	−	+	+
30	−	−	−	+	+	+	−	+	−	−
31	−	−	−	+	−	−	+	−	+	+
32	+	+	+	+	+	+	+	+	+	+

Table A.7 2^6 Full Factorial (*Continued*)

	Treatments									
RUN	ABC	ABD	ABE	ABF	ACD	ACE	ACF	ADE	ADF	AEF
33	−	−	−	+	−	−	+	−	+	+
34	+	+	+	−	+	+	−	+	−	−
35	+	+	+	−	−	−	+	−	+	+
36	−	−	−	+	+	+	−	+	−	−
37	+	−	−	+	+	+	−	−	+	+
38	−	+	+	−	−	−	+	+	−	−
39	−	+	+	−	+	+	−	−	+	+
40	+	−	−	+	−	−	+	+	−	−
41	−	+	−	+	+	−	+	+	−	+
42	+	−	+	−	−	+	−	−	+	−
43	+	−	+	−	+	−	+	+	−	+
44	−	+	−	+	−	+	−	−	+	−
45	+	+	−	+	−	+	−	+	−	+
46	−	−	+	−	+	−	+	−	+	−
47	−	−	+	−	−	+	−	+	−	+
48	+	+	−	+	+	−	+	−	+	−
49	−	−	+	+	−	+	+	+	+	−
50	+	+	−	−	+	−	−	−	−	+
51	+	+	−	−	−	+	+	+	+	−
52	−	−	+	+	+	−	−	−	−	+
53	+	−	+	+	+	−	−	+	+	−
54	−	+	−	−	−	+	+	−	−	+
55	−	+	−	−	+	−	−	+	+	−
56	+	−	+	+	−	+	+	−	−	+
57	−	+	+	+	+	+	+	−	−	−
58	+	−	−	−	−	−	−	+	+	+
59	+	−	−	−	+	+	+	−	−	−
60	−	+	+	+	−	−	−	+	+	+
61	+	+	+	+	−	−	−	−	−	−
62	−	−	−	−	+	+	+	+	+	+
63	−	−	−	−	−	−	−	−	−	−
64	+	+	+	+	+	+	+	+	+	+

Table A.7 2^6 Full Factorial (*Continued*)

	Treatments										
RUN	BCD	BCE	BCF	BDE	BDF	BEF	CDE	CDF	CEF	DEF	ABCD
1	–	–	–	–	–	–	–	–	–	–	+
2	–	–	–	–	–	–	–	–	–	–	–
3	+	+	+	+	+	+	–	–	–	–	–
4	+	+	+	+	+	+	–	–	–	–	+
5	+	+	+	–	–	–	+	+	+	–	–
6	+	+	+	–	–	–	+	+	+	–	+
7	–	–	–	+	+	+	+	+	+	–	+
8	–	–	–	+	+	+	+	+	+	–	–
9	+	–	–	+	+	–	+	+	–	+	–
10	+	–	–	+	+	–	+	+	–	+	+
11	–	+	+	–	–	+	+	+	–	+	+
12	–	+	+	–	–	+	+	+	–	+	–
13	–	+	+	+	+	–	–	–	+	+	+
14	–	+	+	+	+	–	–	–	+	+	–
15	+	–	–	–	–	+	–	–	+	+	–
16	+	–	–	–	–	+	–	–	+	+	+
17	–	+	–	+	–	+	+	–	+	+	+
18	–	+	–	+	–	+	+	–	+	+	–
19	+	–	+	–	+	–	+	–	+	+	–
20	+	–	+	–	+	–	+	–	+	+	+
21	+	–	+	+	–	+	–	+	–	+	–
22	+	–	+	+	–	+	–	+	–	+	+
23	–	+	–	–	+	–	–	+	–	+	+
24	–	+	–	–	+	–	–	+	–	+	–
25	+	+	–	–	+	+	–	+	+	–	–
26	+	+	–	–	+	+	–	+	+	–	+
27	–	–	+	+	–	–	–	+	+	–	+
28	–	–	+	+	–	–	–	+	+	–	–
29	–	–	+	–	+	+	+	–	–	–	+
30	–	–	+	–	+	+	+	–	–	–	–
31	+	+	–	+	–	–	+	–	–	–	–
32	+	+	+	+	+	+	+	+	+	+	+

Table A.7 2^6 Full Factorial (*Continued*)

RUN	Treatments										
	BCD	BCE	BCF	BDE	BDF	BEF	CDE	CDF	CEF	DEF	ABCD
33	−	−	+	−	+	+	−	+	+	+	+
34	−	−	+	−	+	+	−	+	+	+	−
35	+	+	−	+	−	−	−	+	+	+	−
36	+	+	−	+	−	−	−	+	+	+	+
37	+	+	−	−	+	+	+	−	−	+	−
38	+	+	−	−	+	+	+	−	−	+	+
39	−	−	+	+	−	−	+	−	−	+	+
40	−	−	+	+	−	−	+	−	−	+	−
41	+	−	+	+	−	+	+	−	+	−	−
42	+	−	+	+	−	+	+	−	+	−	+
43	−	+	−	−	+	−	+	−	+	−	+
44	−	+	−	−	+	−	+	−	+	−	−
45	−	+	−	+	−	+	−	+	−	−	+
46	−	+	−	+	−	+	−	+	−	−	−
47	+	−	+	−	+	−	−	+	−	−	−
48	+	−	+	−	+	−	−	+	−	−	+
49	−	+	+	+	+	−	+	+	−	−	+
50	−	+	+	+	+	−	+	+	−	−	−
51	+	−	−	−	−	+	+	+	−	−	−
52	+	−	−	−	−	+	+	+	−	−	+
53	+	−	−	+	+	−	−	−	+	−	−
54	+	−	−	+	+	−	−	−	+	−	+
55	−	+	+	−	−	+	−	−	+	−	+
56	−	+	+	−	−	+	−	−	+	−	−
57	+	+	+	−	−	−	−	−	−	+	−
58	+	+	+	−	−	−	−	−	−	+	+
59	−	−	−	+	+	+	−	−	−	+	+
60	−	−	−	+	+	+	−	−	−	+	−
61	−	−	−	−	−	−	+	+	+	+	+
62	−	−	−	−	−	−	+	+	+	+	−
63	+	+	+	+	+	+	+	+	+	+	−
64	+	+	+	+	+	+	+	+	+	+	+

Table A.7 2^6 Full Factorial (*Continued*)

	Treatments									
RUN	ABCE	ABCF	ABDE	ABDF	ABEF	ACDE	ACDF	ACEF	ADEF	BCDE
1	+	+	+	+	+	+	+	+	+	+
2	−	−	−	−	−	−	−	−	−	+
3	−	−	−	−	−	+	+	+	+	−
4	+	+	+	+	+	−	−	−	−	−
5	−	−	+	+	+	−	−	−	+	−
6	+	+	−	−	−	+	+	+	−	−
7	+	+	−	−	−	−	−	−	+	+
8	−	−	+	+	+	+	+	+	−	+
9	+	+	−	−	+	−	−	+	−	−
10	−	−	+	+	−	+	+	−	+	−
11	−	−	+	+	−	−	−	+	−	+
12	+	+	−	−	+	+	+	−	+	+
13	−	−	−	−	+	+	+	−	−	+
14	+	+	+	+	−	−	−	+	+	+
15	+	+	+	+	−	+	+	−	−	−
16	−	−	−	−	+	−	−	+	+	−
17	−	+	−	+	−	−	+	−	−	−
18	+	−	+	−	+	+	−	+	+	−
19	+	−	+	−	+	−	+	−	−	+
20	−	+	−	+	−	+	−	+	+	+
21	+	−	−	+	−	+	−	+	−	+
22	−	+	+	−	+	−	+	−	+	+
23	−	+	+	−	+	+	−	+	−	−
24	+	−	−	+	−	−	+	−	+	−
25	−	+	+	−	−	+	−	−	+	+
26	+	−	−	+	+	−	+	+	−	+
27	+	−	−	+	+	+	−	−	+	+
28	−	+	+	−	−	−	+	+	−	−
29	+	−	+	−	−	−	+	+	+	−
30	−	+	−	+	+	+	−	−	−	−
31	−	+	−	+	+	−	+	+	+	−
32	+	+	+	+	+	+	+	+	+	+

Table A.7 2^6 Full Factorial (*Continued*)

	Treatments									
RUN	ABCE	ABCF	ABDE	ABDF	ABEF	ACDE	ACDF	ACEF	ADEF	BCDE
33	+	−	+	−	−	+	−	−	−	+
34	−	+	−	+	+	−	+	+	+	+
35	−	+	−	+	+	+	−	−	−	−
36	+	−	+	−	−	−	+	+	+	−
37	−	+	+	−	−	−	+	+	−	−
38	+	−	−	+	+	+	−	−	+	−
39	+	−	−	+	+	−	+	+	−	+
40	−	+	+	−	−	+	−	−	+	+
41	+	−	−	+	−	−	+	−	+	−
42	−	+	+	−	+	+	−	+	−	−
43	−	+	+	−	+	−	+	−	+	+
44	+	−	−	+	−	+	−	+	−	+
45	−	+	−	+	−	+	−	+	+	+
46	+	−	+	−	+	−	+	−	−	+
47	+	−	+	−	+	+	−	+	+	−
48	−	+	−	+	−	−	+	−	−	−
49	−	−	−	−	+	−	−	+	+	−
50	+	+	+	+	−	+	+	−	−	−
51	+	+	+	+	−	−	−	+	+	+
52	−	−	−	−	+	+	+	−	−	+
53	+	+	−	−	+	+	+	−	+	+
54	−	−	+	+	−	−	−	+	−	+
55	−	−	+	+	−	+	+	−	+	−
56	+	+	−	−	+	−	−	+	−	−
57	−	−	+	+	+	+	+	+	−	+
58	+	+	−	−	−	−	−	−	+	+
59	+	+	−	−	−	+	+	+	−	−
60	−	−	+	+	+	−	−	−	+	−
61	+	+	+	+	+	−	−	−	−	−
62	−	−	−	−	−	+	+	+	+	−
63	−	−	−	−	−	−	−	−	−	+
64	+	+	+	+	+	+	+	+	+	+

Table A.7 2^6 Full Factorial (*Continued*)

	Treatments										
RUN	BCDF	BCEF	BDEF	CDEF	ABCDE	ABCDF	ABCEF	ABDEF	ACDEF	BCDEF	ABCDEF
1	+	+	+	+	−	−	−	−	−	−	+
2	+	+	+	+	+	+	+	+	+	−	−
3	−	−	−	+	+	+	+	+	−	+	−
4	−	−	−	+	−	−	−	−	+	+	+
5	−	−	+	−	+	+	+	−	+	+	−
6	−	−	+	−	−	−	−	+	−	+	+
7	+	+	−	−	−	−	−	+	+	−	+
8	+	+	−	−	+	+	+	−	−	−	−
9	−	+	−	−	+	+	−	+	+	+	−
10	−	+	−	−	−	−	+	−	−	+	+
11	+	−	+	−	−	−	+	−	+	−	+
12	+	−	+	−	+	+	−	+	−	−	−
13	+	−	−	+	−	−	+	+	−	−	+
14	+	−	−	+	+	+	−	−	+	−	−
15	−	+	+	+	+	+	−	−	−	+	−
16	−	+	+	+	−	−	+	+	+	+	+
17	+	−	−	−	+	−	+	+	+	+	−
18	+	−	−	−	−	+	−	−	−	+	+
19	−	+	+	−	−	+	−	−	+	−	+
20	−	+	+	−	+	−	+	+	−	−	−
21	−	+	−	+	−	+	−	+	−	−	+
22	−	+	−	+	+	−	+	−	+	−	−
23	+	−	+	+	+	−	+	−	−	+	−
24	+	−	+	+	−	+	−	+	+	+	+
25	−	−	+	+	−	+	+	−	−	−	+
26	−	−	+	+	+	−	−	+	+	−	−
27	+	+	−	+	+	−	−	+	−	+	−
28	+	+	−	+	−	+	+	−	+	+	+
29	+	+	+	−	+	−	−	−	+	+	−
30	+	+	+	−	−	+	+	+	−	+	−
31	−	−	−	−	−	+	+	+	+	−	+
32	+	+	+	+	+	+	+	+	+	+	+

Table A.7 2^6 Full Factorial (*Continued*)

	Treatments										
RUN	BCDF	BCEF	BDEF	CDEF	ABCDE	ABCDF	ABCEF	ABDEF	ACDEF	BCDEF	ABCDEF
33	−	−	−	−	−	+	+	+	+	+	−
34	−	−	−	−	+	−	−	−	−	+	+
35	+	+	+	−	+	−	−	−	+	−	+
36	+	+	+	−	−	+	+	+	−	−	−
37	+	+	−	+	+	−	−	+	−	−	+
38	+	+	−	+	−	+	+	−	+	−	−
39	−	−	+	+	−	+	+	−	−	+	−
40	−	−	+	+	+	−	−	+	+	+	+
41	+	−	+	+	+	−	+	−	−	−	+
42	+	−	+	+	−	+	−	+	+	−	−
43	−	+	−	+	−	+	−	+	−	+	−
44	−	+	−	+	+	−	+	−	+	+	+
45	−	+	+	−	−	+	−	−	+	+	−
46	−	+	+	−	+	−	+	+	−	+	+
47	+	−	−	−	+	−	+	+	+	−	+
48	+	−	−	−	−	+	−	−	−	−	−
49	−	+	+	+	+	+	−	−	−	−	+
50	−	+	+	+	−	−	+	+	+	−	−
51	+	−	−	+	−	−	+	+	−	+	−
52	+	−	−	+	+	+	−	−	+	+	+
53	+	−	+	−	−	−	+	−	+	+	−
54	+	−	+	−	+	+	−	+	−	+	+
55	−	+	−	−	+	+	−	+	+	−	+
56	−	+	−	−	−	−	+	−	−	−	−
57	+	+	−	−	−	−	−	+	+	+	−
58	+	+	−	−	+	+	+	−	−	+	+
59	−	−	+	−	+	+	+	−	+	−	+
60	−	−	+	−	−	−	−	+	−	−	−
61	−	−	−	+	+	+	+	+	−	−	+
62	−	−	−	+	−	−	−	−	+	−	−
63	+	+	+	+	−	−	−	−	−	+	−
64	+	+	+	+	+	+	+	+	+	+	+

Table A.7 2^6 Full Factorial (*Continued*)

	Factor		
RUN	1	2	3
1	1	1	1
2	1	2	2
3	2	1	2
4	2	2	1

Table A.8 L-4 Taguchi Design

	Factor						
RUN	1	2	3	4	5	6	7
1	1	1	1	1	1	1	1
2	1	1	1	2	2	2	2
3	1	2	2	1	1	1	2
4	1	2	2	2	2	2	1
5	2	1	2	1	2	2	2
6	2	1	2	2	1	1	1
7	2	2	1	1	2	2	1
8	2	2	1	2	1	1	2

Table A.9 L-8 Taguchi Design

	Factor										
RUN	1	2	3	4	5	6	7	8	9	10	11
1	1	1	1	1	1	1	1	1	1	1	1
2	1	1	1	1	1	2	2	2	2	2	2
3	1	1	1	2	2	1	1	1	2	2	2
4	1	2	2	2	2	1	2	2	1	1	2
5	1	2	2	1	2	2	1	2	1	2	1
6	1	2	2	2	1	2	2	1	2	1	1
7	2	1	1	2	1	1	2	2	1	2	1
8	2	1	1	1	2	2	2	1	1	1	2
9	2	1	1	2	2	2	1	2	2	1	1
10	2	2	2	1	1	1	1	2	2	1	2
11	2	2	2	2	1	2	1	1	1	2	2
12	2	2	2	1	2	1	2	1	2	2	1

Table A.10 L-12 Taguchi Design

Run	Factor														
	1	2	3	4	5	6	7	8	9	10	11	12	13	14	15
1	1	1	1	1	1	1	1	1	1	1	1	1	1	1	1
2	1	1	1	1	1	1	1	2	2	2	2	2	2	2	2
3	1	1	1	2	2	2	2	1	1	1	1	2	2	2	2
4	1	1	1	2	2	2	2	2	2	2	2	1	1	1	1
5	1	2	2	1	1	2	2	1	1	2	2	1	2	2	2
6	1	2	2	1	1	2	2	2	2	1	1	2	1	1	1
7	1	2	2	2	2	1	1	1	1	2	2	2	1	1	1
8	1	2	2	2	2	1	1	2	2	1	1	1	2	2	2
9	2	1	2	1	2	1	2	1	2	1	2	2	1	1	2
10	2	1	2	1	2	1	2	2	1	2	1	1	2	2	1
11	2	1	2	2	1	2	1	1	2	1	2	1	2	2	1
12	2	1	2	2	1	2	1	2	1	2	1	2	1	1	2
13	2	2	1	1	2	2	1	1	2	2	1	2	2	2	1
14	2	2	1	1	2	2	1	2	1	1	2	1	1	1	2
15	2	2	1	2	1	1	2	1	2	1	1	1	1	1	2
16	2	2	1	2	1	1	2	2	1	2	2	2	2	2	1

Table A.11 L-16 Taguchi Design

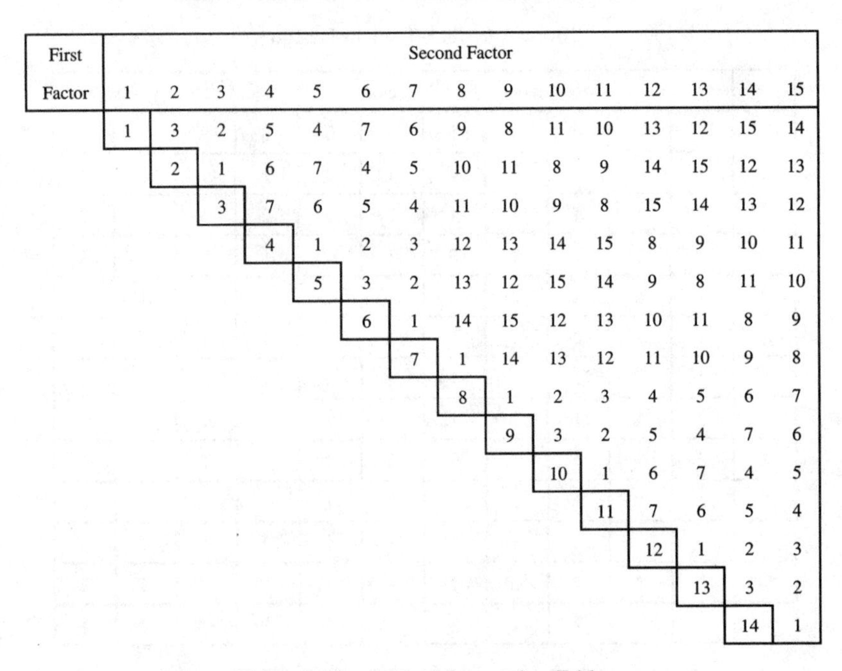

First Factor	Second Factor														
	1	2	3	4	5	6	7	8	9	10	11	12	13	14	15
	1	3	2	5	4	7	6	9	8	11	10	13	12	15	14
		2	1	6	7	4	5	10	11	8	9	14	15	12	13
			3	7	6	5	4	11	10	9	8	15	14	13	12
				4	1	2	3	12	13	14	15	8	9	10	11
					5	3	2	13	12	15	14	9	8	11	10
						6	1	14	15	12	13	10	11	8	9
							7	1	14	13	12	11	10	9	8
								8	1	2	3	4	5	6	7
									9	3	2	5	4	7	6
										10	1	6	7	4	5
											11	7	6	5	4
												12	1	2	3
													13	3	2
														14	1

Table A.12 2-Level Interaction Table

Number of Factors k	Fraction k-p	Number of Runs	Design Generators
3	2^{3-1}	4	C = AB
4	2^{4-1}	8	D = ABC
5	2^{5-1}	16	E = ABCD
	2^{5-2}	8	D = AB
			E = AC
6	2^{6-1}	32	F = ABCDE
	2^{6-2}	16	E = ABC
			F = BCD
	2^{6-3}	8	D = AB
			E = AC
			F = BC
7	2^{7-1}	64	G = ABCDEF
	2^{7-2}	32	F = ABCD
			G = ABDE
	2^{7-3}	16	E = ABC
			F = BCD
			G = ACD
	2^{7-4}	8	D = AB
			E = AC
			F = BC
			G = ABC

Table A.13 Fractional Factorial Generators

(*continues*)

Number of Factors k	Fraction k-p	Number of Runs	Design Generators
8	2^{8-2}	64	G = ABCD
			H = ABEF
	2^{8-3}	32	F = ABC
			G = ABD
			H = BCDE
	2^{8-4}	16	E = BCD
			F = ACD
			G = ABC
			H = ABD
9	2^{9-2}	128	H = ACDFG
			J = DCEFG
	2^{9-3}	64	G = ABCD
			H = ACEF
			J = CDEF
	2^{9-4}	32	F = BCDE
			G = ACDE
			H = ABDE
			J = ABCE
	2^{9-5}	16	E = ABC
			F = BCD
			G = ACD
			H = ABD
			J = ABCD

Table A.13 Fractional Factorial Generators (*Continued*)

Number of Factors k	Fraction k-p	Number of Runs	Design Generators
10	2^{10-3}	128	H = ABCG
			J = ACDE
			K = ACDF
	2^{10-4}	64	G = BCDF
			H = ACDF
			J = ABDE
			K = ABCE
	2^{10-5}	32	F = ABCD
			G = ABCE
			H = ABDE
			J = ACDE
			K = BCDE
	2^{10-6}	16	E = ABC
			F = BCD
			G = ACD
			H = ABD
			J = ABCD
			K = AB

Table A.13 Fractional Factorial Generators (*Continued*)

Number of Factors k	Fraction k-p	Number of Runs	Design Generators
11	2^{11-5}	64	G = CDE
			H = ABCD
			J = ABF
			K = BDEF
			L = ADEF
	2^{11-6}	32	F = ABC
			G = BCD
			H = CDE
			J = ACD
			K = ADE
			L = BDE
	2^{11-7}	16	E = ABC
			F = BCD
			G = ACD
			H = ABD
			J = ABCD
			K = AB
			L = AC

Table A.13 Fractional Factorial Generators (*Continued*)

	Treatments		
Runs	A	B	C
1	−1	−1	1
2	1	−1	−1
3	−1	1	−1
4	1	1	1

Table A.14 2^{3-1}

	Treatments			
Runs	A	B	C	D
1	−1	−1	−1	−1
2	1	−1	−1	1
3	−1	1	−1	1
4	1	1	−1	−1
5	−1	−1	1	1
6	1	−1	1	−1
7	−1	1	1	−1
8	1	1	1	1

Table A.15 2^{4-1}

	Treatments				
Runs	A	B	C	D	E
1	−1	−1	−1	−1	1
2	1	−1	−1	−1	−1
3	−1	1	−1	−1	−1
4	1	1	−1	−1	1
5	−1	−1	1	−1	−1
6	1	−1	1	−1	1
7	−1	1	1	−1	1
8	1	1	1	−1	−1
9	−1	−1	−1	1	−1
10	1	−1	−1	1	1
11	−1	1	−1	1	1
12	1	1	−1	1	−1
13	−1	−1	1	1	1
14	1	−1	1	1	−1
15	−1	1	1	1	−1
16	1	1	1	1	1

Table A.16 2^{5-1}

	Treatments				
Runs	A	B	C	D	E
1	−1	−1	−1	1	1
2	1	−1	−1	−1	−1
3	−1	1	−1	−1	1
4	1	1	−1	1	−1
5	−1	−1	1	1	−1
6	1	−1	1	−1	1
7	−1	1	1	−1	−1
8	1	1	1	1	1

Table A.17 2^{5-2}

	Treatments					
Runs	A	B	C	D	E	F
1	−1	−1	−1	−1	−1	−1
2	1	−1	−1	−1	1	−1
3	−1	1	−1	−1	1	1
4	1	1	−1	−1	−1	1
5	−1	−1	1	−1	1	1
6	1	−1	1	−1	−1	1
7	−1	1	1	−1	−1	−1
8	1	1	1	−1	1	−1
9	−1	−1	−1	1	−1	1
10	1	−1	−1	1	1	1
11	−1	1	−1	1	1	−1
12	1	1	−1	1	−1	−1
13	−1	−1	1	1	1	−1
14	1	−1	1	1	−1	−1
15	−1	1	1	1	−1	1
16	1	1	1	1	1	1

Table A.18 2^{6-2}

	Treatments					
Runs	A	B	C	D	E	F
1	−1	−1	−1	1	1	1
2	1	−1	−1	−1	−1	1
3	−1	1	−1	−1	1	−1
4	1	1	−1	1	−1	−1
5	−1	−1	1	1	−1	−1
6	1	−1	1	−1	1	−1
7	−1	1	1	−1	−1	1
8	1	1	1	1	1	1

Table A.19 2^{6-3}

	Treatments						
Runs	A	B	C	D	E	F	G
1	−1	−1	−1	−1	−1	1	1
2	1	−1	−1	−1	−1	−1	−1
3	−1	1	−1	−1	−1	−1	−1
4	1	1	−1	−1	−1	1	1
5	−1	−1	1	−1	−1	−1	1
6	1	−1	1	−1	−1	1	−1
7	−1	1	1	−1	−1	1	−1
8	1	1	1	−1	−1	−1	1
9	−1	−1	−1	1	−1	−1	−1
10	1	−1	−1	1	−1	1	1
11	−1	1	−1	1	−1	1	1
12	1	1	−1	1	−1	−1	−1
13	−1	−1	1	1	−1	1	−1
14	1	−1	1	1	−1	−1	1
15	−1	1	1	1	−1	−1	1
16	1	1	1	1	−1	1	−1
17	−1	−1	−1	−1	1	1	−1
18	1	−1	−1	−1	1	−1	1
19	−1	1	−1	−1	1	−1	1
20	1	1	−1	−1	1	1	−1
21	−1	−1	1	−1	1	−1	−1
22	1	−1	1	−1	1	1	1
23	−1	1	1	−1	1	1	1
24	1	1	1	−1	1	−1	−1
25	−1	−1	−1	1	1	−1	1
26	1	−1	−1	1	1	1	−1
27	−1	1	−1	1	1	1	−1
28	1	1	−1	1	1	−1	1
29	−1	−1	1	1	1	1	1
30	1	−1	1	1	1	−1	−1
31	−1	1	1	1	1	−1	−1
32	1	1	1	1	1	1	1

Table A.20 2^{7-2}

	Treatments						
Runs	A	B	C	D	E	F	G
1	−1	−1	−1	−1	−1	−1	−1
2	1	−1	−1	−1	1	−1	1
3	−1	1	−1	−1	1	1	−1
4	1	1	−1	−1	−1	1	1
5	−1	−1	1	−1	1	1	1
6	1	−1	1	−1	−1	1	−1
7	−1	1	1	−1	−1	−1	1
8	1	1	1	−1	1	−1	−1
9	−1	−1	−1	1	−1	1	1
10	1	−1	−1	1	1	1	−1
11	−1	1	−1	1	1	−1	1
12	1	1	−1	1	−1	−1	−1
13	−1	−1	1	1	1	−1	−1
14	1	−1	1	1	−1	−1	1
15	−1	1	1	1	−1	1	−1
16	1	1	1	1	1	1	1

Table A.21 2^{7-3}

	Treatments						
Runs	A	B	C	D	E	F	G
1	−1	−1	−1	1	1	1	−1
2	1	−1	−1	−1	−1	1	1
3	−1	1	−1	−1	1	−1	1
4	1	1	−1	1	−1	−1	−1
5	−1	−1	1	1	−1	−1	1
6	1	−1	1	−1	1	−1	−1
7	−1	1	1	−1	−1	1	−1
8	1	1	1	1	1	1	1

Table A.22 2^{7-4}

	Treatments							
Runs	A	B	C	D	E	F	G	H
1	−1	−1	−1	−1	−1	−1	−1	1
2	1	−1	−1	−1	−1	1	1	1
3	−1	1	−1	−1	−1	1	1	−1
4	1	1	−1	−1	−1	−1	−1	−1
5	−1	−1	1	−1	−1	1	−1	−1
6	1	−1	1	−1	−1	−1	1	−1
7	−1	1	1	−1	−1	−1	1	1
8	1	1	1	−1	−1	1	−1	1
9	−1	−1	−1	1	−1	−1	1	−1
10	1	−1	−1	1	−1	1	−1	−1
11	−1	1	−1	1	−1	1	−1	1
12	1	1	−1	1	−1	−1	1	1
13	−1	−1	1	1	−1	1	1	1
14	1	−1	1	1	−1	−1	−1	1
15	−1	1	1	1	−1	−1	−1	−1
16	1	1	1	1	−1	1	1	−1
17	−1	−1	−1	−1	1	−1	−1	−1
18	1	−1	−1	−1	1	1	1	−1
19	−1	1	−1	−1	1	1	1	1
20	1	1	−1	−1	1	−1	−1	1
21	−1	−1	1	−1	1	1	−1	1
22	1	−1	1	−1	1	−1	1	1
23	−1	1	1	−1	1	−1	1	−1
24	1	1	1	−1	1	1	−1	−1
25	−1	−1	−1	1	1	−1	1	1
26	1	−1	−1	1	1	1	−1	1
27	−1	1	−1	1	1	1	−1	−1
28	1	1	−1	1	1	−1	1	−1
29	−1	−1	1	1	1	1	1	−1
30	1	−1	1	1	1	−1	−1	−1
31	−1	1	1	1	1	−1	−1	1
32	1	1	1	1	1	1	1	1

Table A.23 2^{8-3}

	Treatments							
Runs	A	B	C	D	E	F	G	H
1	−1	−1	−1	−1	−1	−1	−1	−1
2	1	−1	−1	−1	−1	1	1	1
3	−1	1	−1	−1	1	−1	1	1
4	1	1	−1	−1	1	1	−1	−1
5	−1	−1	1	−1	1	1	1	−1
6	1	−1	1	−1	1	−1	−1	1
7	−1	1	1	−1	−1	1	−1	1
8	1	1	1	−1	−1	−1	1	−1
9	−1	−1	−1	1	1	1	−1	1
10	1	−1	−1	1	1	−1	1	−1
11	−1	1	−1	1	−1	1	1	−1
12	1	1	−1	1	−1	−1	−1	1
13	−1	−1	1	1	−1	−1	1	1
14	1	−1	1	1	−1	1	−1	−1
15	−1	1	1	1	1	−1	−1	−1
16	1	1	1	1	1	1	1	1

Table A.24 2^{8-4}

	Treatments								
Runs	A	B	C	D	E	F	G	H	J
1	−1	−1	−1	−1	−1	1	1	1	1
2	1	−1	−1	−1	−1	1	−1	−1	−1
3	−1	1	−1	−1	−1	−1	1	−1	−1
4	1	1	−1	−1	−1	−1	−1	1	1
5	−1	−1	1	−1	−1	−1	−1	1	−1
6	1	−1	1	−1	−1	−1	1	−1	1
7	−1	1	1	−1	−1	1	−1	−1	1
8	1	1	1	−1	−1	1	1	1	−1
9	−1	−1	−1	1	−1	−1	−1	−1	1
10	1	−1	−1	1	−1	−1	1	1	−1
11	−1	1	−1	1	−1	1	−1	1	−1
12	1	1	−1	1	−1	1	1	−1	1
13	−1	−1	1	1	−1	1	1	−1	−1
14	1	−1	1	1	−1	1	−1	1	1
15	−1	1	1	1	−1	−1	1	1	1
16	1	1	1	1	−1	−1	−1	−1	−1
17	−1	−1	−1	−1	1	−1	−1	−1	−1
18	1	−1	−1	−1	1	−1	1	1	1
19	−1	1	−1	−1	1	1	−1	1	1
20	1	1	−1	−1	1	1	1	−1	−1
21	−1	−1	1	−1	1	1	1	−1	1
22	1	−1	1	−1	1	1	−1	1	−1
23	−1	1	1	−1	1	−1	1	1	−1
24	1	1	1	−1	1	−1	−1	−1	1
25	−1	−1	−1	1	1	1	1	1	−1
26	1	−1	−1	1	1	1	−1	−1	1
27	−1	1	−1	1	1	−1	1	−1	1
28	1	1	−1	1	1	−1	−1	1	−1
29	−1	−1	1	1	1	−1	−1	1	1
30	1	−1	1	1	1	−1	1	−1	−1
31	−1	1	1	1	1	1	−1	−1	−1
32	1	1	1	1	1	1	1	1	1

Table A.25 2^{9-4}

	Treatments								
Runs	A	B	C	D	E	F	G	H	J
1	−1	−1	−1	−1	−1	−1	−1	−1	1
2	1	−1	−1	−1	1	−1	1	1	−1
3	−1	1	−1	−1	1	1	−1	1	−1
4	1	1	−1	−1	−1	1	1	−1	1
5	−1	−1	1	−1	1	1	1	−1	−1
6	1	−1	1	−1	−1	1	−1	1	1
7	−1	1	1	−1	−1	−1	1	1	1
8	1	1	1	−1	1	−1	−1	−1	−1
9	−1	−1	−1	1	−1	1	1	1	−1
10	1	−1	−1	1	1	1	−1	−1	1
11	−1	1	−1	1	1	−1	1	−1	1
12	1	1	−1	1	−1	−1	−1	1	−1
13	−1	−1	1	1	1	−1	−1	1	1
14	1	−1	1	1	−1	−1	1	−1	−1
15	−1	1	1	1	−1	1	−1	−1	−1
16	1	1	1	1	1	1	1	1	1

Table A.26 2^{9-5}

	Treatments									
Runs	A	B	C	D	E	F	G	H	J	K
1	–1	–1	–1	–1	–1	1	1	1	1	1
2	1	–1	–1	–1	–1	–1	–1	–1	–1	1
3	–1	1	–1	–1	–1	–1	–1	–1	1	–1
4	1	1	–1	–1	–1	1	1	1	–1	–1
5	–1	–1	1	–1	–1	–1	–1	1	–1	–1
6	1	–1	1	–1	–1	1	1	–1	1	–1
7	–1	1	1	–1	–1	1	1	–1	–1	1
8	1	1	1	–1	–1	–1	–1	1	1	1
9	–1	–1	–1	1	–1	–1	1	–1	–1	–1
10	1	–1	–1	1	–1	1	–1	1	1	–1
11	–1	1	–1	1	–1	1	–1	1	–1	1
12	1	1	–1	1	–1	–1	1	–1	1	1
13	–1	–1	1	1	–1	1	–1	–1	1	1
14	1	–1	1	1	–1	–1	1	1	–1	1
15	–1	1	1	1	–1	–1	1	1	1	–1
16	1	1	1	1	–1	1	–1	–1	–1	–1
17	–1	–1	–1	–1	1	1	–1	–1	–1	–1
18	1	–1	–1	–1	1	–1	1	1	1	–1
19	–1	1	–1	–1	1	–1	1	1	–1	1
20	1	1	–1	–1	1	1	–1	–1	1	1
21	–1	–1	1	–1	1	–1	1	–1	1	1
22	1	–1	1	–1	1	1	–1	1	–1	1
23	–1	1	1	–1	1	1	–1	1	1	–1
24	1	1	1	–1	1	–1	1	–1	–1	–1
25	–1	–1	–1	1	1	–1	–1	1	1	1
26	1	–1	–1	1	1	1	1	–1	–1	1
27	–1	1	–1	1	1	1	1	–1	1	–1
28	1	1	–1	1	1	–1	–1	1	–1	–1
29	–1	–1	1	1	1	1	1	1	–1	–1
30	1	–1	1	1	1	–1	–1	–1	1	–1
31	–1	1	1	1	1	–1	–1	–1	–1	1
32	1	1	1	1	1	1	1	1	1	1

Table A.27 2^{10-5}

	Treatments									
Runs	A	B	C	D	E	F	G	H	J	K
1	–1	–1	–1	–1	–1	–1	–1	–1	1	1
2	1	–1	–1	–1	1	–1	1	1	–1	–1
3	–1	1	–1	–1	1	1	–1	1	–1	–1
4	1	1	–1	–1	–1	1	1	–1	1	1
5	–1	–1	1	–1	1	1	1	–1	–1	1
6	1	–1	1	–1	–1	1	–1	1	1	–1
7	–1	1	1	–1	–1	–1	1	1	1	–1
8	1	1	1	–1	1	–1	–1	–1	–1	1
9	–1	–1	–1	1	–1	1	1	1	–1	1
10	1	–1	–1	1	1	1	–1	–1	1	–1
11	–1	1	–1	1	1	–1	1	–1	1	–1
12	1	1	–1	1	–1	–1	–1	1	–1	1
13	–1	–1	1	1	1	–1	–1	1	1	1
14	1	–1	1	1	–1	–1	1	–1	–1	–1
15	–1	1	1	1	–1	1	–1	–1	–1	–1
16	1	1	1	1	1	1	1	1	1	1

Table A.28 2^{10-6}

Number of Factors k	Fraction k-p	Number of Runs
3	2^{3-1}	4
	Confounding and Aliases	
A = BC	B = AC	C = AB
4	2^{4-1}	8
	Confounding and Aliases	
A = BCD	B = ACD	C = ABD
D = ABC	AB = CD	AC = BD
	AD = BC	
5	2^{5-1}	16
	Confounding and Aliases	
A = BD = CE	B = AD = ADE	C = AE = BDE
D = AB = BCE	E = AC = BCD	BC = DE = ACD = ABE
	CD = BE = ABC = ADE	
5	2^{5-2}	8
	Confounding and Aliases	
A = BD = CE	B = AD = CDE	C = AE = BDE
D = AB = BCE	E = AC = BCD	BC = DE = ACD = ABE
	CD = BE = ABC = ADE	
6	2^{6-1}	32
	Confounding and Aliases	
ABC = DEF	ABD = CEF	ABE = CDF
ABF = CDE	ACD = BEF	ACE = BDF
ACF = BDE	ADE = BCF	ADF = BCE
	AEF = BCD	

Table A.29 Fractional Factorial Confounding and Aliases

(*continues*)

Number of Factors k	Fraction k-p	Number of Runs
6	2^{6-2}	16
	Confounding and Aliases	
A = BCE = DEF	B = ACE = CDF	C = ABE = BDF
D = BCF = AEF	E = ABC = ADF	F = BCD = ADE
AB = CE	AC = BE	AD = EF
AE = BC = DF	AF = DE	BD = CF
ABD = CDE = ACF = BEF	ACD = BDE = ABF = CEF	
6	2^{6-3}	8
	Confounding and Aliases	
	A = BD = CE = CDF = BEF	
	B = AD = CF = CDE = AEF	
	C = AE = BF = BDE = ADF	
	D = AB = EF = BCE = ACF	
	E = AC = DF = BCD = ABF	
	F = BD = DE = ACD = ABE	
	CD = BE = AF = ABC = ADE = BDF = CEF	
7	2^{7-2}	32
	Confounding and Aliases	
A =	AF = BCD	CG = EF
B =	AG = BDE	DE = ABG
C = EFG	BC = ADF	DF = ABC
D =	BD = ACF = AEG	DG = ABE
E = CFG	BE = ADG	ACE = AFG
F = CEG	BF = ACD	ACG = AEF
G = CEF	BG = ADE	BCE = BFG
AB = CDF = DEG	CD = ABF	BCG = BEF
AC = BDF	CE = FG	CDE = DFG
AD = BCF = BEG	CF = ABD = EG	CDG = DEF
	AE = BDG	

Table A.29 Fractional Factorial Confounding and Aliases (*Continued*)

Number of Factors *k*	Fraction *k-p*	Number of Runs
7	2^{7-3}	16
	Confounding and Aliases	
A = BCE = DEF = CDG = BFG		AB = DE = FG
B = ACE = CDF = DEG = AFG		AC = BE = DG
C = ABE = BDF = ADG = EFG		AD = EF = CG
D = BCF = AEF = ACG = BEG		AE = BC = DF
E = ABC = ADF = BDG = CFG		AF = DE = BG
F = BCD = ADE = ABG = CEG		AG = CD = BF
G = ACD = BDE = ABF = CEF		BD = CF = EG
	ABD = CDE = ACF = BEF = BCG = AEG = DFG	
7	2^{7-4}	8
	Confounding and Aliases	
	A = BD = CE = FG = CDF = BEF = BCG = DEG	
	B = AD = CF = EG = CDE = AEF = ACG = DFG	
	C = AE = BF = DG = BDE = ADF = ABG = EFG	
	D = AB = EF = CG = BCE = ACF = AEG = BFG	
	E = AC = DF = BG = BCD = ABF = ADG = CFG	
	F = BC = DE = AG = ACD = ABE = BDG = CEG	
	G = CD = BE = AF = ABC = ADE = BDF = CEF	

Table A.29 Fractional Factorial Confounding and Aliases (*Continued*)

Number of Factors k	Fraction k-p	Number of Runs
8	2^{8-3}	32
	Confounding and Aliases	
A = BCF = BDG	AE = DFH = CGH	DE = BCH = AFH
B = ACF = ADG	AF = BC = DEH	DH = BCE = AEF
C = ABF = DFG	AG = BD = CEH	EF = ADH = BGH
D = ABG = CFG	AH = DEF = CEG	EG = ACH = BFH
E =	BE = CDH = FGH	FH = ADE = BEG
F = ABC = CDG	BH = CDE = EFG	GH = ACE = BEF
G = ABD = CDF	CD = FG = BEH	ABE = CEF = DEG
H =	CE = BDH = AGH	ABH = CFH = DGH
AB = CF = DG	CG = DF = AEH	AD = BG = EFH
AC = BF = EGH	CH = BDE = AEG	
	ACD = BDF = BCG = AFG	
	EH = BCD = ADF = ACG = BFG	
8	2^{8-4}	16
	Confounding and Aliases	
	A = CDF = BEF = BCG = DEG = BDH = CEH = FGH	
	B = CDE = AEF = ACG = DFC = ADH = CFH = EGH	
	C = BDE = ADF = ABG = EFG = AEH = BFH = DGH	
	D = BCE = ACF = AEG = BFG = ABH = EFH = CGH	
	E = BCD = ABF = ADG = CFG = ACH = DFH = BGH	
	F = ACD = ABE = BDG = CEG = BCH = DEH = AGH	
	G = ABC = ADE = BDF = CEF = CDH = BEH = AFH	
	H = ABD = ACE = BCF = DEF = CDG = DEG = AFG	
	AB = EF = CG = DH	
	AC = DF = BG = EH	
	AD = CF = EG = BH	
	AE = BF = DG = CH	
	AF = CD = BE = GH	
	AG = BC = DE = FH	
	AH = BD = CE = FG	

Table A.29 Fractional Factorial Confounding and Aliases (*Continued*)

Number of Factors k	Fraction k-p	Number of Runs
9	2^{9-4}	32
	Confounding and Aliases	
A = BGF = CFH = DFJ		AD = CEG = BEH = FJ
B = AFG = CGH = DGJ		CD = BEF = AEG = HJ
C = AFH = BGH = DHJ		AF = BG = CH = DJ
D = AFJ = BGJ = CHJ		AG = CDE = BF = EHJ
E =		AH = BDE = CF = EGJ
F = ABG = ACH = ADJ		AJ = BCE = DF = EGH
G = ABF = BCH = BDJ		BC = DEF = GH = AEJ
H = ACF = BCG = CDJ		BD = CEF = AEH = GJ
J = ADF = BDG = CDH		CJ = ABE = EFG = DH
AB = FG = DEH = CEJ		BH = ADECGEFJ
AC = DEG = FH = BEJ		
	BJ = ACE = DG = EFH	
	CE = BDF = ADG = ADJ = FGJ	
	BE = CDF = ADH = ACJ = FHJ	
	DE = BCF = ACG = ABH = FGH	
	EF = BDC = DGH = CGJ = BHJ	
	EG = ACD = DFH = CFJ = AHJ	
	EH = ABD = DFG = BFJ = AGJ	
	EJ = ABC = CFG = BFH = AGH	
	AEF = BEG = CEH = DEJ	
	AE = CDG = BDH = BCJ = GHJ	

Table A.29 Fractional Factorial Confounding and Aliases (*Continued*)

Number of Factors k	Fraction k-p	Number of Runs
9	2^{9-5}	16

Confounding and Aliases

A = FJ = BCE = DEF = CDG = BFG = BDH = CFH = EGH

B = GJ = ACE = CDF = DEG = AFG = ADH = EFH = CGH

C = HJ = ABE = BDF = ADG = EFG = DEH = AFH = BGH

D = EJ = BCF = AEF = ACG = BEG = ABH = CEH = FGH

E = DJ = ABC = ADF = BDG = CFG = CDH = BFH = AGH

F = AJ = BCD = ADE = ABG = CEG = ACH = AEH = DGH

G = BJ = ACD = BDE = ABF = CEF = BCH = AEH = DFH

H = CJ = ABD = CDE = ACF = BEF = BCG = AEG = DFG

J = DE = AF = BG = CH

AB = CE = FG = DH = CDJ = BFJ = AGJ = EHJ

AC = BE = DG = FH = BDJ = CFJ = EGJ = AHJ

AD = EF = CG = BH = BCJ = AEJ = DFJ = GHJ

AE = BC = DF = GH = ADJ = EFJ = CGJ = EHJ

AG = CD = BF = BH = ABJ = CEJ = FGJ = DHJ

AH = BD = CF = EG = ACJ = BEJ = DGJ = FHJ

Table A.29 Fractional Factorial Confounding and Aliases (*Continued*)

Number of Factors k	Fraction k-p	Number of Runs
10	2^{10-5}	32

Confounding and Aliases

A = EFK = DGK = CHK = BJK
B = EFJ = DGJ = CHJ = AJK =
C = EFH = DGH = BHJ = AHK
D = EFG = CGH = BGJ = AGK
E = DFG = CFH = BFJ = AFK
F = DEG = CEH = BEJ = AEK
G = DEF = CDH = BDJ = ADK
H = CEF = CDG = BCJ = ACK
J = BEF = BDG = BCH = ABK
K = AEF = ADG = ACH = ABJ
AB = CDF = CEG = DEH = FGH = JK
AC = BDF = BEG = DEJ = FGJ = HK
AD = BCF = BEH = CEJ = FHJ = GK
AE = BCJ = BDH = CDJ = GHJ = FK
AF = BCD = BGH = CGJ = DHJ = EK
AG = BCE = BFH = CFJ = EHJ = DK
AH = BDE = BFG = DFJ = EGJ = CK
AJ = CDE = CFG = DFH = EJH = BK
AK = EF = DG = CH = BJ
BC = ADF = AEG = HJ = DEK = FGK
BD = ACF = AEH = GJ = CEK = FHK
BE = ACG = ADH = FJ = CDK = CHK
BF = ACD = AGH = EJ = CGK = DHK
BG = ACW = AFH = DJCFK = EHK
BH = ADE = AFG = CJ = DFK = EGK
CD = ABF = GH = AEJ = BEK = FJK
CE = ABG = FH = ADJ = BDK = GJK
CF = ABD = EH = AGJ = BGK = DJK
CG = ADE = DH = AFJ = BFK = EJK
DE = FG = ABH = ACJ = BCK = HJK
DF = ABD = EG = AHJ = BHK = CJK

Table A.29 Fractional Factorial Confounding and Aliases (*Continued*)

$\alpha/2$	0.1	0.05	0.025	0.0125	0.005
α	0.2	0.1	0.05	0.025	0.01
df					
1	3.078	6.314	12.706	25.452	63.656
2	1.886	2.92	4.303	6.205	9.925
3	1.638	2.353	3.182	4.177	5.841
4	1.533	2.132	2.776	3.495	4.604
5	1.476	2.015	2.571	3.163	4.032
6	1.44	1.943	2.447	2.969	3.707
7	1.415	1.895	2.365	2.841	3.499
8	1.397	1.86	2.306	2.752	3.355
9	1.383	1.833	2.262	2.685	3.25
10	1.372	1.812	2.228	2.634	3.169
11	1.363	1.796	2.201	2.593	3.106
12	1.356	1.782	2.179	2.56	3.055
13	1.35	1.771	2.16	2.533	3.012
14	1.345	1.761	2.145	2.51	2.977
15	1.341	1.753	2.131	2.49	2.947
16	1.337	1.746	2.12	2.473	2.921
17	1.333	1.74	2.11	2.458	2.898
18	1.33	1.734	2.101	2.445	2.878
19	1.328	1.729	2.093	2.433	2.861
20	1.325	1.725	2.086	2.423	2.845
21	1.323	1.721	2.08	2.414	2.831
22	1.321	1.717	2.074	2.405	2.819
23	1.319	1.714	2.069	2.398	2.807
24	1.318	1.711	2.064	2.391	2.797
25	1.316	1.708	2.06	2.385	2.787
26	1.315	1.706	2.056	2.379	2.779
27	1.314	1.703	2.052	2.373	2.771
28	1.313	1.701	2.048	2.368	2.763
29	1.311	1.699	2.045	2.364	2.756
30	1.31	1.697	2.042	2.36	2.75
32	1.309	1.694	2.037	2.352	2.738
34	1.307	1.691	2.032	2.345	2.728
36	1.306	1.688	2.028	2.339	2.719
38	1.304	1.686	2.024	2.334	2.712
40	1.303	1.684	2.021	2.329	2.704
45	1.301	1.679	2.014	2.319	2.69
50	1.299	1.676	2.009	2.311	2.678
60	1.296	1.671	2	2.299	2.66
75	1.293	1.665	1.992	2.287	2.643
100	1.29	1.66	1.984	2.276	2.626
150	1.287	1.655	1.976	2.264	2.609
500	1.283	1.648	1.965	2.248	2.586
∞	1.282	1.645	1.96	2.241	2.576

Table A.30 *t* Distribution

	n_x													
n_y	2	3	4	5	6	7	8	9	10	11	12	13	14	15
4			10											
5		6	11	17										
6		7	12	18	26									
7		7	13	20	27	36								
8	3	8	14	21	29	38	49							
9	3	8	15	22	31	40	51	63						
10	3	9	15	23	32	42	53	65	78					
11	4	9	16	24	34	44	55	68	81	96				
12	4	10	17	26	35	46	58	71	85	99	115			
13	4	10	18	27	37	48	60	73	88	103	119	137		
14	4	11	19	28	38	50	63	76	91	106	123	141	160	
15	4	11	20	29	40	52	65	79	94	110	127	145	164	185
16	4	12	21	31	42	54	67	82	97	114	131	150	169	
17	5	12	21	32	43	56	70	84	100	117	135	154		
18	5	13	22	33	45	58	72	87	103	121	139			
19	5	13	23	34	46	60	74	90	107	124				
20	5	14	24	35	48	62	77	93	110					
21	6	14	25	37	50	64	79	95						
22	6	15	26	38	51	66	82							
23	6	15	27	39	53	68								
24	6	16	28	40	55									
25	6	16	28	42										
26	7	17	29											
27	7	17												
28	7													

For n_x and n_y both greater than 8, you can use the normal approximation and z score test with

$$\mu_R = \frac{n_x(n_x + n_y + 1)}{2} \quad \text{and} \quad \sigma_R^2 = \frac{n_x n_y (n_x + n_y + 1)}{12}$$

Table A.31 Critical Values for Wilcoxon Rank-Sum Test at the α=.05 Level

ratio	n	90%	95%	99%
$r = \frac{x_n - x_{n-1}}{x_n - x_1}$	3	0.886	0.941	0.988
	4	0.679	0.765	0.889
	5	0.557	0.642	0.780
	6	0.482	0.560	0.698
	7	0.434	0.507	0.637
$r = \frac{x_n - x_{n-1}}{x_n - x_2}$	8	0.479	0.554	0.683
	9	0.441	0.512	0.635
	10	0.409	0.477	0.597
$r = \frac{x_n - x_{n-2}}{x_n - x_2}$	11	0.517	0.576	0.679
	12	0.490	0.546	0.642
	13	0.467	0.521	0.615
$r = \frac{x_n - x_{n-2}}{x_n - x_3}$	14	0.492	0.546	0.641
	15	0.472	0.525	0.616
	16	0.454	0.507	0.595
	17	0.438	0.490	0.577
	18	0.424	0.475	0.561
	19	0.412	0.462	0.547
	20	0.401	0.450	0.535
	21	0.391	0.440	0.524
	22	0.382	0.430	0.514
	23	0.374	0.421	0.505
	24	0.367	0.413	0.497
	25	0.360	0.406	0.489

Table A.32 Critical Values for Dixon's Test

B

STATISTICAL SOFTWARE

Applications of Taguchi Methods and Taguchi Software

The American Supplier Institute (ASI), 17333 Federal Drive, Suit 220, Allen Park, MI 48101; (313) 336-8877; Fax (313) 336-3187.

Taguchi Certification of Experts A comprehensive certification program offered by ASI that covers the entire spectrum of Taguchi Methods in an intensive modular 16-day program. Those completing the program will be certified by ASI, Dr. Taguchi, and Shin Taguchi as a "Taguchi Expert."

Robust Technology Development Provides you with the tools and skills needed to begin developing robust, generic technology that can be applied efficiently across a broad range, or family, of present and future products. On-site training and implementation assistance is available.

Taguchi Methods for Problem-Solving For those whose designs are already known or in use, this approach will provide a means for improving an existing design, as well as determining the process parameters that will most effectively produce the improved design. Various problem-solving methods are explored. Public workshops as well as on-site training and implementation assistance are available.

Taguchi Methods for Manufacturing Process Control Helps to cost-effectively control production processes in order to maintain improvements identified in the design phase. Public workshops are available, as are on-site training and implementation assistance.

ANOVA -TM

ANOVA -TM FOR WINDOWS is an easy-to-use, Design of Experiments (DOE) software application for Windows (compatible with Windows 3.1, 95, and NT) based on Dr. Genichi Taguchi's methods. It is an orderly, methodical, and meaningful tool for systematically assessing experimental results (traditional or Taguchi) when a variety of factors dynamically affect a specific quality characteristic. This popular software combines vast capabilities and an easy-to-use Graphical User Interface (GUI) in Microsoft Windows.

Benefits of using ANOVA-TM:

- Has all the power and flexibility for complex or basic experiments.
- Utilizes a complete set of orthogonal arrays all the way to an L108, or customize your own arrays.
- Prints to any Windows printer.
- Supports OLE (Object, Linking, and Embedding).
- S/N calculations for dynamic characteristics.
- Cut, copy, or paste data to and from your favorite Windows program.

An enhanced version, ANOVA-TM Professional, is also available. Call ASI for more information.

Technical Support for ANOVA-TM for Windows is provided by Advanced Systems and Design Inc Item Number: TM000670.

STN

STN for Windows is the first software designed exclusively to support Dr. Genichi Taguchi's methods. With more than 20 new features, engineers can concentrate on the important task of technology improvement, while leaving the computations to STN for Windows.

Features include:

- Icon-based button bar provides quick access to frequent menu functions.
- Worksheet editor permits easy data entry.
- Imports data from common spreadsheet formats (Lotus, Excel, etc.) and ASCII delimited files.
- Data transformation feature converts raw data with user-defined equations.
- Orthogonal array modification helps you develop specialized, efficient experimental designs.
- Various analysis cases for dynamic characteristics, including zero point proportional, reference point proportional, and linear equations, double signals, dynamic operating window for chemical reactions, moving signal factor levels, and user-defined equations.
- Analysis for nondynamic characteristics.

- Analysis for attributes.
- Output signal-to-noise ratio, raw data mean (nondynamic), and linear coefficient and sensitivity (dynamic).

Technical Support for STN for Windows is provided by InfoMate Item Number TM000671.

DOE-PC IV

Quality America Inc., 7650 E. Broadway, Suite 208, Tucson AZ 85710; (800) 722-6154 or (520) 722-6154.

Quality America's DOE-PC IV software product is designed for all users of designed experiments. The software enables the novice and the expert to efficiently isolate process parameters that have primary influence on a process. The Windows environment provides an interface for the software user that enhances usability while improving analysis features. Using the built-in help system as a guide, a novice user can attain advice as to how to generate a design, gather data, and interpret results from the analysis. The expert user will see advantages in the ability to import data from other applications, copy and paste charts and analysis from DOE-PC IV to presentation software, and overlay analysis plots to gain additional information from the data.

DOE-PC IV is designed with the goal of enabling the user to learn more about and improve their process. This is done without the user having to know the many classes of experimental designs that DOE-PC IV uses, or how to compare these experimental designs. The software only asks the user to specify the parameters of the process on which they wish to do an experiment. These parameters can be entered as main factors, subsidiary (noise) factors for Taguchi designs, blocking factors, or causal factors. The factors can have from 2 to 30 levels, and can be entered as quantitative or qualitative values. These factors and their levels are used to define the type of design that will be applicable to your process. This design selection is based off of the interactions that you choose to investigate using the parameters you defined. The designs are generated internally using proven design generation methods. The user can investigate main factors only, main and two-factor interactions, or main, two-, and three-factor interactions, with all of these factor settings capable of being analyzed in linear an/or quadratic forms. There is also a selection for response surface interactions if a response surface design is necessary.

After a design is generated, and the experiment has been performed, DOE-PC IV has analysis tools which help the user get as much information out of their experimental design as possible. The tools are graphical and analytic in nature. The analytic tools include ANOVA and Regression summaries for the data and Response Surface analysis for RSM models. The Response Surface analysis is a canonical analysis of the system to determine the optimal operating region. With respect to graphical tools, DOE-PC IV provides point-line charts, box-and-whisker plots, half-normal probability plots, interaction plots, square/cube plots, and contour plots. These tools enable the user to locate contributing effects, check for model accuracy, develop a model for the system, graphically portray the systems output over specified regions, etc.

DOE-PC IV incorporates models for classical factorial designs, Taguchi designs, and response surface models. The designs are constructed to be free of confounding within the specified interactions. This enables all of the requested interactions to be estimated. All designs generated in DOE-PC IV can be reduced to a D-optimal design using DOE-PC IV's built-in design reduction feature. This enables the user to reduce a design in run size without having to worry about inducing confounding amongst the factors. If the user enters their own design and confounding is present, DOE-PC IV will warn the user. At this point the user can have DOE-PC IV extend the design, automatically add runs, to remove the confounding. These features enable the production of designs that have no confounding yet are minimal in run size.

Using DOE-PC IV software, the experimentalist can quickly develop a design for their processes and extract information through design analysis. Using this information, the process can be investigated, and improvement methods found. Using DOE-PC IV, optimal operating conditions for a process can be found which enable an increase in output and quality while producing a decrease in cost.

DOE-PC IV requires a 386/16 or better CPU (486 with math co-processor strongly recommended), 4 Megabytes of RAM, approximately 3 Megabytes of free hard drive space, VGA graphics adapter or better, Microsoft Windows 3.1 or higher, MS Windows 95, MS Windows NT, IBM OS/2 2.1 or higher.

MINI TAB

Minitab Inc., 3081 Enterprise Drive, State College, PA 16801-3008. Toll-free: (800) 448-3555 (814) 238-3280; Fax: (814) 238-4383.

MINITAB was developed in 1972 by leading university statisticians (Barbara F. Ryan, Brian L. Joiner, and Thomas A Ryan, Jr.) and has been a leader in statistical analysis in research and education ever since. Over 2000 colleges and universities use MINITAB, and over 300 statistics texts support it. MINITAB is the tool of choice for businesses in more than 60 countries.

MINITAB runs on mainframe, minicomputer, workstation, and microcomputer systems—and the list of supported computers is continually growing. A few of the many operating systems MINITAB works under include: Windows, DOS, Macintosh, UNIX, Open VMS. This wide range of supported computers makes MINITAB'S power accessible to you regardless of your computer type.

MINITAB is incredibly easy to learn and use, right out of the box. Pull-down menus and dialog boxes give you easy prompts every step of the way. There is no lengthy learning process and no need to struggle with unwieldy manuals. MINITAB lets you spend more time exploring your data, less time telling the computer what you want it to do.

Entering data in MINITAB'S data window is a breeze, as simple as using your favorite spreadsheet package. And now you have the option of importing data directly from a variety of file formats. In the Windows version, you can import Lotus, Excel, Symphony, Quattro Pro, dBase, and text ASCII files. The Macintosh version allows you to easily import Excel, Lotus, dBase, FoxPro and text files. Both versions make importing data effortless.

Now you can ensure that your analyses will attract the attention they deserve. With MINITAB'S new graphics features you have the power of a drawing package right at your fingertips. It's easy to create and customize a wide range of graphs. And you can make your graphs look just the way you want them to . . . add text. change line and fill styles, vary the format, and experiment with color. These easy-to-use features help you visualize and explore your data graphically with just one or two clicks of the mouse.

You can build a macro to consolidate repetitive tasks, and run complex analyses with just a single keystroke or click of the mouse. MINITAB'S macros let you create your own custom operations, designed specifically for your applications. DO LOOPS, IF THEN ELSE statements, and GOTO statements give you unlimited power when designing your analyses.

Since Release 10Xtra is available for both Macintosh and Windows machines, you have complete flexibility in selecting your working environment. You can transparently transition from the Macintosh version of MINITAB to the Windows version (and vice versa) since output and screens are nearly identical. Training people new to statistics becomes easier because the trainee can now select the environment in which she or he feels most comfortable and the trainer does not need to differentiate operations by platform.

Windows 32-bit Version: 8 MB RAM; 22 MB hard disk space: 386 machines and above; MS-DOS 5.0 or later; Windows version 3.1 or greater; VGA or SVGA monitor: 3.5" high density disk drive; minimum 8MB Windows swap file; mouse required for some capabilities; math coprocessor strongly recommended.

Windows 16-bit Version: for users with limited RAM; other requirements same as above, except 4MB RAM are required, although 6MB are recommended.

Macintosh Version: Macintosh 68020, 68040; System 7 or greater; 8 MB memory available for MINITAB; 19 MB hard disk space; total memory can include RAM from virtual memory or RAM doubling software.

Power Macintosh Version: 8MB of memory; a minimum of 12 MB of hard disk space required; 21 MB required to install all optional files.

SYSTAT Design

SPSS Inc., (800) 543-2185; Fax: (312) 329-3668.

For performing experiments or surveys where there are trials, observations, or questionnaires, use SYSTAT Design to determine your sample size requirements and to test the power of your design. Using power analysis, expected mean squares and a host of other analytical tools, SYSTAT Design helps you find differences—if they exist—in your data.

Flexible power and sample size analyses SYSTAT Design's flexibility helps you get the right answers from the information you have. If you provide the power, SYSTAT Design will give you the sample size required to obtain that power. Or, if you provide the sample size, SYSTAT Design will give you the power achieved using that sample. You can estimate sample size for analyses like *t*-tests for independent samples, paired *t*-tests, single- and two-factor designs.

The program creates plots that help you visualize the trade-off between power and sample size so you can pick the most appropriate combination. Use a complete range of tools: SYSTAT Design gives you all the tools you need for constructing your research appropriately. For instance, you can:

- Generate randomization plans for up to 20 treatments to control your study. Select equal or unequal subjects per treatment and choose whether or not to change the order of treatments within a subject. Choose a blocking variable of up to 20 categories with up to 400 subjects per block. Or, if your study prematurely ends, SYSTAT Design ensures against serious imbalance by carrying out randomization within subblocks.
- Compute tables of expected mean squares for balanced experiments. SYSTAT Design automatically generates a table showing you the expected size of difference by degrees of freedom based on your particular data set. Rather than manually looking up a set of tables, you can let SYSTAT Design quickly and easily create tables specific to your study. SYSTAT Design uses the Cornfield-Tukey algorithm to construct tables of expected mean squares for balanced experiments involving up to nine fixed or random factors that are either crossed or nested.
- Calculate orthogonal polynomials. This tells you which contrasts you need to analyze to find differences in your data. SYSTAT Design figures out all the possible combinations of values of the variable and advises which comparisons to include in your study.

If you need to structure your analysis quickly and easily with power analysis and expected mean squares, you need SYSTAT Design.

Analysis Sample size estimation

- Determines sample size requirements for
  - Single-factor designs
  - Two-factor designs
- Paired *t*-tests
- Specify effects
- Specify range
- Specify standardized average squared effect
- Correlation coefficients and proportions
  - Individual effects
  - A range
  - Standardized average squared effect
- Correlation coefficients
- Proportions—equal and unequal group sizes
- Two sample *t*-tests
- Two period cross-over designs

- Compare a signle sample to a known standard
- Power of specific sample sizes
- Plots of power vs. sample size
- Noncentrality parameters
  - Single-factor experiment
  - Two factor experiment
  - Randomized block designs
  - Simple linear regression
  - Quadratic regression
- Tables of Expected Mean Squares
  - Based on Cornfield and Tukey (1956)
  - Specify a model
  - Handles interactions and nested effects
- Randomization Plans
- Equal or unequal number of subjects per experiment
- Number of subjects per treatment prespecified or not
- Permute treatments within subjects
- Random-number generator (Wichmann and Hill, McLeod)
- Orthogonal polynomials
  - Specify metric of 3.20 distinct values
  - Double precision of Cooper (1968)

System Requirements Requires SYSTAT data file, but does not require SYSTAT Base System CBMS/COPY PLUS is available to translate files from other applications into SYSTAT data file format.

For DOS

- DOS version 3.3 or higher
- 286-based and better personal computers
- 4MB RAM minimum
- 6MB available hard disk space
- Mouse recommended
- Math coprocessor recommended

For the Macintosh

- System 6.0.2 or higher
- 400K RAM
- 200K hard disk space
- Coprocessor not required

STAT-EASE

Hennepin Square, Suite 191, 2021 E. Hennepin Avenue, Minneapolis, MN 55413-2723.

DESIGN-EASE SOFTWARE identifies the breakthrough factors for process or product improvement. It helps you set up and analyze two-level factorial, fractional factorial (up to 15 variables) and Plackett-Burman designs (up to 31 variables). These designs screen for critical factors and their interactions.

DESIGN-EXPERT SOFTWARE leads you to the peak of performance for your process or formulation. The program provides in-depth analysis for as many as 10 process factors or 12 components (screening designs extend to 24 components). DESIGN-EXPERT provides rotatable 3D plots so you can visualize your response surface. Explore the 2D contours with your mouse, setting flags along the way to identify coordinates and predict responses. The ultimate peak of performance can be found via the program's numerical optimization function, which finds the most desirable factor settings for up to 12 responses simultaneously.

Put the power of experiment design to work.

Capabilities Stat-ease Corporation makes it easy to use statistical design of experiments. We offer:

- DESIGN-EASE and
- DESIGN-EXPERT

These two programs place the power of DOE at your fingertips. DESIGN-EASE offers two-level factorial screening designs to find the critical factors that lead to breakthrough improvements. Then move up to DESIGN-EXPERT for an in-depth exploration. Use response surface methods to optimize your process or mixture. Display peak performance with 3D plots.

Requirements

DESIGN-EASE:	Macintosh:	Mac Classic or better, 1.5 MB RAM 3 MB disk space, System 7 or later
	Windows:	286/higher, 4 MB RAM recommended Hard disk with 3 MB available EGA or higher resolution MS-DOS version 3.3 or higher Windows version 3.1 or higher
DESIGN-EXPERT:	DOS	286 or higher Needs CGA, EGA, VGA or Hercules graphics 2MB hard disk space 640 conventional memory (EMS utilized)

Availability

Network Usage: Software may be installed on a network. If more than one person wishes to use the software concurrently, you must buy a separate software package for each additional user.

Within the United States:

Stat-Ease Corporation

2021 East Hennepin Avenue, Suite 191

Minneapolis, MN 55413

Call Toll-free 800/325-9816 (612/378-9449 direct); or Fax 612/378-2152.

BIBLIOGRAPHY

Barnett, V., *Sample Survey Principles and Methods*, Second Edition, Oxford (1991).

Box, George E. P., William G. Hunter, and J. Stuart Hunter, *Statistics for Experimenters*, Wiley (1978).

Brassard, Michael, *The Memory Jogger Plus+*, GOAL/QPC (1989).

Bruning, James L., and B. L. Kintz, *Computational Handbook of Statistics, Second Edition*, Scott, Foresman and Co. (1977) [medical].

Burrington, Richard S., and Donald C. May, *Handbook of Probability and Statistics with Tables, Second Edition*, McGraw-Hill (1970).

Cheney, Ward, and David Kincaid, *Numerical Mathematics and Computing, Second Edition*, Brooks/Cole (1985) [random numbers].

Chow, Alan F., Variability Reduction: The Shainin Way, Hughes Aircraft Co., Quality Directorate Training Course. Bound copies of viewgraphs.

Deming, W. E., *Sample Design in Business Research*, Wiley (1960).

Dixon, W. J., "Processing Data for Outliers," *Biometrics*, vol. 9, pp. 74–89, 1953.

Dixon, Wilfrid, and Frank J. Massey, Jr., *Introduction to Statistical Analysis, Second Edition*, McGraw-Hill (1957).

Farnum, Nicholas R., and LaVerne W. Stanton, *Quantitative Forecasting Methods*, PWS-Kent (1989).

Fasal, John H. (ed.), *Practical Value Analysis Methods*, Hayden Book Co. (1972).

Hahn, Gerald J., and Samuel S. Shapiro, *Statistical Models in Engineering*, Wiley (1967).

Hicks, Charles R., *Fundamental Concepts in the Design of Experiments, Third Edition*, Holt, Rinehart and Winston (1982)

Hines, William W,. and Douglas C. Montgomery, *Probability and Statistics in Engineering and Management Science, Third Edition*, Wiley (1990).

Holman, J. P., *Experimental Methods for Engineers, Fifth Edition*, McGraw-Hill (1989).

Ishikawa, Kaoru, *Introduction to Quality Control*, 3A Corporation, Japan (1990).

Jackson, Harry, and Normand L. Frigon, *Management 2000*, Wiley (1996).

Mandel, John, *The Statistical Analysis of Experimental Data*, Dover (1984); Wiley (1964).

Milliken, George A., and Dallas E. Johnson, *Analysis of Messy Data, Volume I: Designed Experiments*, Van Nostrand Reinhold (1992).

Montgomery, Douglas C., *Design and Analysis of Experiments, Third Edition*, Wiley (1991).

Noether, Gottfried E., *Introduction to Statistics: A Nonparametric Approach, Second Edition*, Houghton Mifflin (1976).

Phadke, Madhav S., *Quality Engineering Using Robust Design*, Prentice-Hall (1989) [Taguchi].

Rickmers, Albert D., and Hollis N. Todd, *Statistics: An Introduction*, McGraw-Hill (1967).

Ryan, Thomas P., *Statistical Methods for Quality Improvement*, Wiley (1989).

Scholtes, Peter R., *The Team Handbook*, Joiner Associates (1988).

L. A. Seder, "Diagnostics with Diagrams, Part I," *Industrial Quality Control*, 6 (4), pp. 11–19 (Jan. 1950) [Multi-vari].

Snedecor, George W., *Statistical Methods, Fifth Edition*, Iowa State U. Press (1956) [agriculture; added chapter on sampling by William G. Cochran].

Spence, Janet T., John W. Cotton, Benton J. Underwood, and Carl P. Duncan, *Elementary Statistics, Fifth Edition*, Prentice-Hall (1990).

Steel, Robert G. D., and James H. Torrie, *Principles and Procedures of Statistics*, McGraw-Hill (1960) [agricultural, biological].

Taguchi, Genichi, *Taguchi Methods*, American Supplier Institute (1992).

Taguchi, Genichi, Elsayed A. Elsayed, and Thomas Hsiang, *Quality Engineering in Production Systems*, McGraw-Hill (1989).

Walpole, Ronald E., and Raymond H. Myers, *Probability and Statistics for Engineers and Scientists, Second Edition*, Macmillan (1978).

Winer, B. J., Donald R. Brown, Kenneth M. Michels, *Statistical Principles in Experimental Design, Third Edition*, McGraw-Hill (1991).

Wright, Susan E., *Social Science Statistics*, Allyn and Bacon (1986).

INDEX